PROGRESS IN SOLID STATE CHEMISTRY RESEARCH

PROGRESS IN SOLID STATE CHEMISTRY RESEARCH

RONALD W. BUCKLEY
EDITOR

Nova Science Publishers, Inc.
New York

For permission to use material from this book please contact us:
Telephone 631-231-7269; Fax 631-231-8175
Web Site: http://www.novapublishers.com

Library of Congress Cataloging-in-Publication Data

Progress in solid state chemistry / Arte M. Newman (editor).
p. cm.
Includes bibliographical references and index.
ISBN 13 978-1-60021-313-7
ISBN 10 1-60021-313-8
1. Solid state chemistry--Research. I. Newman, Arte M.
QD478.P78 2006
541'.0421--dc22 2006019702

Published by Nova Science Publishers, Inc. ✢New York

CONTENTS

PREFACE

The new book on solid state chemistry presents studies of chemical, structural, thermodynamic, electronic, magnetic, and optical properties and processes in solids. Research areas include: bonding in solids, crystal chemistry, crystal growth mechanisms, diffusion epitaxy, high-pressure processes, magnetic properties of materials, optical characterization of materials, order-disorder, phase equilibria and transformation mechanisms, reactions at surfaces, statistical mechanics of defect interactions, structural studies and transport phenomena.

In Chapter one, Layered oxide materials, which have been widely studied in the past couple of decades for their very unique and highly applicable chemical and physical properties are discussed The preparation, characterization and the application of several series of inorganic layered oxides in different fields are of common interest. Among those materials are mainly titanates, niobates, and tantalates. These promising materials find applications in several fields such as ion-exchange, electrochemistry, photoelectrochemistry and photoluminescence. Their ability to modify easily makes them applicable in these fields, and their properties are highly dependent on the modification and interlayer materials. The synthesis of layered oxides has been generally carried out with a soft chemistry method, which does not require any abrupt conditions and allows low temperature modification and design of layered oxides in a broad range of shape and structure.

The acidic properties of the layers make them available to easily exfoliate, and exfoliation brings about the easy processing of the layered oxides for intercalation and thin-film preparation. The development of the thin-films from layered oxides with novel characteristics is inevitable for the advancement in thin film applications in a wide range of fields. Production of nanoscale thin films with higher ion exchange and modification capability, thermal stability and appropriate electrochemical properties contributes to the utilization of the films as functional materials with advanced features.

Modification of the interlayer by the introduction of various materials has been previously performed by applying several methods. In our study, we have been mostly focused on the Electrostatic self-assembly deposition (ESD), which was found in our laboratory, and Layer-by-layer assembly (LBL) methods. Both methods are based on the basic electrostatic principles, where cations (guests) combine with the negatively charged host layers right after coming into contact.

The chapter will focus on the synthesis and properties of semiconductor layered oxide materials as well as their applications in the field of electrochemistry and photoelectrochemistry. The properties of the layered oxides will be discussed by giving some experimental data supported by the theoretical approach. Among those properties and applications are intercalation behavior when in contact with complex molecules, water splitting or photocatalytic capability, and photoluminescence characteristics of Lanthanide cation intercalated films. We believe that the chapter will fill a gap in the solid state chemistry of layered oxide materials, since there are few numbers of publications but most of them are on layered silicates.

Chapter two states that quantum dots of several metal chalcogenide semiconducting compounds were deposited in thin film form by chemical and sonochemical methods. The studied systems included ZnSe, SnSe, Bi_2S_3, In_2S_3, $AgBiS_2$ and $CuInS_2$. All of the developed methods were based on controlled colloidal precipitation reactions in the presence or absence of 20 kHz ultrasonic irradiation, applied using a direct immersion ultrasonic horn. The predominant crystal growth mechanism during the deposition was studied by coherent light scattering experiments. The synthesized materials in thin film form as well as bulk precipitates were identified by X-ray diffraction method. Besides for identification, the XRD data were used to calculate the average crystal size of the synthesized materials within the Debye-Scherrer approach. On the basis of XRD patterns, a refinement of the unit cell parameter values was also performed employing multiple regression analysis techniques and further conclusions concerning the existence and magnitude of lattice strain in the samples were derived. The dislocation densities were calculated as well. Optical absorption measurements in the UV-VIS-NIR spectral region were used to calculate the band gap energies of the studied systems within the framework of Fermi's golden rule for interband electronic transitions employing the parabolic approximation for the dispersion relation. Blue-shifted band gap values with respect to the corresponding bulk specimen, which exhibit red shifts upon thermal annealing of the thin film samples, were observed for all synthesized materials, further confirming their low-dimensional character. The size quantization effects were rationalized within the framework of the models of Brus, Nosaka and the hyperbolic band model. In the case of semiconductors in which valence band is split due to spin-orbit coupling, the evolution of splitting energy upon average crystal size increase was followed and discussed. On the basis of temperature dependence of thin films' resistivity, thermal band gap energies and impurity levels ionization energies were calculated. The changes in intercrystalline barriers to electronic transport properties in the case of synthesized nanocrystalline semiconductor films upon thermal treatment were studied as well. In the case of systems exhibiting photoconduction properties, the spectral dependence of stationary nonequilibrium conductivity was measured by the constant field method. These photoconductivity spectral response data were used to calculate the parameters relevant to the band structures of the studied semiconducting films. Photoconductivity relaxation dynamics upon interaction of the films with light impulses was studied by the oscilloscopic method. On the basis of data concerning the photoconductivity time-decay, conclusions about the kinetics of nonequilibrium charge carriers' recombination processes were derived and also their average lifetimes were calculated. Further conclusions on the photoconduction dynamics were derived on the basis of measured lux-ampere characteristics of the synthesized nanocrystalline thin films.

Chapter three informs us that Nb-containing multimetallic oxide materials have generated considerable interest because of their attractive physical properties such as catalysis, ferroelectricity, ion conductivity and luminescence. Most of the time, the preparation method is the classical ceramic route; because of the serious limitations of this high-temperature route, alternatives based on precursors have also been considered. The development of such processes in the case of niobium-containing materials constitutes a great challenge because of the cruel lack of appropriate sources of niobium(V). Many efforts have been made recently on the syntheses of novel precursors that display versatile properties.

The purpose of this review is to cover the open literature concerning precursors routes for the preparation of Nb-containing multimetallic oxides. It provides a classification of the different kinds of such methods developed up to now and offers a broad view of the possible ways to form niobium-based oxide materials.

Chapter four shows that synthesis of nano-sized magnetic materials has become a particularly important area of research and is attracting a growing interest because of their potential applications in ferrofluids, advanced magnetic materials, catalysts, colored pigments, high-density magnetic recording media, and medical diagnostic etc. In this chapter, sol-gel method, polymer-pyrolysis method, template method, hydrothermal method and refluxing method are introduced for synthesis of nano-sized spinel ferrites (or composites) and metal oxides including $NiFe_2O_4/SiO_2$, $CoFe_2O_4/BiFeO_4$, $CoFe_2O_4$, $MnFe_2O_4$, $ZnFe_2O_4$, $NiFe_2O_4$, α-Fe_2O_3, NiO, β-MnO_2 etc, respectively. These nanosized magnetic materials prepared using the above methods are characterized using X-ray diffraction (XRD), transmission electron microscopy (TEM), electron diffraction (ED), scanning electron microscopy (SEM), electron dispersed X-ray diffraction (EDX), X-ray photoelectron spectroscopy (XPS) and vibrating sample magnetometer (VSM) measurements, respectively. Influencing parameters such as reaction conditions etc on their magnetic properties are discussed.

In Chapter five, the mild hydrothermal synthesis is an adequate method to obtain novel microporous materials with occluding protonated diamines, which act as templating agents directing the synthesis towards a determinate product. Among these materials, iron phosphates occupy the first position because they increment a rich structural chemistry. Nevertheless, the iron arsenates are scarce. In the last years we have synthesize organically templated iron(III) arsenates with different crystalline dimensionality (1D, 2D and 3D) using the ethylenediamine, piperazine, 1,4-diazobicyclooctane diamines,... as templating agents. The synthesis was carried out in aqueous solution using as starting reagents arsenic(V) oxide, iron(III) salt, hydrofluoric acid, if the final phase contains the fluoride anions connecting the iron centers, and the diamine that regulates the pH of the initial reaction mixture. The syntheses were performed inside poly(tetrafluoroethylene) containers under autogeneous pressure at 170 °C for five days.

On this synthetic basis several iron(III) arsenates were obtained. $(C_2H_{10}N_2)[Fe(HAsO_4)_2(H_2AsO_4)].H_2O$ (1) consists of $[Fe(HAsO_4)_2(H_2AsO_4)]^{2-}$ anionic *chains*. The FeO_6 octahedra are isolated and bridged by two $HAsO_4$ and one H_2AsO_4 tetrahedra. The $(C_4H_{12}N_2)_{1.5}[Fe_3(HAsO_4)_{1.2}(HPO4)_{0.98}(AsO_4)_{0.88}(PO4)_{0.12}F_5]$ (7) *layered* phase is constructed from $[Fe_3(HAsO_4)_{1.2}(HPO4)_{0.98}(AsO_4)_{0.88}(PO4)_{0.12}F_5]^{3-}$ inorganic sheets. The layers have cavities where the piperazinium dications are placed. $(C_3H_{12}N_2)_{0.75}[Fe^{II}_{1.5}Fe^{III}_{1.5}(AsO_4)F_6]$ (12) is a three-dimensional mixed valent compound. The structure of $(C_6H_{14}N_2)[Fe_3(HAsO_4)_2(AsO_4)F_4].0.5H_2O$ (15) and the isomorphous mixed arsenate-phosphate compound consist of a *three-dimensional* framework formed by $[Fe_3(HAsO_4)_2(AsO_4)F_4]^{2-}$ inorganic layers interconnected by hydrogenarsenate groups.

These compounds exhibit interesting magnetic behaviors. The thermal evolution of χ_m for (1) is indicative of *antiferromagnetic* exchange couplings. The magnetic data were fitted to the equation given by Fischer and Dingle for chains of spin S= 5/2. The best fit is obtained for a value of the J-exchange parameter of J= -0.835K, where g= 2.00 was obtained from ESR spectroscopy. The magnetic susceptibility for (7) increases with decreasing temperature and reaches a maximum at 8.5K. After that, the susceptibility slightly decreases until 5K. However, in the $\chi_m T$ *vs.* T curve at low temperatures an abrupt increase is observed with a maximum at 13.4K. After this temperature the $\chi_m T$ curve decreases down to 5K. This result indicates the existence of a *ferromagnetic* behavior, which can take place due to the decompensation of the magnetic moments of the different sublattices of the FeO_6 octahedra. Compound (12) is a three-dimensional antiferromagnet with a weak hysteresis loop at low temperatures, being the remnant magnetization of 400 emu/mol and the coercive field of 1180 Gauss. In the case of compound (15) magnetic susceptibility measurements indicate the presence of antiferromagnetic interactions. In order to obtain the J exchange parameter the system was considered as a cubic network. The equation developed by Rooshbroke and Wood for a cubic antiferromagnetic system with S= 5/2 was used. Only the low temperature magnetic data, in the proximity of the maximum, can be reasonably fitted with a value of |J/K|= 0.41. The isomorphous arsenate-phosphate phase [16] is an intermediate 3D-2D antiferromagnetic system.

The more important characteristic of some organically templated iron(III) arsenates found in the literature will be studied in this revision.

Chapter six states that the near-infrared spectra of hydrothermally synthesized beidellites with increasing layer charge, pyrophyllite and paragonite have been compared and interpreted based on their associated mid-infrared spectra. The 6000-7500 cm^{-1} region for beidellite is dominated by the first overtone of the adsorbed H_2O stretching mode around 6900 cm^{-1}, while the 7100 and 7150 cm^{-1} bands are the first overtones of the two Al-OH stretching modes. Pyrophyllite is characterised by a sharp band at 7172 cm^{-1} while paragonite shows a weaker and broader band at 7147 cm^{-1}. The 5000-6000 cm^{-1} region is characterised by combination bands of interlayer H_2O stretching modes with the adsorbed H_2O bending mode around 5240 and 5140 cm^{-1}. Furthermore, the combination modes are broad as expected based on the broad nature of the H_2O bands in the mid-infrared region. Minor bands in the pyrophyllite spectrum are due to the presence of a small amount of beidellite. Bands in the 4000-5500 cm^{-1} region at 4545 and 4595 cm^{-1} are attributed to the combination bands of the stretching and libration modes of the Al_2OH units in the beidellite structure. Pyrophyllite exhibits a single sharp band around 4615 cm^{-1}. A similar but broader band is observed for paragonite at 4575 cm^{-1}.

In: Progress in Solid State Chemistry Research
Editor: Ronald W. Buckley, pp. 1-54
ISBN 1-60021-313-8

Chapter 1

SEMICONDUCTOR LAYERED OXIDES: SYNTHESIS, PROPERTIES AND APPLICATIONS

Ugur Unal**, Yasumichi Matsumoto and Shintaro Ida
Department of Nano Science and Technology, Faculty of Engineering
Kumamoto University, Kurokami 2-39-1 860-8555 Kumamoto, Japan

ABSTRACT

Layered oxide materials have been widely studied in the past couple of decades, because of their very unique, highly applicable chemical and physical properties. The preparation, characterization and the application of several series of inorganic layered oxides in different fields are of common interest. Among those materials are mainly titanates, niobates, and tantalates. These promising materials find applications in several fields such as ion-exchange, electrochemistry, photoelectrochemistry and photoluminescence. Their ability to be modified easily makes them applicable in these fields and their properties are highly dependent on the modification and interlayer materials. The synthesis of layered oxides has been generally carried out with soft chemistry method, which does not require any abrupt conditions and allows low temperature modification and design of layered oxides in a broad range of shape and structure.

The acidic properties of the layers make them available to be easily exfoliated and exfoliation brings about the easy processing of the layered oxides for intercalation and thin-film preparation. The development of the thin-films from layered oxides with novel characteristics is inevitable for the advancement in thin film applications in a wide range of fields. Production of nanoscale thin films with higher ion exchange and modification capability, thermal stability and appropriate electrochemical properties contributes to the utilization of the films as functional materials with advanced features.

Modification of the interlayer by introducing various materials has been previously performed by applying several methods. In our study, we have been mostly focused on the Electrostatic self-assembly deposition (ESD), which was found in our laboratory and Layer-by-layer assembly (LBL) methods. Both methods are based on the basic

* E-mail: ugurunal@chem.kumamoto-u.ac.jp. TEL: +81-96-342-3658; FAX: +81-96-342-3679

electrostatic principles, where cations (guests) combine with the negatively charged host layers right after coming into the contact.

The chapter will be focused on the synthesis and properties of semiconductor layered oxide materials as well as their applications in the field of electrochemistry and photoelectrochemistry. The properties of the layered oxides will be discussed by giving some experimental data supported by the theoretical approach. Among those properties and applications are intercalation behavior when in contact with complex molecules, water splitting or photocatalytic capability, and photoluminescence characteristics of Lanthanide cation intercalated films. We believe that the chapter will fill a gap in the solid state chemistry of layered oxide materials, since there are few numbers of publications but most of them are on layered silicates.

1. INTRODUCTION

Layered materials have been extensively studied in the past couple of decades. The word "layered" gives the structural concept of the material as the layered materials are composed of two dimensional units, which have a thickness in nanoscale, with or without ionic species in-between. These two dimensional units are piled up one over another to form the bulk phase of the materials, (Figure 1). Layers are hold together with weak interatomic forces, such as electrostatic or Van der Waals. For example, Cs^+ cations in $Cs_xTi_{(2-x/4)}\square_{x/4}O_4$ act as a counter ion between negatively charged TiO_6 layers to hold the layer together according to the electrostatic principles. On the other hand, graphite possesses no species between layers, which are hold together with weak interatomic forces. The range of layered materials is quite large. Aluminosilicates, clays, and semiconductive layered oxides such as titanates can be counted in the class of layered materials. The subject of this chapter will, however, be semiconductive layered oxides, as there are many collective work for other type of layered materials but just few for semiconductive layered oxides. In this chapter, thus, we will try to discuss the processing, properties and applications of semiconductive layered oxides by representing our work and reviewing other studies in the field.

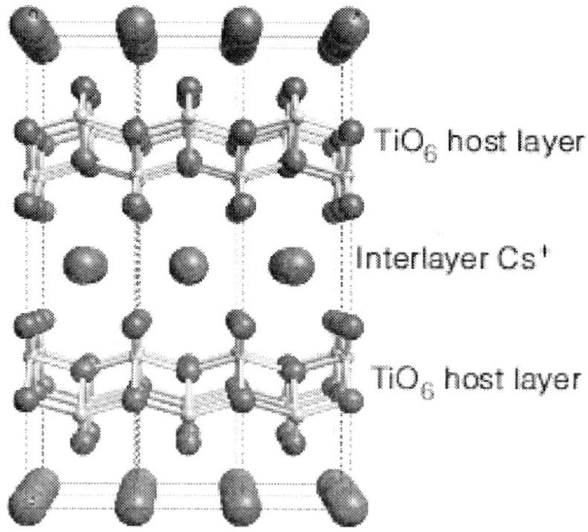

Figure 1. Structure of a layered oxide, $Cs_xTi_{(2-x/4)}\square_{x/4}O_4$.

Synthesis and modification of layered oxides using soft chemistry (chimie douce) is a leading method [1,2], since it does not need any abrupt conditions such as high temperature, high pressure and expensive experimental setups. Soft chemistry has many benefits that the crystal and bond structure of the oxide framework is not affected during the chemical processing. It allows processing at temperatures low enough to keep the main structure intact. In addition, the reaction media is mostly aqueous, which makes the method relatively environmentally friendly. One can design a consistent path for the fabrication of a metastable phase of a layered oxide starting from the thermodynamically stable phase synthesised by conventional solid state reactions. Ion exchange, intercalation or exfoliation reactions carried out with soft chemistry method only give the desired new material without changing the main structure of the starting material. Products of soft chemistry method show structural similarity to the parent material. Soft chemistry allows fabrication of new materials that can not be prepared by conventional methods. For instance, a layered perovskite might easily be converted into three-dimensional perovskite structure by simply removing the interlayer H^+ of protonated two dimensional layered perovskite structures or a new form of layered oxide can be easily prepared by simply exchanging the interlayer guest cations with a new one in an aqueous solution. Most of the processes given in this chapter are based on soft chemistry method.

Layered oxides can be utilized in many applications because of their unique chemical and physical properties. For instance, superconducting properties of Cu and Co layered oxides are resulted mainly from the host oxide layer with a planar structure [3-5]. The semiconductive properties can be utilized in photocatalytic applications such as water splitting. Electrochemical intercalation for the positive electrode in the Li cell is a well known behavior of layered oxides with both properties of electrical conductivity of host layer and high ionic conductivity of the guest ion in the interlayer [6]. Semiconductive layered oxides offer a wide range of chemical and physical properties that can be easily modified and these properties can be utilized in a variety of applications. The space between layers possesses metal cations initially and ion-exchanging in acidic environment protonates the interlayer to increase the reactivity and make it more susceptible to ion-exchange and intercalation reactions. The acidic interlayer might experience expansion when in contact with the large basic molecules or even exfoliation to single nanosize sheets. Exfoliation brings about many advantages like easier processing of layered oxide for the purpose of interest. For instance, the introduction of guest molecules into interlayer might face with difficulties due to high charge density of layers. The difficulty might be alternatively overcome by the exfoliation of layers. In addition, exfoliation makes fabrication of very thin film possible in a controllable manner.

Nanosize sheets of layered oxides exhibit diverse behavior when compared with their bulk forms, especially in the field of photochemistry. General application of layered oxide materials is in the bulk form in the field of photocatalyst. The photocatalytic and photoelectrochemical properties of this type of materials have been widely studied. Their thin films in nanosize, however, have not been deeply examined in terms of electrochemical and photoelectrochemical properties. In addition, utilization of the quantum size effect leads oxide films to be available as functional materials and disclosure of their electrochemical and photoelectrochemical properties, especially in aqueous solutions, will aid in clarifying and evaluation of the catalytic and photocatalytic mechanisms.

Some cations and materials existing in the interlayer highly affect the physical, electrochemical and photoelectrochemical properties, and also their photochemical behavior.

The relative energy level of interlayer cations or molecules determines whether interlayer compounds act as an electron donor, acceptor or recombination center. It is quite important to know the relative energy levels of host and guest in order to develop efficient photocatalysts intercalated with stable compounds. Briefly, we will try to discuss all above concepts of semiconductive layered oxides in this chapter.

2. STRUCTURES OF LAYERED OXIDES

Semiconductive layered oxides are composed of repeating units of two dimensional metal oxide layers. Alkali metal cations are generally sandwiched between these layers and these cations have two purposes in the interlayer: (1) to keep the charge balance, and (2) to hold the layers together. As semiconductive metal oxide layers are negatively charged, the cations between these layers keep them together according to the electrostatic principles. Metal oxide layers are composed of MO_6 (M: transition metal, i.e. Ti, Nb, or Ta) octahedral units connected to each other via the edge or corner sharing. Figure 2 shows the structure of some layered oxides. For example, lepidocrocite type $Cs_xTi_{(2-x/4)}\square_{x/4}O_4$ ($\square$=vacancy, x≈0.7) (Figure 2a) has layers composed of TiO_6 octahedral units connected to each other via edge sharing in the layer and Cs^+ cations are sandwiched between these layers. The crystal has a body centered orthorhombic symmetry, *Immm*, with lattice parameters given in Table 1. The negative charge of the layer of this type of layered oxide arises from the disordered titanium vacancies [7,8]. The thickness of a layer was estimated to be 0.75 nm [9-11].

Table 1. Structural parameters of some layered oxides

Layered Oxide	Space group	a (Å)	b (Å)	c (Å)
$Cs_xTi_{(2-x/4)}\square_{x/4}O_4$	Immm	3.83	17.00	2.96
$K_2Ti_4O_9$	C2/m	18.25	3.79	12.01
$K_4Nb_6O_{17}$	P222	7.85	37.67	6.46
$K_{0.5}MnO_2$	R3m	2.88	-	19.07
$KCa_2Nb_3O_{10}$	Cmcm	3.88	29.51	7.71

There are also other types of titanate layered oxides, such as $K_2Ti_4O_9$ [12], $Cs_2Ti_5O_{11}$ [13,14], generally called layered hydrous titanium oxides. In the structure, four or five TiO_6 octahedral units are edge shared to form a block and these blocks line in a zig-zag form by combining through their edges one above another. The structure is given in detail in Figure 2b and the structural parameters were tabulated in Table 1. The protonated forms of these types of titanates are known to have strong acidity because of the dangling oxygen atoms at the free edges of blocks as marked with arrows in Figure 2b.

On the other hand, NbO_6 octahedral units in $K_4Nb_6O_{17}$ and niobates with similar crystal structure [15,16] are connected to each other via corner sharing as seen in Figure 2c and $K_4Nb_6O_{17}$ has two alternative interlayers in contrast to layered titanate oxides. This distinctive property brings about the high ion-exchange capability and unique photocatalytic properties. Interlayer I hosts potassium with water molecules, while only dehydrated potassium ions exist in the interlayer II under ambient conditions. The alternative layers are different from each other in their properties of hydration and ion exchange. Interlayer II can accommodate monovalent cations only, while interlayer I can be host to both monovalent and divalent molecules or other cations [17-19]. The crystal structure belongs to orthorhombic system. In the ambient conditions with a humidity of higher than 25%, $K_4Nb_6O_{17}$ exists in its hydrated from, $K_4Nb_6O_{17}.3H_2O$ and the hydrated structure can be easily determined by XRD pattern, where (020) peak appears at $2\theta=5^o$.

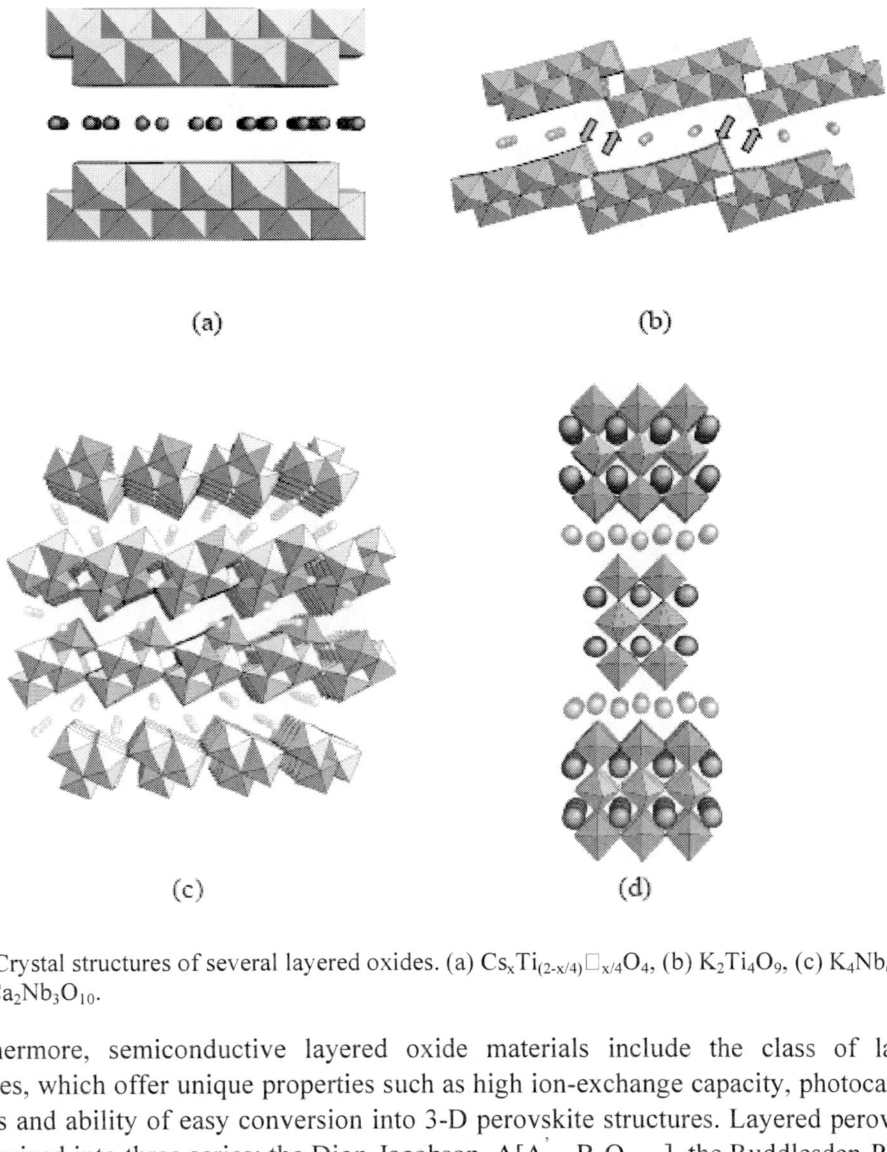

Figure 2. Crystal structures of several layered oxides. (a) $Cs_xTi_{(2-x/4)}\square_{x/4}O_4$, (b) $K_2Ti_4O_9$, (c) $K_4Nb_6O_{17}$ and (d)$KCa_2Nb_3O_{10}$.

Furthermore, semiconductive layered oxide materials include the class of layered perovskites, which offer unique properties such as high ion-exchange capacity, photocatalytic properties and ability of easy conversion into 3-D perovskite structures. Layered perovskites are categorized into three series: the Dion-Jacobson, $A[A'_{n-1}B_nO_{3n+1}]$, the Ruddlesden-Popper, $A_2[A'_{n-1}B_nO_{3n+1}]$, and Aurivillius, $(Bi_2O_2)[A_{m-1}(B)_mO_{3m+1}]$ series. A represents the exchangeable interlayer alkali metal cation and A' is alkali earth metal cation located in the cavities among the octahedral BO_6 (B: Ti, Nb, or Ta) units where smaller B cations are placed at the center of octahedra. BO_6 octahedral units are corner shared to form perovskite blocks and these blocks in the crystal structure are bound together by cations as seen in Figure 2d. $KCa_2Nb_3O_{10}$ is a well known example for Dion-Jacobson phase. $[Ca_2Nb_3O_{10}]^-$ perovskite blocks are separated by interlayer K^+ cations and are aligned in such a way that adjacent blocks are dislocated one-half unit cell distance relative to each other [20,21]. Dion-Jacobson phase layered perovskites possess one interlayer cation per unit cell while Ruddlesden-Popper phases have two [22,23]. In Aurivillius phases, two perovskite slabs are connected to each other via $Bi_2O_2^{2+}$ covalently [24].

3. Ion Exchange and Intercalation

Layered structure and readily available ion-exchange sites in the interlayer receive attraction for the host-guest chemistry, electrochemical, catalytic and photocatalytic applications. The high reactivity of the host layer and large inner surface area enable ion exchange reactions, especially, of alkaline metal cations, which can enhance the photocatalytic properties of layered oxides. Another advantage of the ion exchange ability is its functionality in preparation of different type layered oxides that cannot be prepared by high temperature solid state reactions or chemical methods. Ion-exchange ability lets us to prepare different phases of a layered oxide with a different metal cation in the interlayer with a soft chemical approach. Intercalation of bulky molecules is also possible in this manner [25-28]. Introduction of a complex or photoactive molecule into the interlayer of a layered oxide adds certain properties, such as visible light response when intercalated with a photoactive molecule.

3.1. Layered Oxides: Solid Acids

Layered oxide materials are generally known to be highly acidic as they possess oxide sites open to interlayer and these sites make bond with protons in the interlayer to make strongly acidic OH groups fixed to the nanosize sheets metal atoms (i.e. Ti-OH) (denoted with arrows in Figure 2b) as in tetratitanate with zigzag structure or exchangeable H^+ or H_3O^+ in the interlayer are the source of acidity as in $H_xTi_{(2-x/4)}\square_{x/4}O_4$ when the interlayer is protonated [25,29-33]. Bronsted acidity is highly expressed for the layered oxides. High acidity is obtained only in the proton exchanged form, which can be obtained by treating the layered oxide with an acidic solution. In an acidic solution, interlayer alkali cations are exchanged with protons and protonation, which occurs without destroying the original structure, extends interlayer to subsequent ion-exchange, intercalation and exfoliation processes. Products possess distinctive solid acidic property, which leads to the ability of hosting various types of cations, metal complexes, and organic bases or to distinguished catalytic properties.

The acidic properties of proton exchanged layered oxides have been widely studied in order to reveal the acidic sites and functionality of the layers. Interlayer proton, hydronium ions or positions of oxide sites on the layer as well as the transition metal cation forming the host layer regulate the acidity. The order of the electronegativity in the periodic table gives higher acidity in niobates than that in titanates [2,8]. On the other hand, structural features are also important. For example, tantalates are expected to be less acidic than niobates according to their position in the periodic table. However, extensive distortion in niobates compared to tantalates in the MO_6 octahedra results in shorter terminal M-O bonds, which leads to less acidic nonterminal, bridging oxygen atoms in niobates than in tantalates [34]. These oxygen atoms closely coordinate with H^+ in the structure to form highly acidic bridging hydroxyl groups. The necessity of bridging hydroxyl groups was also reported by Domen's group [31-33]. They suggested that the high acidity corresponding to 90% H_2SO_4 can be assigned to Ti(OH)Nb or Ti(OH)Ta sites on nanosize sheets of $HTiMO_5$ and HTi_2NbO_5, where TiO_6 and NbO_6 octahedral units were alternately arranged in the crystal structure. These sites were found to be source of the acidity, although these nanosize sheets also possess Ti-OH sites. Moreover, Ti(OH)Ta sites were found to be more acidic than Ti(OH)Nb sites. For HNb_3O_8, the Bronsted acidity was similarly assigned to Nb(OH)Nb sites possessing bridging oxygen atoms [33]. On the other hand, terminal oxygen atoms are responsible for the acidity in a series of di, tri, tetra and penta titanates of which layered structure possess zigzagged blocks of octahedral units (Figure 2b). Unshared oxygen atoms at the edge of each block (shown by an arrow) are highly basic and protonated to form hydroxyls under acid leaching and act as acidic sites on the nanosize sheets. $H_xTi_{(2-x/4)}\square_{x/4}O_4$, however, does not contain any oxygen atoms to be protonated and its acidity comes from the exchangeable interlayer H_3O^+ [30]. The acidity of $H_xTi_{(2-x/4)}\square_{x/4}O_4$ was found to be the highest among other titanates (tri, tetra and pentatitanates) as it was able to intercalate pyridine but others did not undergo an intercalation reaction [25].

Strong solid acidity of layered oxides allows preparation of new materials that cannot be prepared by conventional methods. Exchanging interlayer protons or hydronium ions with different alkali cations leads to new materials with unique properties. Another advantage of layered solid acids is readily available interlayer space for acid-base reactions, which results in the production of organic-inorganic composites. Moreover, they can be used as solid acid catalysts [31-33]. Exfoliated $HTiNbO_5$, HTi_2NbO_7, and $HTiTaO_5$ layered oxides showed high catalytic activity in various reactions such as esterification of acetic acid, cracking of cumene and dehydration of 2-propanol.

3.2. Ion Exchange Reactions

Ion exchange reactions involving layered oxide materials have been widely studied. Development of new materials as well as the modification of interlayer and increasing the reactivity of the layered oxide materials put new dimensions into the research on the application areas of these kinds of materials. The most fundamental ion exchange reaction and also the very initial step in modification of the layered oxide is the protonation of the interlayer (Figure 3), which is exchanging the interlayer alkali metal cations with protons or hydronium ions in an acidic solution. The process does not alter the structural features of layered oxides but gives more functionality and reactivity into the interlayer. In most cases, it was observed that the interlayer has expanded after proton exchange reaction as a result of the presence of water molecules in the interlayer. Proton exchange process is a part of a soft solution processing of layered oxides. Interlayer proton or hydronium ions can be easily replaced by other alkali metal cations or organic molecules in an aqueous solution. Otherwise, parent layered oxide with an alkali cation in the interlayer needs to be treated at harsh conditions, such as high pressure and temperature [35] and this might result in the structural change.

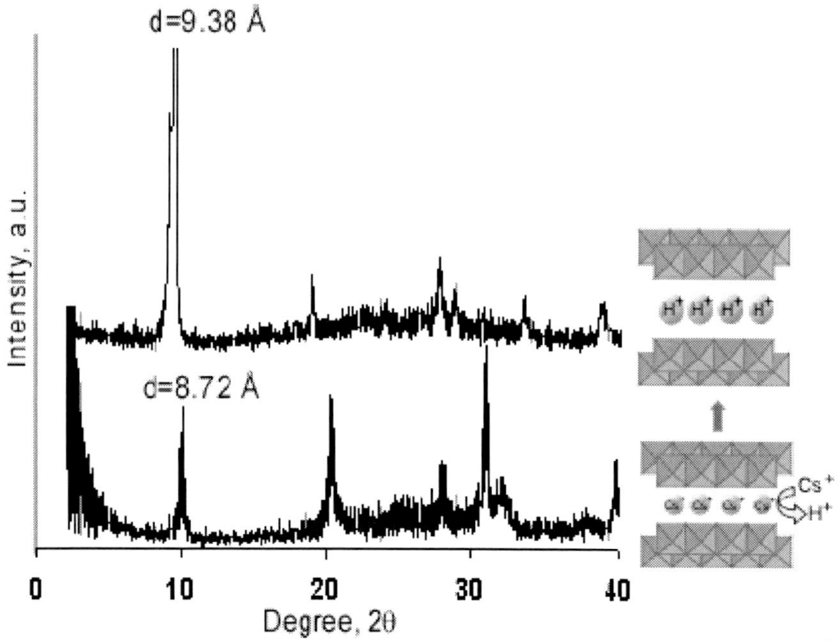

Figure 3. XRD patterns of $Cs_xTi_{(2-x/4)}\square_{x/4}O_4$ before and after H^+ exchange in an acidic solution.

Sasaki et al examined the introduction of several cations into the protonated interlayer of a series of titanates [12,30,36,37]. In a long term process, cations exchanged the interlayer protons in a gradual manner. The formation of different phases at various cation loadings was examined with XRD patterns. Proton exchanged pentatitanate was reproduced by Cs^+ cation exchange and its K^+ phase was also formed by introducing K^+ cation into the interlayer [37]. Same phenomenon was observed for other titanate series. Introduction of Rb^+ and Cs^+ into the protonated interlayer of tetratitanate produced different phases at different cation loadings. Cation loading over ¾ of the theoretical limit resulted in mixed phases of tetra and hexatitanates for Rb^+ loading and tetra and pentatitanates for Cs^+ loading [36]. Depending on cation loading level, cations and water molecules are placed in the interlayer in mono or bi-layer arrangement. Thin films of layered oxides showed similar behavior to their bulk from in terms of ion exchange, but replacing interlayer cations in an aqueous solution needed a very long period of time, yet stoichiometric ratios were below ion exchange capacity [38-40].

Layered perovskites are the most widely studied materials among layered oxides in terms of ion exchange. Layered perovskites can go into many structural changes under ion exchange reactions. Schaak and Mallouk have summarized the modification and transition between layered perovskite structures or to three dimensional perovskites in an extensive review [2]. Taking a layered perovskite as a precursor, one can fabricate various types of layered perovskites using ion exchange reactions under appropriate conditions. Transformation between layered perovskite phases is one of the applications. As given in Figure 4, Dion-Jacobson phase can be prepared by exchanging two interlayer alkali cations of a Ruddlesden-Popper phase with a divalent cation [41].

As other layered oxides, layered perovskites have a solid acid property, when interlayer is protonated after treatment with an acidic solution. During protonation, interlayer cations are replaced by H^+ and layered perovskite gains acidic properties, which will make it available for acid-base reactions, such as intercalation of organic molecules [42-47].

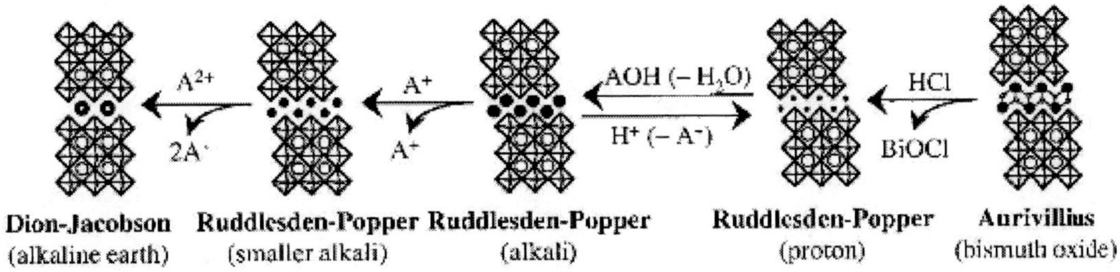

Figure 4. Ion exchange reactions of layered perovskites that involve phase changes [Reprinted with permission from *Chem. Mater.* 2002, 14, 1455-1471. Copyright 2002 Am. Chem. Soc.].

3.3. Intercalation

Intercalation reaction of layered oxides is mostly referred to acid-base reactions involving large alkali molecules, alcohols, metal complexes or organic molecules. Intercalation of these types of molecules into the interlayer of a layered oxide serves for various goals. Large alkali molecules may delaminate layers of a proton exchanged layered oxide into single nanosize sheets after acid-base reaction in an aqueous solution. Photoactive organic molecules introduced into the interlayer may make use of layered oxide as a photocatalyst effective in visible light range. For such aims, an extensive range of molecules has been examined as an intercalation material to modify and increase the functionality of layered oxides. Intercalation of guest molecules into the interlayer is mainly governed by the acidity of layered oxide. Another factor is the negative charge density of layers. Higher charge density hinders the penetration of guest molecules into interlayer more than lower charge density does. Thus, using layered oxides with lower charge density is more favourable in terms of intercalation. For instance, intercalation of pyridine into $H_xTi_{(2-x/4)}\square_{x/4}O_4$ is found to be the most feasible in a series of layered titanates as it has the lowest layer charge density (1/32.2 $Å^{-2}$) [25]. Other titanates were tri, tetra and pentatitanates with layered charge densities of 1/17.2, 1/22.5, and 1/28.2 $Å^{-2}$, respectively.

Intercalation is a long term process and introduction of large molecules into the interlayer of layered oxide takes time in order to obtain complete intercalation. In most cases, an amount of intercalated molecules is less than the real capacity of the layered oxide and is not enough to compensate the negative charge of layer. Hence, intercalation of H_2O and H_3O^+ takes place in order to do so. Sasaki et al assigned partial intercalation of pyridine molecules to the steric effect, that a pyridine molecule extends out of an intercalation sites and obstacles the occupation of two neighbouring sites [25]. Loading rate was found to be 30% of the intercalation sites and the remaining part was occupied by H_2O molecules or H_3O^+ ions. They have suggested that the pyridine molecule was inclined 64^0 with respect to the layer. Gopalakrishnan et al reported another pyridine intercalated layered oxide, $HLaNb_2O_7$ that could be prepared after 20 days treatment [44]. Intercalation of some large molecules may not be favourable as interlayer of layered oxide is not large enough to accommodate. In addition, high charge densities of layers obstacles the introduction of bulky ionic molecules into interlayer. An approach to these problems is to open up the interlayer gallery with pre-intercalated large alkali molecules by taking advantage of the acid-base reactions that proton exchanged layered oxides can undergo (Figure 5).

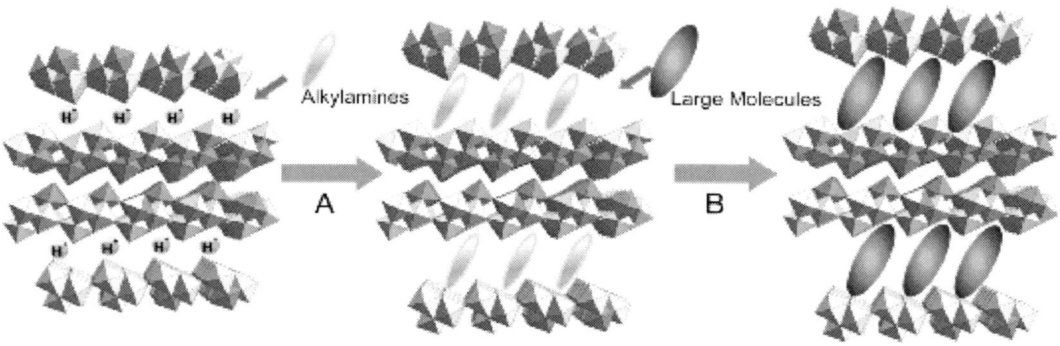

Figure 5. Interlayer is opened up initially by preintercalation of alkylamines (A) to ease the intercalation of large molecules (B), such as dyes.

Alkali molecules are substituted by interlayer protons or hydronium ions and further treatment of layered oxide with an aqueous solution of a desired molecule produces target intercalated layered oxide by replacement of pre-intercalated alkali molecules (Figure 5). Intercalation of various molecules such as $Ru(bpy)_3^{2+}$ [39,40,48-53], phenols [54,55], porphyrin [26,56,57], and dye molecules [58-63] could be achieved into $K_4Nb_6O_{17}$ or titanates pre-intercalated with alkylamine molecules in a two-step reaction. Alternatively, intercalation of target molecule can be achieved in a one-step reaction [64]. For instance, methylene blue could be intercalated into the n-propylamine pre-intercalated interlayers of the single crystal $K_4Nb_6O_{17}$. The reaction was carried out in an aqueous mixture of n-propylamine and methylene blue, where $K_4Nb_6O_{17}$ preferably intercalates n-propylamine molecules in advance and then methylene blue in the same solution replaces these molecules.

$K_4Nb_6O_{17}$ has been extensively studied in terms of intercalation of various kinds of molecules such as dyes, organometallics, and photoactive complexes. It receives the attraction because of its unique structure with two different layers and high photocatalytic ability. In most cases, however, the intercalation of these molecules was below the capacity of niobate layered oxides. There occasionally remained nonexchangable K^+ or alkylamine ions in the cavities. The unique, two structurally distinguishable interlayer property of $K_4Nb_6O_{17}$ results in two different products after intercalation of any molecule depending on whether both layers are intercalated or only one [48,49,53,65]. Intercalation into Interlayer I is more favourable than intercalation into Interlayer II. Type A corresponds to intercalation of molecules into Interlayer I only. When both interlayers were intercalated, the product is Type B intercalation as seen in Figure 6. The type of the product is deduced from the XRD pattern by comparing the d value and the size of the intercalated molecule. Type A is formed under milder conditions and shorter reaction time, while, on the other hand, Type B may require better intercalation-driving conditions, such as higher temperature and longer reaction time [48], because K^+ in interlayer II is relatively more difficult to remove.

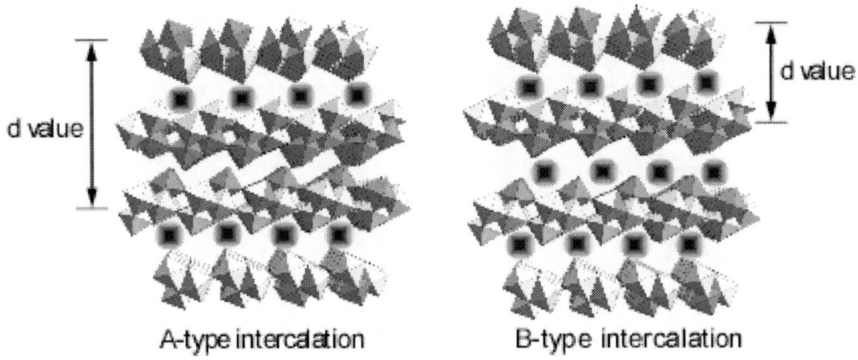

Figure 6. A and B type intercalation models for $K_4Nb_6O_{17}$

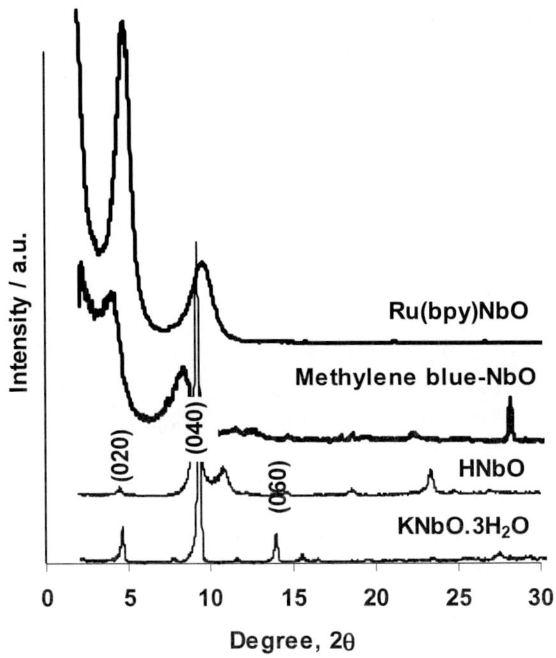

Figure 7. XRD patterns of $K_4Nb_6O_{17}.3H_2O$, $H_4Nb_6O_{17}$, $Ru(bpy)_3^{2+}$ intercalated and Methylene blue intercalated samples, respectively, from bottom to top [Reproduced from J. Solid State Chem. 2006, 179, 33-40]

We have achieved intercalation of $Ru(bpy)_3^{2+}$ and Methylene blue into both layers of $H_4Nb_6O_{17}$. The intercalation compound was B-type intercalation [53]. The layer distances (d values) were calculated as 18.5 Å (2θ=4.78°) and 21.4 Å (2θ=4.12°) from corresponding peaks on XRD patterns (Figure 7) for $Ru(bpy)_3^{2+}$ and Methylene Blue intercalation compounds, respectively. Intercalation models were drawn in Figure 8. Taking account of the size of $Ru(bpy)_3^{2+}$ and Methylene blue molecules, the intercalation type was assigned to B-type intercalation. $Ru(bpy)_3^{2+}$ is placed in the interlayer in a single layer arrangement form. On the other hand, the planar structure of Methylene blue proposes two possible types of arrangement in the interlayer. If molecules lie in the interlayer parallel to $[Nb_6O_{17}]^{4-}$ layers with their long axis, a double layer arrangement is formed in the interlayer. Otherwise, molecules form a single layer arrangement, however in this case, slightly inclined to the host layer with the long axis (Figure 8).

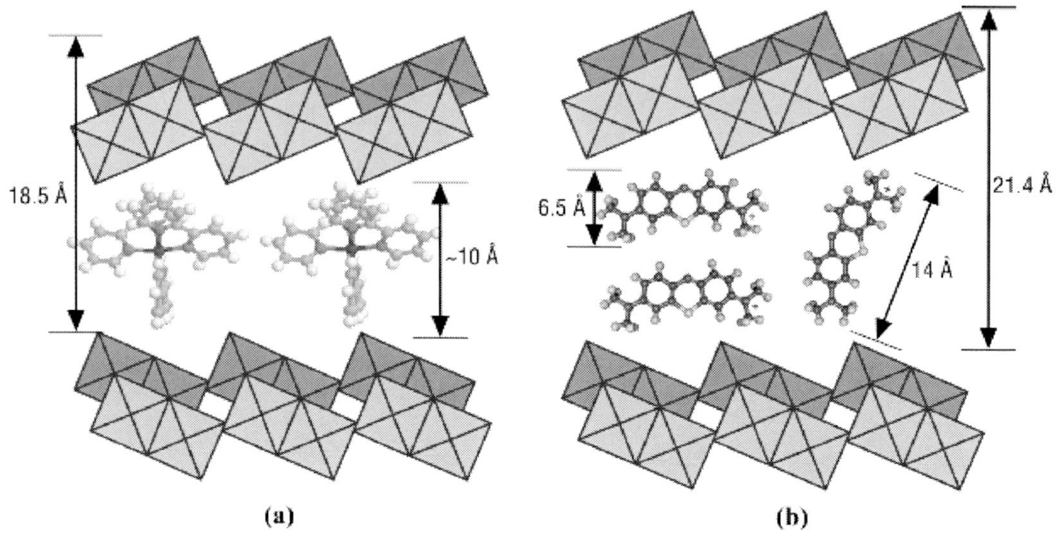

Figure 8. Intercalation models of: (a) $Ru(bpy)_3^{2+}$ and (b) Methylene blue intercalated niobate layered oxides [Reproduced from J. Solid State Chem. 2006, 179, 33-40]

$Cs_xTi_{2-x/4}\square_{x/4}O_4$ is easier to intercalate with large molecules because of its lower layer charge density. Intercalation of pyridine into $H_xTi_{2-x/4}\square_{x/4}O_4$ was reported by Sasaki et al [25]. The reaction was also carried out for other titanates but they failed to intercalate pyridine because of their higher charge densities. Sumida et al has reported intercalation of several dyes into the interlayer of $H_xTi_{2-x/4}\square_{x/4}O_4$ thin film [40]. Thin film preintercalated with alkylamine was kept in an aqueous solution of dye molecule over a long period of time to achieve intercalation. Airoldi et al attained the intercalation of n-alkyldiamine series into $H_2Ti_4O_9$ over 30 days of reaction with the aqueous solution of n-alkyldiamines [27].

In our research, we have observed the intercalation of alcohol and alkylamine molecules into the interlayer of $Cs_xTi_{2-x/4}\square_{x/4}O_4$ during the electrophoretic deposition of as-prepared powder in various alcohol solutions or $H_xTi_{2-x/4}\square_{x/4}O_4$ in an aqueous solution of an alkylamine [66]. Intercalation of alcohol and alkylamine molecules into the interlayer was possible only under electrical bias. XRD patterns of the films revealed highly oriented films deposited in alcohol and alkylamine solutions and interlayer distances of the intercalated titanate films were large enough to accommodate alcohol or alkylamine molecules according to XRD patterns. It was also deduced from the thermal and FTIR data that water molecules were co-intercalated. In addition, the atomic ratio of Cs^+ in the film was less than that in the starting material, which shows that alcohol molecules replaced the interlayer Cs^+ to some extent. On the other hand, proton exchanged form of $Cs_xTi_{2-x/4}\square_{x/4}O_4$, that is $H_xTi_{2-x/4}\square_{x/4}O_4$, had a trace amount of Cs^+ and resulted in better orientated films in (020) direction in alkylamine solutions. Another report for the forced intercalation of alcohol molecules into a layered titanate shows that the intercalation is possible under applied pressure [67].

3.4. Pillaring

Another method for the introduction of molecules, especially bulky inorganic species, into layered oxides is pillaring process. Pillaring is widely used for preparation of microporous materials that show zeolitic properties and results in reasonably large pores to provide a basis for molecular sieving and shape-selective catalysis for relatively high molecular compounds. Specific surface area as high as 460 m^2/g was reported for pillared layered titanates [68-70]. In addition, photocatalytic activity might increase due to the interlayer semiconductor pillars. Pillared layered materials are utilized in the field of catalysis and adsorption. One of the disadvantages in catalytic application is the thermal stability of the catalysis. The development of thermally stable catalysts is possible with the introduction of cross-linking inorganic molecules into the interlayer of layered materials. Smectite clays can be easily pillared in aqueous solutions as a result of their high swelling ability. However, titanates or niobates do not have such swelling ability because of their higher layer charge density. In spite of this, Anderson et. al. was able to pillar layered potassium titanate powders with Keggin cations directly in an aqueous solution [71]. Thus, prior to the pillaring process, the interlayer was expanded or exfoliated by the intercalation of bulky amine molecules and large inorganic cations were exchanged into the interlayer [72-74]. Alternatively, Yanagisawa et al. used hydrothermal synthesis for the fabrication of iron oxide pillared tetratitanate [75]. They have investigated the synthesis temperature on the photocatalytic activity of the pillared titanate. Semiconducting CdS pillars were constructed in the interlayer of titanate or niobate by reaction of the intercalated Cd^{2+} cation with H_2S to CdS [76-78].

4. EXFOLIATION

Layered oxides have a potential of breaking apart into their nanosize sheets under appropriate reaction conditions. The process is generally called exfoliation or in some cases delamination. After realizing the functionality and distinct property of nanosize sheets, exfoliation of layered oxides has been researched extensively. Single sheets after exfoliation can be thought of as the smallest element of parent layered oxide reflecting the bulk properties as well as opening new functionalities. Altering the ionic strength or adding counter ions into the colloidal solution may result in producing new materials. Taking advantage of the negative charge of the layered oxide sheet and its nanosize two dimensional structure, one can fabricate ultrathin films or put exfoliated sheets into new shapes by combining with counterions.

Exfoliation is a further stage of intercalation and occurs as a result of acid-base reactions taking place between the layered oxide and bulky solvent molecules, such as ethylamine or tetrabutylamine. The reaction can be achieved if interlayer possesses solid acid characteristic. Factors determining solid acid characteristics of layered oxides play an important role in this manner as described in Section 3.1. When an acidic layered oxide comes in contact with bulky alkali molecules, molecules initially penetrate into the interlayer and extend the interlayer gallery (Figure 9). If the concentration goes beyond a certain limit, weak attraction forces between layers are overcome and repulsive forces come into existence [9,34,79-85]. As a result, layers composing the bulk layered structure come apart in the solvent to form stable colloidal solution of single nanosize sheets without losing their crystallinity. At very high concentrations osmotic swelling occurs instead of exfoliation (Figure 10). Concentration limits of intercalation, exfoliation and osmotic swelling were well determined for $H_xTi_{2-x/4}\square_{x/4}O_4$ [80].

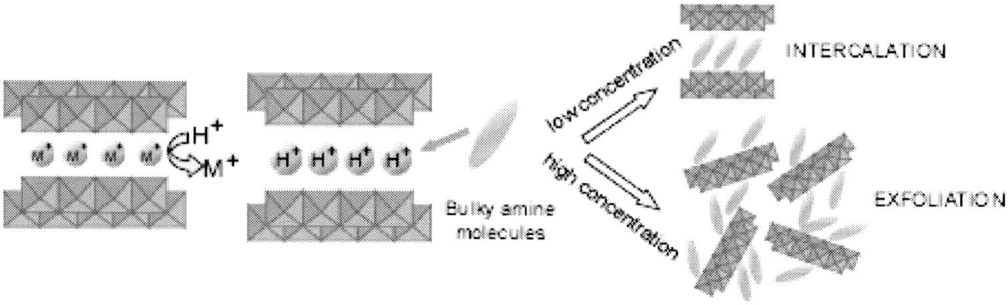

Figure 9. Exfoliation occurs as a result of excessive amount of bulky alkali molecule loading into acidic interlayer of a layered oxide

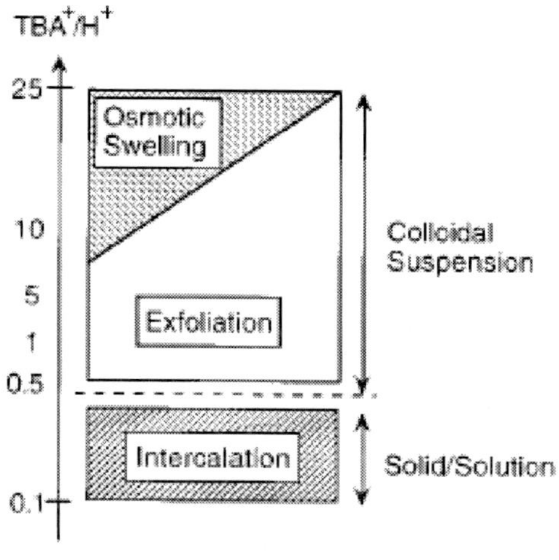

Figure 10. Schematic diagram of the osmotic swelling and exfoliation process [Reprinted with permission from *J. Am. Chem. Soc.* 1998, *120,* 4682-4689. Copyright 1998 American Chemical Society]

On the other hand, not every organic alkali molecule can exfoliate a layered oxide. Organic alkali molecules with long chain may stand in the interlayer making paraffin-like arrangement, which results in strong van der Waals forces to form stable organic-inorganic composite structure. Alkali molecules with shorter chains or bulky organic alkali molecules with steric effect like ethylamine or tetrabutylamine are bound on layer only loosely so that the result is highly swollen interlayers eventually separated into single parts.

Mallouk et al has examined the exfoliation behavior of α-zirconium phosphate by controlling the pH of an exfoliation solution. The titration curve of α-zirconium phosphate with tetrabutylamine revealed the exfoliation point [84,86]. As concentration increases, the arrangement of tetrabutylamine molecules in the interlayer changes from single layer arrangement to multilayer arrangement and eventually α-zirconium phosphate is exfoliated into a colloid solution. Kim et al proposed that stabilization of single nanosize sheets in an aqueous solution is not likely to occur even if molecules cover whole surface of each sheet and as long as the cation-cation contact between layers still exists [86]. Same phenomenon was also reported for $H_xTi_{2-x/4}\square_{x/4}O_4$ and $H_{0.13}MnO_2.0.7H_2O$ for which intercalation occurs at lower tetrabutylamine concentrations [80-81]. In order to break cation-cation contact, an enough number of protons should be ionized by binding more tetrabutylamine molecules on each side of nanosize sheets. At higher concentrations, intercalation phase of the layered oxide is obtained and when the concentration of tetrabutylamine exceeds the threshold point exfoliation occurs to form colloidal suspension of nanosize sheets. For $H_xTi_{2-x/4}\square_{x/4}O_4$, the threshold concentration corresponds to approximately 50% equivalent of the exchangeable proton concentration at which the whole surface of a layer can be covered with tetrabutylamine molecules (Figure 10). Complete delamination, however, occurs when an equivalent amount of tetrabutylamine molecules is 5 times higher than the total concentration of the exchangeable protons. For other layered oxides such as $HCa_2Nb_3O_{10}$, replacement of as low as 15% of interlayer protons with alkylamines was found to be necessary for exfoliation [34].

One of the main factors affecting the exfoliation behavior of layered oxides is the charge density of the layers. Similar to intercalation behavior, charge density rules the penetration of molecules into the interlayer. For example, in a series of titanates, exfoliation of $H_xTi_{2-x/4}\square_{x/4}O_4$ is most likely to occur because of its lowest charge density [9]. Weaker interaction between the layers allows the penetration of tetrabutylamine or ethylamine molecules into the interlayer without much difficulty. Treating $H_xTi_{2-x/4}\square_{x/4}O_4$ with an aqueous ethylamine or tetrabutylamine solution results in exfoliation and these colloidal solutions can stand stable for over months. Atomic force microscopy is a very useful analysis technique to see whether the exfoliation of the oxide into single sheets has occurred (Figure 11). Nanosize sheets were electrostatically deposited on a polycation (polyethyleneimine) modified mica substrate and subjected to the analysis. Height profile analysis gives a thickness of 1.02 nm for a single sheet of $H_xTi_{2-x/4}\square_{x/4}O_4$ as seen in the Figure 11. The value is slightly larger than the theoretical thickness of a single sheet (0.75 nm). The difference comes from H_2O or alkylamine molecules adsorbed on the surface. The figure also shows AFM photographs of various other layered oxides.

We have observed partial exfoliation of $H_xTi_{2-x/4}\square_{x/4}O_4$ in alcohol solutions during the electrophoretic deposition on a platinum substrate [66]. The exfoliation occurs immediately as a result of the forced intercalation of protic alcohol molecules into the interlayer under an electrical bias. Films deposited in polar protic solvents like ethanol or methanol has smoother surface than the films deposited in polar aprotic solvents such as acetone. This could be explained by the partial exfoliation of the layered oxide in protic solvents. Films deposited in alcohol solution gave better crystallinity as seen XRD patterns in Figure 12. Furthermore, electrophoretic deposition in the presence of an aqueous ethylamine or n-butylamine solution gave very smooth films (Figure 12). Films were highly crystalline oriented in (0k0) direction. Both XRD patterns and SEM photographs in Figure 12 revealed that the exfoliated layers with an approximate lateral size of 1 μm were deposited parallel to the substrate surface. On the other hand, films deposited in an acetone solution had no orientation and very rough surface on the films, which shows that powders were deposited without any exfoliation and intercalation (Figure 12).

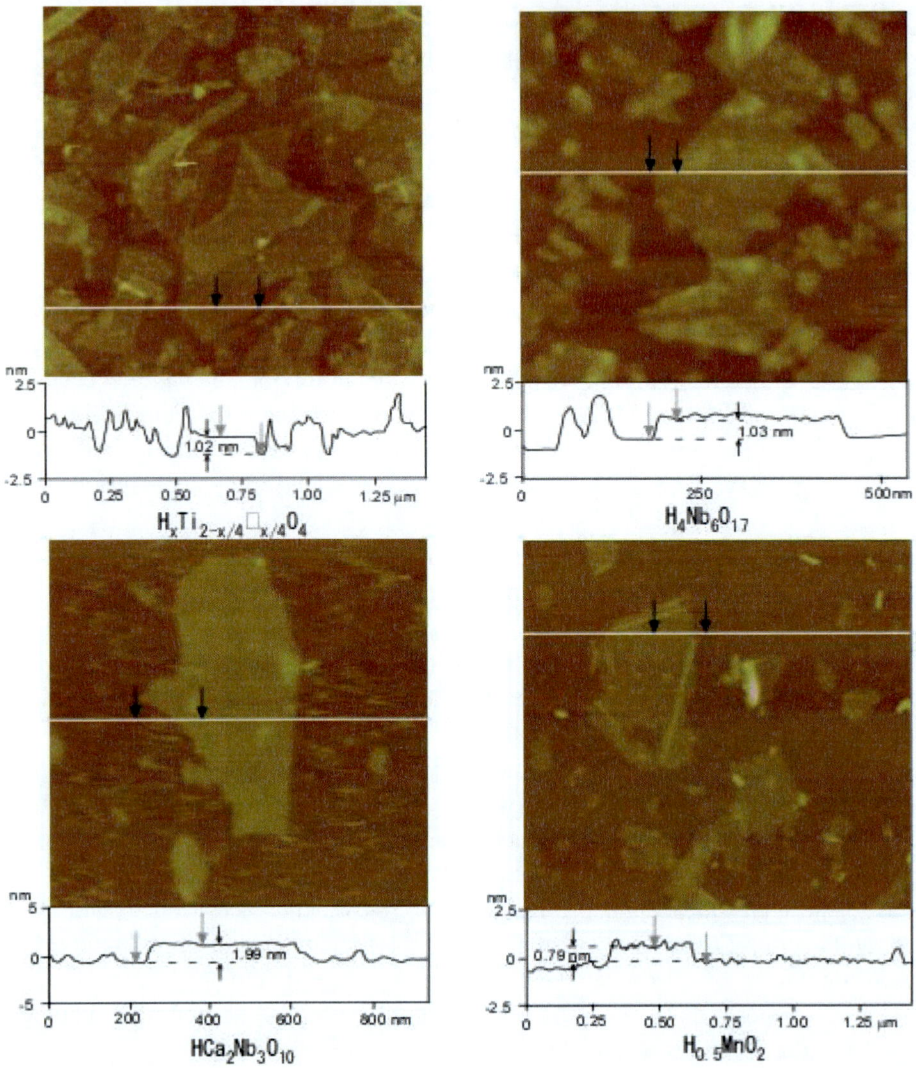

Figure 11. AFM figures of various layered oxides with height profiles

The distinctive two interlayer structure of $K_4Nb_6O_{17}$ and similar layered oxides faces with difficulties of exfoliation because of the different ion-exchange characteristics of two layers and high surface charge density. Interlayer I can be given acidic properties to an entire extent by exchanging K^+ with protons. Interlayer II, however, can be partially protonated and K^+ remains in the interlayer. Thus, interlayer I can be easily exfoliated as a result of intercalation of tetrabutylamine according to acid-base reactions, while intercalation through interlayer II is a difficult process. Abe et al reported exfoliation in the interlayer I only [87]. Exfoliated sheets were found to be consisting of even number of single sheets as a result of exfoliation in Interlayer I. On the other hand, our research showed that exfoliation into single nanosheets at both layers is possible. Proton exchanged niobate layers could be exfoliated into single sheets in an aqueous tetrabutylamine solution but reaction in ethylamine solution resulted in only exfoliation into bilayers [33]. The height profile in AFM images of the exfoliated sheets clearly indicates the thickness of a single layer (Figure 11). Measured thicknesses are in close agreement with the theoretical values. Theoretical and estimated thicknesses of exfoliated sheets were tabulated in Table 3.

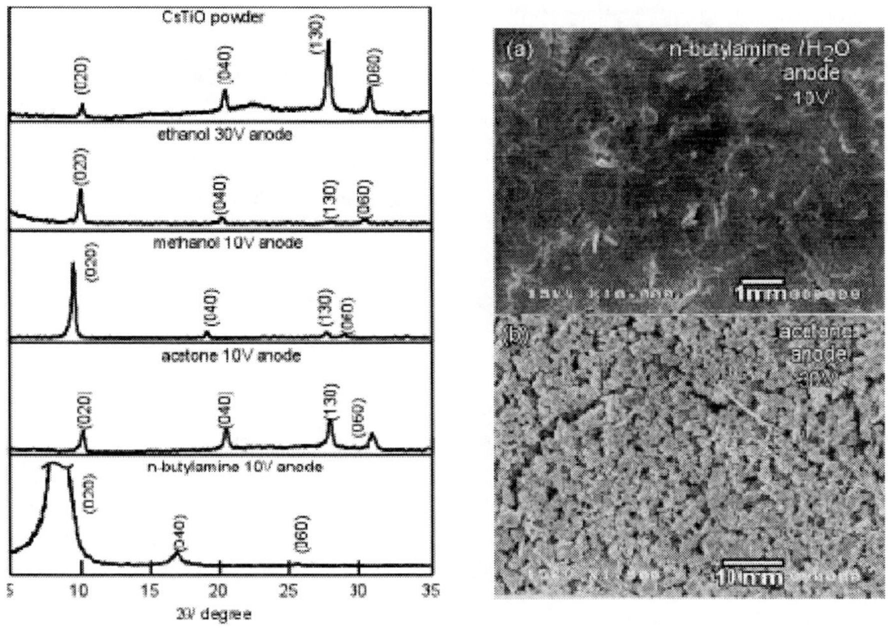

Figure 12. XRD (left image) and SEM (right image) of $Cs_xTi_{2-x/4}\square_{x/4}O_4$ films deposited electrophoretically in different solutions. [Reproduced with permission from *J. Phys. Chem. B* 2001, 105, 10893-10899. Copyright 2001 Am. Chem. Soc.]

Table 3. Dimensions of exfoliated layered oxide sheets

Layered Oxide	Lateral size (nm)	Observed thickness (nm)	Theoretical thickness (nm)
$Ti_{2-x/4}\square_{x/4}O_4$	50-1000	1.02	0.75
Nb_6O_{17}	20-200	1.03	1.00
$Ca_2Nb_3O_{10}$	20-200	1.99	1.47
MnO_2	50-800	0.79	0.52
Ti_4O_9	20-600	1.18	1.08

In further investigations on exfoliation of the niobate layered oxide, exfoliated niobate sheets showed very interesting behavior such as rolling and forming tubes [34,85,88]. Single niobate sheets were able to be obtained under controlled reaction conditions. The aqueous suspension of proton exchanged niobate was titrated with tetrabutylamine until the pH is stabilized around 10. The reaction initially gives niobate sheets exfoliated to bilayers, which are unstable in the beginning. These bilayers are, then, transformed irreversibly into unrolled or rolled single sheets as seen in Figure 13. A similar behavior was also observed by Bizeto et al [89,90]. Exfoliation was performed with butylamine and the yield was the highest at butylamine/niobate ratio of 0.5. The research shows that pH or ionic strength of the colloidal solution influences the rolling behavior of the niobate sheets. Removal of tetrabutylamine and K^+ ions from niobate sheets at very low colloidal concentrations gives rise to electrostatic repulsion forces between sheets and as a result, unrolled organization of sheets in the colloidal solution is observed. Formation of tubules takes place when the pH of colloidal solution is lowered.

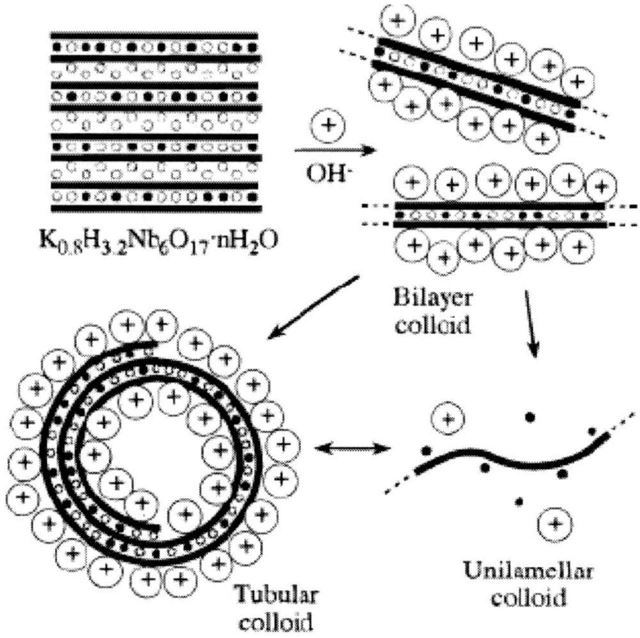

Figure 13. Progressive formation of $K_{0.8}H_{3.2}Nb_6O_{17}$ to bilayer sheets, single sheet tubules, and unrolled single sheet colloids. Dark lines represent the niobium oxide sheets, filled circles represent K^+ ions, and open circles represent H^+ ions. [Reprinted with permission from *Chem. Mater.* 2000, 12, 1556-1562. Copyright 2000 American Chemical Society.]

As a result of the increase in the H^+ concentration, the negatively charged surface of sheets is attracted to H^+ and K^+ more to roll around these ions. Formation of single sheets in the colloidal solution in either unrolled or rolled form was found to be irreversible and entropy driven [85]. These states of the single sheets were well defined by Atomic Force or Transmission Electron Microscopy. High Resolution microscopic analyses showed that sheets were rolled along [100] direction [88].

Although alkali organic molecules have been extensively used as an exfoliation agent, some alternative methods have also been used. Choy et al studied the exfoliation of Dion-Jacobson phase perovskite layered oxide, $HCa_2Nb_3O_{10}$, with an amino acid, 6-aminoundecanoic acid [91]. Amino acids are zwitterions having a negative charge at alkali and positive charge at acidic solutions. Intercalation of the amino acid into the layered oxide in an acidic solution gives a stable intercalation material. Deprotonation of interlayer amino acid groups by increasing pH disturbs the stability and finally negatively charged carboxyls on the amino acid repels negatively charged oxide layers in the bulk solid resulting in exfoliation into single sheets. Another alternative method is in-situ polymerization between the layers [92]. Tetramethylammonium intercalated $HTiNbO_5$, which is a perovskite layered oxide, was subjected to a polymerization of hexamethylcyclotrisiloxane in an aqueous solution. Emulsion polymerization of hexamethylcyclotrisiloxane in the interlayer separated the layers and as a result, exfoliation was achieved in the polymer matrix. In another approach, perovskite type layered $HLaNb_2O_7$ and $HCa_2Nb_3O_{10}$ were exfoliated during a sol-gel process [93]. n-decoxy intercalated layered oxides were treated with an aqueous mixture of poly(dimethylsiloxane) and tetramethoxysilane under hydrolysis with HCl. Development of polysiloxane network in and around layers during the hydrolysis reaction resulted in exfoliation of bulk solid.

5. Nanosize Sheets as Functional Macromolecules

The phenomena that layered oxides can be separable into their two dimensional smallest units, nanosize metal-oxide sheets, opened a wide point of view into the research and utilization of these units in various shapes and applications. It was revealed that these units have the functionality that bulk layered solids do not have. Properties are unique and different than those of bulk solid. After exfoliation, layered oxides behave in a distinct way when compared with the bulk form. Separating bulk solids into two dimensional nanosize sheets put many advantageous in utilizing layered oxides in various applications, such as thin film preparation, fabrication of new structures, producing organic-inorganic composite materials, and getting higher efficiency in catalytic and photocatalytic applications. As a result of quantum size effects, exfoliated nanosize sheets show typical physicochemical properties such as larger band gap energy [91,94,95]. Exfoliated titanium niobate, titanium tantalate and niobate layered oxides possess considerably high catalytic activity to the contrary to their activity in their bulk form [31-33]. In addition, exfoliation also helped in overcoming difficulties in synthetic methods, such as intercalation and ion-exchange. Taking advantage of nanosize of exfoliated sheets and the two dimensional feature, one can prepare very thin films in a controllable manner. Understanding the properties of these nanosize units leads into fabrication of more advanced materials with enhanced properties.

Intercalation into bulk layered oxides is a difficult task. In particular, it takes a long time and efficiency is generally low as discussed in Section 3. Since exfoliated layered oxide sheets are restacked and flocculate when the ionic strength of the colloidal solution is changed or pH is lowered, an intercalated form of layered oxide can be readily formed using the same approach by adding a cation solution into the colloidal solution. Electrostatic affinity of negatively charged sheets to positively charged molecules or cations is taken as an advantage in preparing different types of intercalated layered oxides in flocculated form. Exfoliated single sheets are restacked into their H^+ intercalated form when acid is added into the colloidal solution. Using the same phenomena, we were able to intercalate various types of complex molecules and cations into the interlayer of exfoliated $H_xTi_{2-x/4}\square_{x/4}O_4.H_2O$ and $H_4Nb_6O_{17}$ [52,53,96]. Addition of an aqueous solution of a complex molecule results in an immediate precipitation. The process is called as Electrostatic self-assembly deposition (ESD). The mechanism was pictured in Figure 14. Providing that pH values of both solutions are in a reasonable range, intercalation reaction occurs to intercalate at optimum concentration. pH value of exfoliation solution is at alkali range over 10 and that of an aqueous solution of a metal complex or cation is generally below 5. Titration curves of these solutions revealed that complex molecules will precipitate as hydroxides at higher pH values and exfoliated sheets will be restacked in the proton form of layered oxide at lower pH values. Thus, adjustment of pH to an intermediate value is necessary.

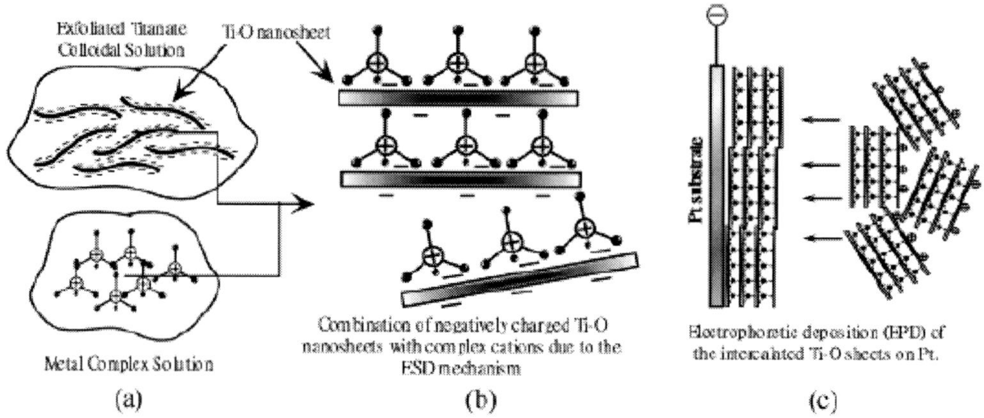

Figure 14. The pictorial view of electrostatic self-assembly deposition (ESD) and electrophoretic deposition (EPD) of intercalated layered oxides [Reproduced with permission from *J. Phys. Chem. B* 2003, 107, 12680-12689. Copyright 2003 Am. Chem. Soc.]

When both solutions are mixed, electrostatic forces govern the reaction and exfoliated sheets precipitate as a single phase intercalated structure. Electrostatic self assembly technique uses the advantage of negatively charged single nanosheets readily having high affinity to cationic molecules. In the colloidal solution of layered oxide sheets, the molecule used for exfoliation, such as tetrabutylamine or ethylamine is bound on the surface of sheets but the binding is loose and when sheets come in contact with a cationic molecule with higher affinity, these surface molecules are readily substituted and colloidal sheets are combined with cationic molecules to precipitate in an intercalated form. Precipitates are in the form of clusters. SEM micrographs of films deposited by electrophoretic deposition revealed that these clusters are formed from around 10 piled-up sheets with interlayer complex cations. XRD diffraction patterns of the films prepared from these clusters revealed highly crystalline structures oriented in (020) directions (Figure 15). Interlayer distance of each intercalated complex molecule was determined by the size of corresponding intercalated molecule, which shows that complex molecules are placed in the interlayer in a single layer arrangement form as given in Figure 16. Layer distances and atomic ratios together with the estimated compositions of each intercalation compound were given in Table 4.

Table 4. Layer distances and atomic ratios of the complex intercalated Ti-O films. The compositions were calculated from the atomic ratio (ICP and XPS) and water content (TG/DTA). [Reproduced with permission from *J. Phys. Chem. B* 2003, 107, 12680-12689. Copyright 2003 Am. Chem. Soc.]

Intercalated Metal complex	Layer Distance (Å)	M/Ti Atomic Ratio	Composition
$Co(NH_3)_6^{3+}$	11.41	0.12	$[Co(NH_3)_6^{3+}]_{0.22}Ti_{1.81}O_4{\cdot}2.4H_2O$
$Ru(NH_3)_6^{3+}$	11.47	0.22	$[Ru(NH_3)_6^{3+}]_{0.19}Ti_{1.81}O_4{\cdot}2.1H_2O$
$Ru(NH_3)_6^{2+}$	11.53	0.16	$[Ru(NH_3)_6^{2+}]_{0.13}Ti_{1.81}O_4{\cdot}4.2H_2O$
$Ni(NH_3)_6^{2+}$	11.41	0.46	$[Ni(NH_3)_6^{2+}]_{0.68}Ti_{1.81}O_4{\cdot}2.9H_2O$
$Ag(NH_3)_2^{+}$	8.99	0.4	$[Ag(NH_3)_2^{+}]_{0.32}Ti_{1.81}O_4{\cdot}0.7H_2O$
$Ru(bpy)_3^{2+}$	16.6	0.04	$[Ru(bpy)_3^{2+}]_{0.07}Ti_{1.81}O_4{\cdot}2.5H_2O$

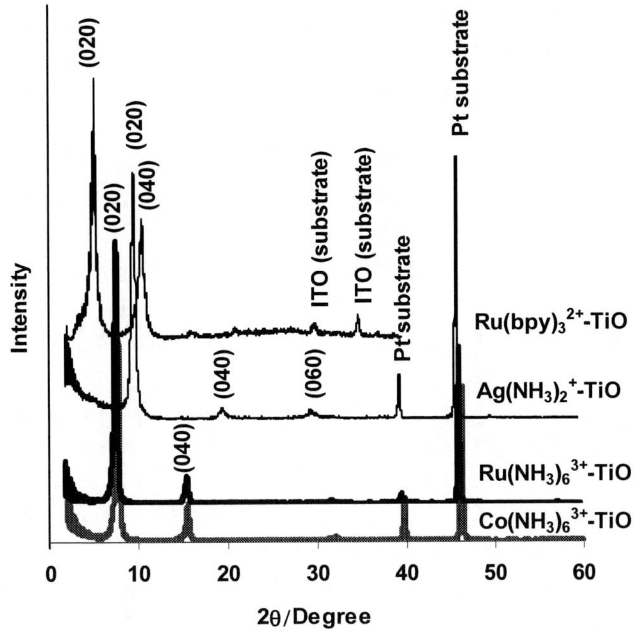

Figure 15. XRD patterns of $H_xTi_{2-x/4}\square_{x/4}O_4$ films intercalated with various complex molecules by ESD technique. [Reproduced with permission from *J. Phys. Chem. B* 2003, 107, 12680-12689. Copyright 2003 Am. Chem. Soc.]

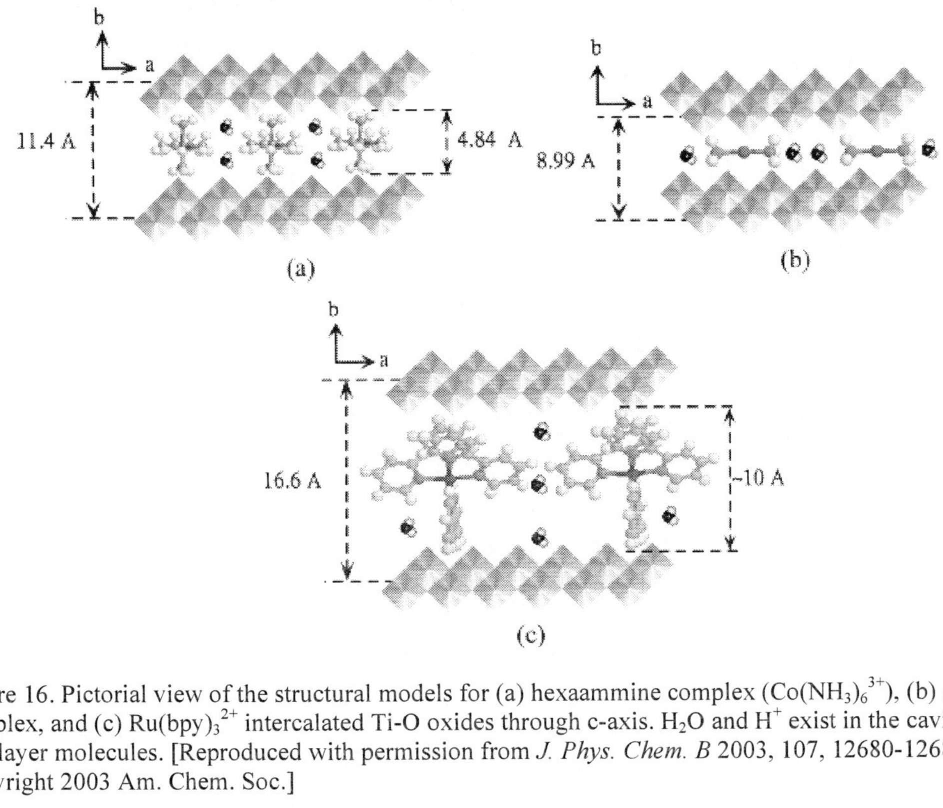

Figure 16. Pictorial view of the structural models for (a) hexaammine complex ($Co(NH_3)_6^{3+}$), (b) silver complex, and (c) $Ru(bpy)_3^{2+}$ intercalated Ti-O oxides through c-axis. H_2O and H^+ exist in the cavities of interlayer molecules. [Reproduced with permission from *J. Phys. Chem. B* 2003, 107, 12680-12689. Copyright 2003 Am. Chem. Soc.]

Charge equilibrium analysis of these samples showed that intercalation has occurred at optimum concentration levels. For $Cs_xTi_{2-x/4}\square_{x/4}O_4.H_2O$ (x=0.76), 0.76 mol of a monovalent cation in the interlayer is required to compensate the negative charge of the layer. Concentration of complex molecules in the interlayer reaches to such a value that maximum possible level of complex molecules in the interlayer is achieved and almost all sites are occupied by the molecules. In addition to complex molecules in the interlayer, co-intercalation of water and a small amount of H^+ occurs according to FTIR and thermal analyses and they will be located in the cavities around intercalated molecules. Intercalation of lanthanide cations such as Eu^{3+} was also achieved by the same method [96]. Concentration of interlayer lanthanide cations were similarly at the highest possible values according to the elemental analysis of samples. Interlayer distances and compositions for intercalated compounds were tabulated in Table 5. Resultant intercalation compounds were highly crystalline oriented in (020) direction as seen in Figure 17. Results revealed that lanthanide cations were intercalated in their aqua complex form, which resulted in very interesting photoluminescent properties [96]. According to the interlayer water content and interlayer distance of samples, lanthanide cations are coordinated with 7-10 water molecules. Another example is intercalation of $Ru(bpy)_3^{2+}$ and Methylene blue into the interlayer of exfoliated $H_4Nb_6O_{17}$. Experimental evidence is clear that intercalation reaction occurs into both interlayers immediately, which actually takes a long time for a bulk solid to intercalate [53].

Table 5. Interlayer distances and compositions of the intercalated layered oxide films prepared by ESD method.[Reproduced with permission from *J. Phys. Chem. B* 2005, 109, 12748-12754 Copyright 2005 Am. Chem. Soc.]

Intercalated Ln ion	Interlayer Distance (Å)	Composition
Y^{3+}	6.43	$Y_{0.21}Ti_{1.81}O_4.2.1H_2O$
Ce^{3+}	6.35	$Ce_{0.21}Ti_{1.81}O_4.1.8H_2O$
Sm^{3+}	6.25	$Sm_{0.31}Ti_{1.81}O_4.2.2H_2O$
Eu^{3+}	6.88	$Eu_{0.31}Ti_{1.81}O_4.2.1H_2O$
Gd^{3+}	6.49	$Gd_{0.27}Ti_{1.81}O_4.2.5H_2O$
Tb^{3+}	6.04	$Tb_{0.25}Ti_{1.81}O_4.2.0H_2O$

In earlier studies, thin films of layered oxides were prepared by spin coating [18,38,39,97] or direct application of powder suspensions onto a substrate. Thin film of exfoliated $H_xTi_{2-x/4}\square_{x/4}O_4$, which was deposited from a colloidal solution by spin coating was reported to have a thickness of around 100 nm [40] and may go as thicker as 4000-5000 nm [97] if the film is deposited from more concentrated exfoliation solutions. These films have quite a rough surface and film thickness is not completely controllable. Exfoliated sheets are deposited parallel to the surface of a substrate with intercalated water and alkali organic molecules used for exfoliation such as ethylamine or tetrabutylamine. Although spin coating might produce well oriented films in (0k0) direction, good quality in crystallinity of films cannot always be ensured. In addition, the stability of the film on substrate is questionable as there is a very weak mechanical adhesion between the film and substrate.

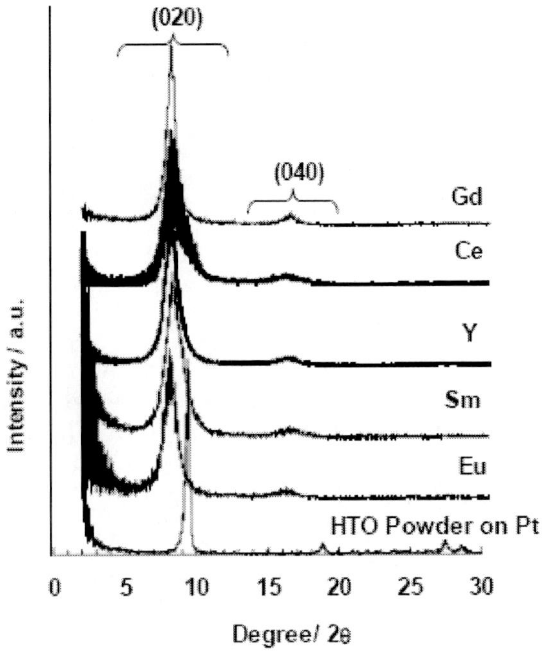

Figure 17. XRD patterns of the deposited films from the $H_xTi_{2-x/4}\square_{x/4}O_4$ starting material (HTO), its ethylamine (EA) exfoliation form, and layered oxides intercalated with different lanthanide(III) cations by the ESD method, respectively from bottom to top. [Reproduced with permission from *J. Phys. Chem. B* 2005, 109, 12748-12754 Copyright 2005 Am. Chem. Soc.]

Langmuir-Blodgett method provides a good technique to fabricate very thin layer of films from exfoliated sheets of layered oxides [98,99] but it requires special equipments and has very limited applications. Monolayers of exfoliated sheets are dispersed uniformly on water and transformed on a substrate by an automated system. There are several parameters in order to obtain good quality films. Substrate shape is an important issue as well as the surface structure. The reaction involves covalent and coordination chemistry, which can only be provided by a certain number of chemicals.

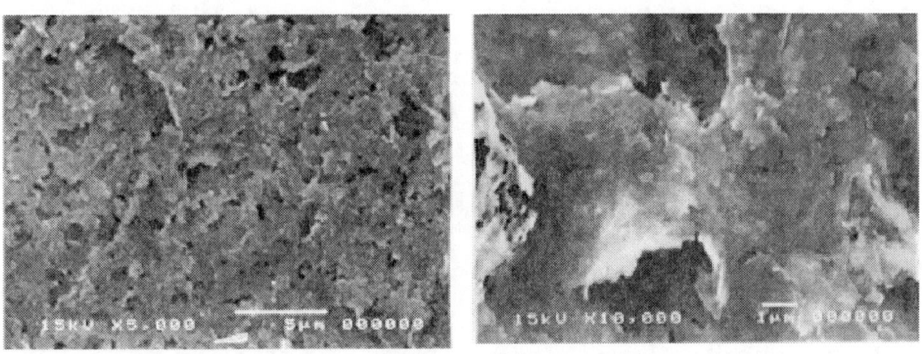

Figure 18. SEM microphotographs of (a) $Ru(NH_3)_6^{3+}$ and (b) $Ni(NH_3)_6^{2+}$ intercalated $H_xTi_{2-x/4}\square_{x/4}O_4$ films prepared by EPD technique. Intercalation of the complexes was carried out with ESD method.

We have obtained well oriented, highly crystalline films of exfoliated $H_xTi_{2-x/4}\square_{x/4}O_4.H_2O$ on platinum substrates by Electrophoretic Deposition Method (EPD) (Figure 14c). Films had good stability and mosaic like surfaces as seen in Figure 18. Applied voltage and deposition time control the thickness of the films, but the thickness is still in micrometer level [66]. Electrophoretic deposition occurred on the cathodic electrode, which shows that surface of these clusters were positively charged. Films were highly crystalline and well oriented according to XRD patterns of the films (Figure 15). As seen in the SEM photographs of films in Figure 18, each cluster can be easily distinguished and their size is virtually uniform, around 1μm in width and about 0.1 μm in thickness, which corresponds to approximately 10 piled-up sheets as depicted in Figure 14c.

A way to prepare thin films of a layered oxide in a controllable manner is depositing exfoliated single sheets on a substrate in a sequential process. The method is initially developed for the fabrication of multilayer polymer films [100,101]. Adsorption of oppositely charged polymers on a substrate is carried out alternatively to obtain ultrathin multilayer films. Then, the method was implemented to the preparation of ultrathin films of layered oxides. The sequential Layer-by-Layer assembly of single sheets of layered oxides allows the fabrication of nanostructured ultrathin films with a desired number of layers and it is one of the simplest and most effective methods for the preparation of thin films of charged layered oxides. There are theoretically no limitations on the shape, dimensions and surface morphology of substrate as long as it provides sufficient space and structure for colloidal sheets to be adsorbed on. Two dimensional nanosize sheets of layered oxides as building blocks are assembled on a substrate by using various polycations or any cationic molecule as a counter ion (Figure 19). A wide range of layered oxides has been studied for the preparation of thin films with Layer-by-Layer assembly method [34,102-115]. One of the first examples of Layer-by-Layer Assembly method for layered oxides is adsorption of zirconium phosphate on several substrates by using poly(allylamine hydrochloride) as a counter ion [86,102,103]. The idea of adsorbing one layer over another by Layer-by-Layer assembly method puts a new trend in the fabrication of intercalated layered oxides. The method produces layered oxides intercalated with desired molecules at optimal concentrations.

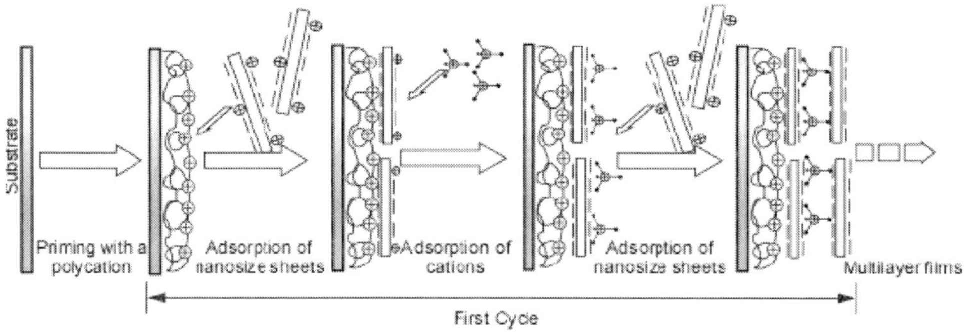

Figure 19. Reaction sequence of Layer-by-Layer film preparation method.

The process was pictured in Figure 19 and involves modification of the substrate surface with a polycation to make the surface positively charged. Then, negatively charged nanosize sheets are adsorbed on the substrate from a colloidal solution according to the electrostatic principles. In the following step, any molecule with a positive charge can be adsorbed on the deposited sheets. Following the sequence, an ultrathin film can be fabricated by adsorbing any number of layers alternatively one over another. Layer-by-layer assembly method is carried out in an aqueous environment to whole extent. Hence, it is environmentally friendly process. In addition, it does not require any complicated, technologically advanced system as it can simply be carried out in an ordinary laboratory beaker. One disadvantage of the method might be in terms of time as it requires tens of minutes for whole coverage of surface with a monolayer. Surface tetrabutylamine or ethylamine molecules on the surface of exfoliated sheets can be readily replaced by counterions in an aqueous solution during Layer-by-Layer assembly because of their loosely bound nature. Adsorption occurs only in a monolayer form because the reaction stops at a point where all surface charge is satisfied with adsorbed counterpart. Condition of a substrate surface is a very important factor to obtain a smooth film and homogenous adsorption of oxide sheets. Covalently bonded charged molecules on a substrate should provide a homogenous layer with a sufficient charge density in order to deposit well ordered layer of nanosize sheets on the substrate. pH of the solution of polycation or colloidal sheets is found to be another important factor to obtain a quality thin film [106,107]. At the pH of exfoliation solution, which is over 10 for most of the layered oxides, exfoliated sheets are well dispersed in the colloidal solution but when the polycation adsorbed on the film is subjected to a colloidal solution with this pH value, polycation loses most of its protons. In this case, the attraction force between the polycation and colloidal sheets is weak and coverage is only limited. When the pH is lowered, sufficient amount of positive charge on the surface provides enough attraction force to adsorb colloidal sheets on any position on the film, yet sheets might be adsorbed in the form of clusters on the film instead of monolayer as a result of the excessive amount of protons in the solution at very low pH values. Flocculation occurs on the surface of the substrate. Therefore, solutions are adjusted to an intermediate value, around 8-9, to provide optimum conditions for reaction to take place.

Fabrication of multilayer ultrathin films of layered oxides provides new materials with distinctive optical, photocatalytic, electrochemical properties. Properties of these thin films can be tuned by incorporating any positively charged species between nanosize sheets by Layer-by-Layer Assembly. Alternatively, thermal processing, condensation or topochemical methods may lead into the formation of new nanostructured films. Fabrication of thin films with Layer-by-Layer Assembly is mostly carried out by utilizing a polymer, such as Polydiallyldimethylammonium or Poly(allylamine hydrochloride) but we have examined and achieved to prepare thin films of various titanate, niobate and manganese layered oxides intercalated with dye molecules, silver species and lanthanide cations [96] by layer-by-layer self assembly method and these films gave distinctive electrochemical and photoluminescence results. The films were deposited on a Pt substrate modified with Polyethylenimine. For example, $Ag(NH_3)_2^+$ was intercalated into several layered oxides by Layer-by-Layer Assembly method. As given in Figure 20, adsorption of layers was monitored by UV-vis spectra and XRD patterns. Increase in the absorbance with the number of layers clearly shows that preparation of the film in Layer-by-Layer Assembly manner is successful. In addition, corresponding peak on XRD patterns of films gets sharper at higher number of layers, which shows that better crystal order and reproducibility of layered structure is obtained at adsorption of more layers. Lanthanide cation intercalated films were also attained by Layer-by-Layer assembly methods and these films showed higher photoluminescence at higher number of layers.

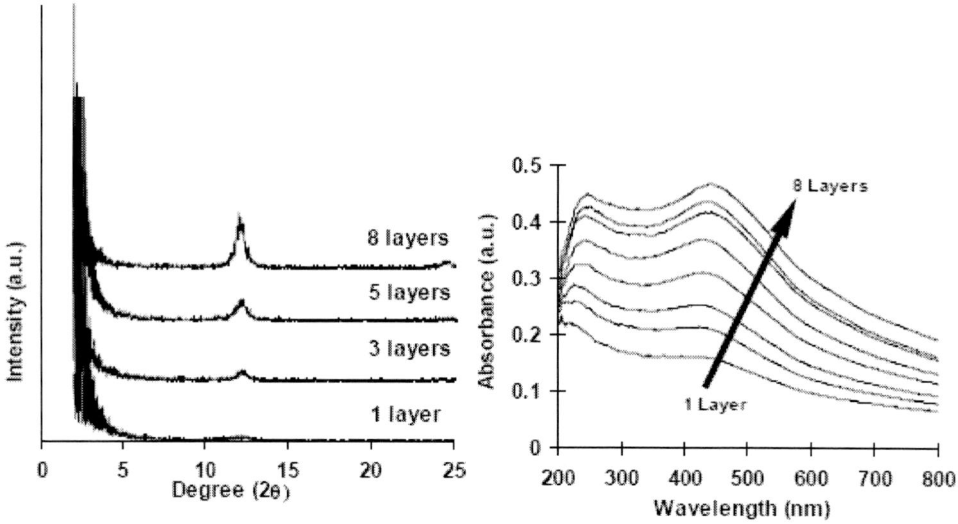

Figure 20. XRD pattern and UV-vis spectra of MnO_2 thin film intercalated with $Ag(NH_3)_2^+$ by sequential adsorption of each layer.

6. Electrochemistry of Semiconductive Layered Oxides

In the electrochemical applications of layered oxides, electrochemical Li intercalation in Co or Mn layered oxides as a positive electrode is most widely known application [116]. High electrical conductivity of the host layer and high ionic conductivity of Li in the interlayer are key properties of these kind of materials making them very good candidate in battery applications. Along with Li intercalated Mn or Co layered oxides, there are several other layered oxides examined for their electrochemical properties. Semiconductive layered oxides consisting of Ti, Ta, or Nb, however, have not received much attention in the field of electrochemistry. They have been used widespread in photoelectrochemical applications but the electrochemical mechanisms particularly in an aqueous solution are not well known. Revealing electrochemical behavior of layered oxides will ease the design and creation of new systems for advanced applications.

One of the main interests of our group is the electrochemical properties of layered oxides consisting of Ti, Nb or Ta. Our report on the electrochemical properties of the thin films of $Cs_xTi_{(2-x/4)}\square_{x/4}O_4$ has revealed various interesting properties in aqueous solutions [66]. Cyclic voltammograms are given in Figure 21. These films produced a reduction peak around 0.2-0.3 V and its peak potential found to be pH dependent. The pH dependency of the peak shows that interlayer proton level highly affects the electrochemical behavior of films. Diffusion and intercalation of electrolyte species, such as H^+, K^+ [66] or Li^+ [117] directly affect the nature of electrochemical properties of films. Electrochemical intercalation/deintercalation of proton or other species during electrochemical analysis in an electrolyte is a typical behavior of layered oxides. X-ray photoelectron spectroscopy shows evidence for the intercalation of electrolyte species under potential difference. Intercalation takes place without destroying the layered structure of film and controls the electrochemical reactions taking in or between layers. Accordingly, we have suggested the following reaction resulting in the reduction peak appearing in the range of 0.2-0.3 V:

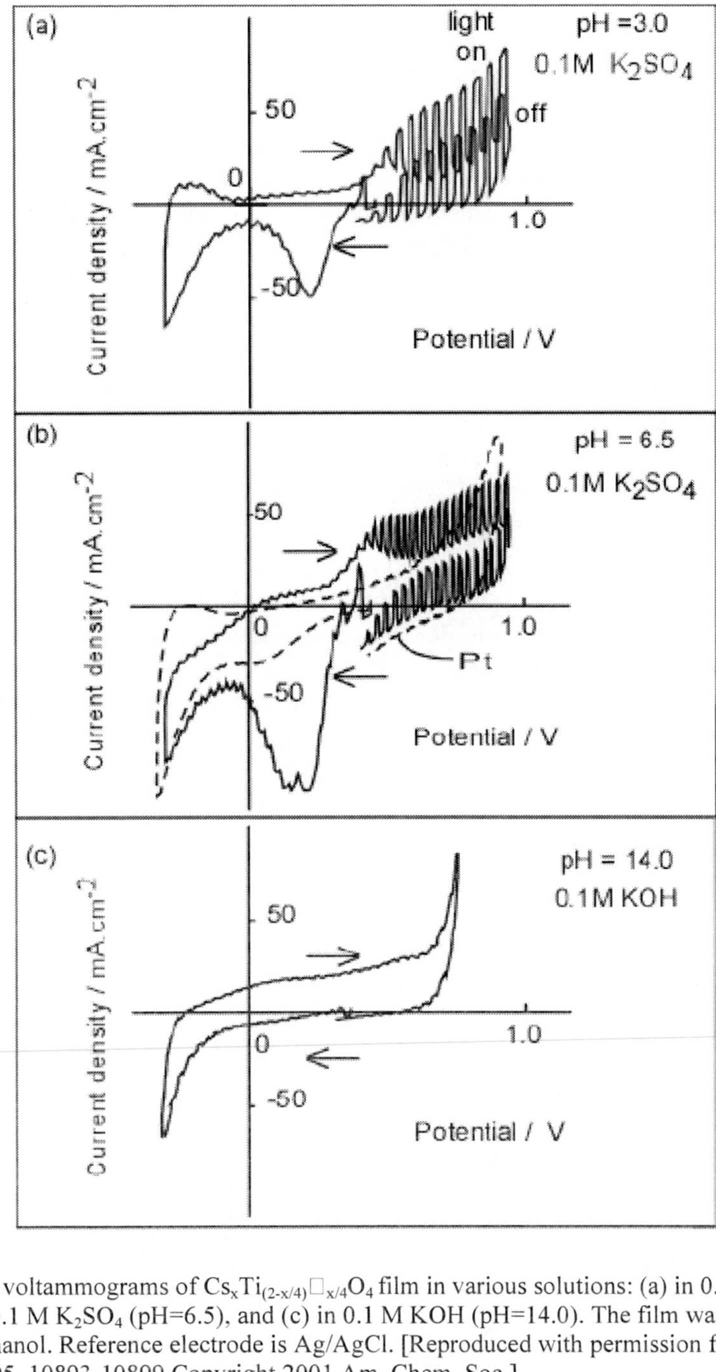

Figure 21. Cyclic voltammograms of $Cs_xTi_{(2-x/4)}\square_{x/4}O_4$ film in various solutions: (a) in 0.1 M K_2SO_4 (pH=3.0), (b) in 0.1 M K_2SO_4 (pH=6.5), and (c) in 0.1 M KOH (pH=14.0). The film was prepared at 30 V for 5 min in ethanol. Reference electrode is Ag/AgCl. [Reproduced with permission from *J. Phys. Chem. B* 2001, 105, 10893-10899 Copyright 2001 Am. Chem. Soc.].

$$xH_3O + K_{0.36}Ti_{1.91}O_4.nH_2O + xe^- \rightarrow H_xK_{0.36}Ti_{1.91}O_4.(n+x)H_2O \text{ (reduction current at low pH)}$$

According to this reaction, intercalation takes places after the reduction of interlayer H_3O^+. Repeating electrochemical experiments in different aqueous electrolytes such as SO_4^- and OH^- salts of Li^+ or Na^+ did not make any alteration in the cyclic behavior and similar intercalation/deintercalation sequence was observed in these electrolytes. Sakai et al carried out an electrochemical analysis of thin films of the same layered oxide in a nonaqueous, $LiClO_4$ containing propylene carbonate solution [117]. Films were prepared by layer-by-layer assembly method from exfoliated single sheets. Comparing to our films, these films were thinner and better controlled in terms of monolayer arrangement. Cyclic voltammograms of these films exhibited reduction and oxidation peaks at -1.2 to -0.9 range, which were assigned to redox reaction involving oxide sheets associated with intercalation/deintercalation reaction of Li^+ in the interlayer. Diffusion of Li^+ was found to be the rate determining step in the redox reaction of Ti^{4+}/Ti^{3+} redox couple. Electrochemical measurements are a powerful step in determining the electrical properties of thin films. Flat band potential, band gap energy or energy levels of conduction and valence band can be resolved by electrochemical experiments in conjunction with optical measurements. Cyclic voltammograms under illumination produced anodic photocurrents on $Cs_xTi_{(2-x/4)}\square_{x/4}O_4$ films. Similarly, we have observed photocurrents from $K_4Nb_6O_{17}$ films deposited by electrophoretic deposition. Band gaps were estimated around 3.5 eV for the layered oxide films, which is in agreement with the band gap estimated from the spectrum of photocurrent at 0.5 V for $H_xTi_{(2-x/4)}\square_{x/4}O_4$ as seen in Figure 22. Observed photocurrent was pH dependent and decreased with increasing alkalinity of electrolyte solution. The pH dependency shows that anodic photocurrent is based on the intercalation of H^+ and H_3O^+ into the interlayer and was assigned to the oxidation of interlayer H_2O to O_2 photochemically. Thus, following equation was proposed for the reaction taking place in the interlayer of the film:

$$H_xK_{0.36}Ti_{1.91}O_4.(n+x)H_2O + 2yh^+ \rightarrow H_xK_{0.36}Ti_{1.91}O_4.(n+x-y)H_2O+2yH^+ +(y/2)O_2 \text{ (photocurrent)}$$

This reaction clearly indicates that interlayer space is photochemically active and these results show that films act as an n-type semiconductor [66,118]. Flat band potential of $H_xTi_{(2-x/4)}\square_{x/4}O_4$ film was estimated to be -0.4 V and pH dependence of the flat band potential was about 60 mV/pH. In addition, the conduction band edge of the film is found to be near the hydrogen equilibrium potential (-0.61 V vs Ag/AgCl at pH=6.5) [66]. The flat band potential of $K_4Nb_6O_{17}$ film was estimated as -0.6 V [118]. On the other hand, $H_xTi_{(2-x/4)}\square_{x/4}O_4$ shows different characteristics when exfoliated to its nanosize layers according to the results reported by Sakai et [117]. Photoelectrochemical experiments revealed that a single titanate layer has a flat band potential of -1.3 V and band gap energy of about 3.8 eV. These results were obtained from electrochemical analyses carried out in a nonaqueous solution. Same group has also reported that a monolayer film of manganese oxide deposited from its exfoliated form has band gap energy of 2.23 eV [119]. Different than other semiconductive layered oxides in which a band gap is formed between O2p and M3d (M= transition metal) orbitals, manganese oxide has its band gap energy formed between t_{2g} and e_g states of Mn3d orbital, which serve as valence and conduction bands, respectively.

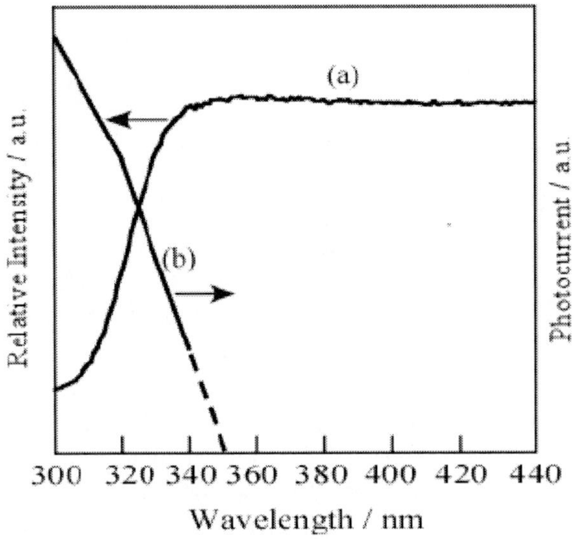

Figure 22. Spectra of (a) absorption and (b) photocurrent at 0.5 V (vs. Ag/AgCl) in 0.1 M K_2SO_4 for $H_xTi_{(2-x/4)}\square_{x/4}O_4$ film. The film was prepared at 30 V for 5 min in ethanol. [Reproduced with permission from *J. Phys. Chem. B* 2001, 105, 10893-10899 Copyright 2001 Am. Chem. Soc.].

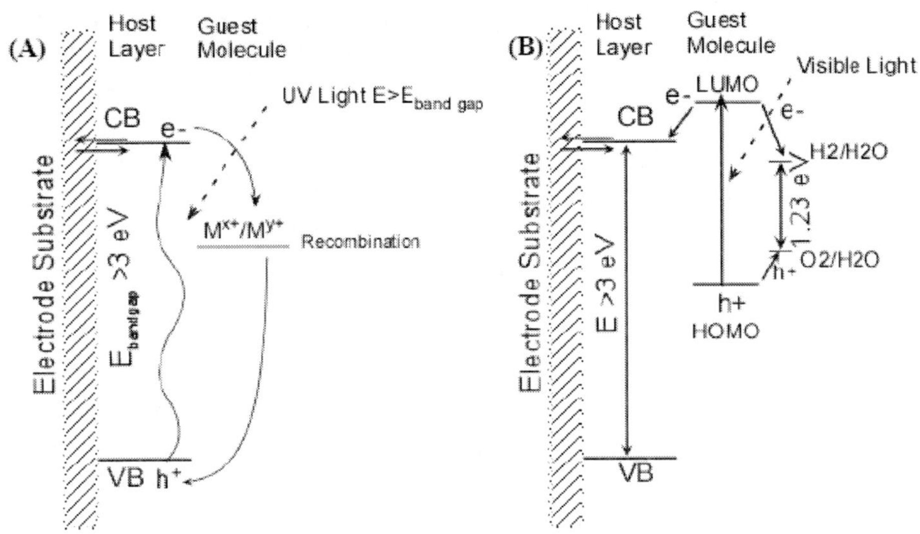

Figure 23. Possible electron transfer mechanisms between the host layer and guest molecules. (A) Guest molecules can act as a recombination center for electron hole pair produced under UV illumination. (B) Photoactive molecules provide electron-hole pair to host layer to take place in photochemical reactions.

Photoelectrochemical experiments were also carried out for $H_xTi_{(2-x/4)}\square_{x/4}O_4$ films intercalated with various transition metal cations such as Fe^{3+}, Cu^{2+}, Co^{2+} and Ni^{2+} and these films produced weaker photocurrents. Redox potential levels of these metal transition metals fall into the band gap of $H_xTi_{(2-x/4)}\square_{x/4}O_4$ film and electron-hole recombination takes place at these redox levels in the band gap [66]. Similar results have been found for $H_xTi_{(2-x/4)}\square_{x/4}O_4$ films intercalated with various metal complexes [52]. These films were prepared from electrophoretic deposition of precipitates obtained from Electrostatic Self-assembly Deposition or layer-by-layer assembly method. Amine complexes of Co^{3+}, Cu^{2+}, Ni^{2+} or Ru^{3+} did not give any electrochemical reaction or photocurrent on the cyclic voltammograms. According to the proposed mechanism, these molecules acted as a recombination center for electron-hole couple produced in the layer under illumination as the redox energy levels of these molecules lies in the band gap of the film (see Figure 23 A). In addition, amine ligands suppress the electron transfer between interlayer complex molecules and host layer.

We have also investigated several systems comprising various photoactive elements with layered oxides such as $Cs_xTi_{(2-x/4)}\square_{x/4}O_4$ or $K_4Nb_6O_{17}$ and suggested the mechanism of energy transfer between interlayer photoactive molecules and host layer as given in Figure 23. Some molecules such as $Ru(bpy)_3^{+}$ and Methylene Blue produced quite a large amount of photocurrent in the interlayer. These molecules can absorb light in the visible light range and yield electrons at their excited state. Visible light photoresponse is a crucial element for efficient photoactivity of layered oxides under readily available daylight and this can only be obtained by modifying interlayer with photoactive molecules since semiconductive layered oxides can absorb in the UV region only. Photoactive molecules can be excited to their Metal to Ligand Charge Transfer (MLCT) state under visible light illumination and electrons excited to the Lowest Unoccupied Molecular Orbital (LUMO) from Highest Occupied Molecular Orbital (HOMO) drifts into the conduction band of the oxide layer, providing that the LUMO level of the interlayer molecule is at more negative potentials than the conduction band energy level of the oxide layer. The mechanism was demonstrated in Figure 23 B. $Ru(bpy)_3^{2+}$ intercalated $H_xTi_{(2-x/4)}\square_{x/4}O_4$ was photochemically active in a visible light range (Figure 24b). However, it was necessary to apply heat treatment in order to get visible light response from $Ru(bpy)_3^{2+}$ intercalated layered oxides. Otherwise, redox reaction of $Ru(bpy)_3^{2+}$ and deintercalation was observed as a result of the instability of the molecules in the interlayer. Instability is caused by water molecules surrounding the complex, which brings about weak interaction between the host and the guest. As seen in figure 24a, as-prepared film produced a large anodic photocurrent after the oxidation to Ru(III) state. However, only very low photocurrents was observed in the region of Ru(II) as Ru(II) species in the interlayer act as recombination centers. Observed photocurrent is due to holes produced in the host layer under UV illumination. Back reaction of electrons, as shown in Figure 23A, leading to recombination of electrons and holes lowers the efficiency of the system.

In most systems, some sacrificial agents such as methanol are required for efficient evaluation of H_2 from water after a photocatalytic reaction [19,53,87,120-123]. Its preferential oxidation by capturing holes produced under illumination hinders the electron-hole recombination in the system and thus promotes photocatalytic H_2 evolution. Intercalation mechanism of alcohols and their role as sacrificial agents in the promotion of H_2 evolution are crucial factors in determining the photocatalytic behavior of layered oxides. Our photoelectrochemical approach to resolve the mechanism for H_2 evolution from layered oxides in the presence of various alcohols revealed that layered oxides produce the highest anodic photocurrent in the presence of methanol among other tested alcohols [124].

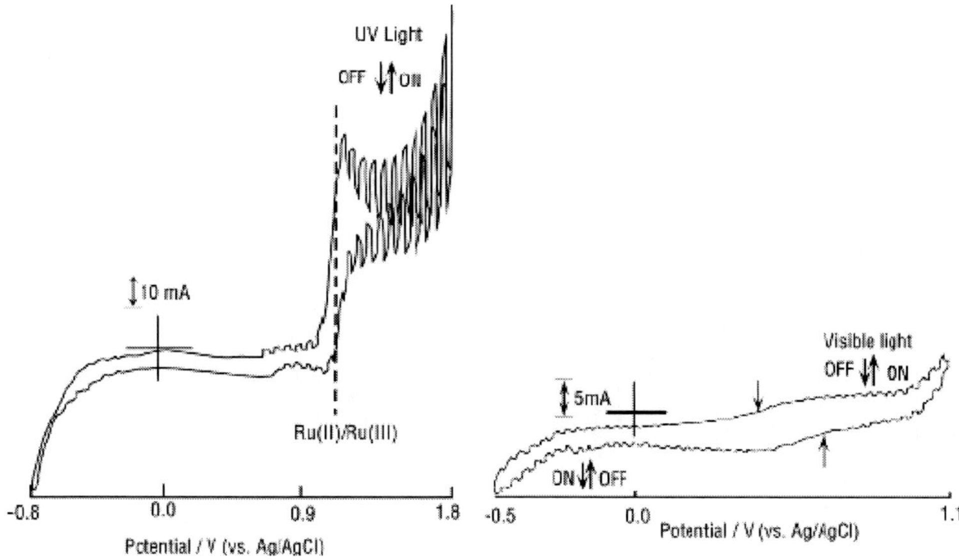

(a) (b)

Figure 24. Cyclic voltammograms of the $Ru(bpy)_3^{2+}$ intercalated $H_xTi_{(2-x/4)}\square_{x/4}O_4$ film with ESD technique: (a) as-deposited; (b) heat-treated at 300 °C. The arrows in figure b denote the changing potentials from anodic to cathodic photocurrents. Cyclic voltammograms were carried out in 0.1 M K_2SO_4 solution.[Reproduced with permission from *J. Phys. Chem. B* 2003, 107, 12680-12689. Copyright 2003 Am. Chem. Soc.].

Cyclic voltammograms of the films prepared by Layer-by-Layer assembly method showed a dramatic increase in the anodic photocurrent in the presence of methanol (Figure 25A). Among tested layered oxides, niobate films ($H_4Nb_6O_{17}$ or $HCa_2Nb_3O_{10}$) produced the highest photocurrent as seen in Figure 25B. The result is thoroughly related with the phenomenon that conduction band edge of niobates lies at more negative potentials than that of titanates, which also causes appearance of photocurrent onset potentials at more negative values for niobates films (Figure 26). Band structures of the photocatalysts in Figure 26 were estimated according to the onset potentials of photocurrents obtained from Figure 25B. Conduction band levels were approximated to flat band potentials of layered oxides because of their n-type semiconductor characteristics. Valence band of semiconductors were assumed to be at the similar energy level because valence bands are formed of O2p orbitals and ionic radius of oxygen atom coordinated to Ti^{4+} and Nb^{5+} is taken similar. Consequently, higher conduction band levels of niobates offer more favorable conditions for water reduction to H_2. In addition, higher photocurrent is related to higher photooxidation rate of methanol by hole in preference to the surface combination rate between electron and hole at the surface [124].

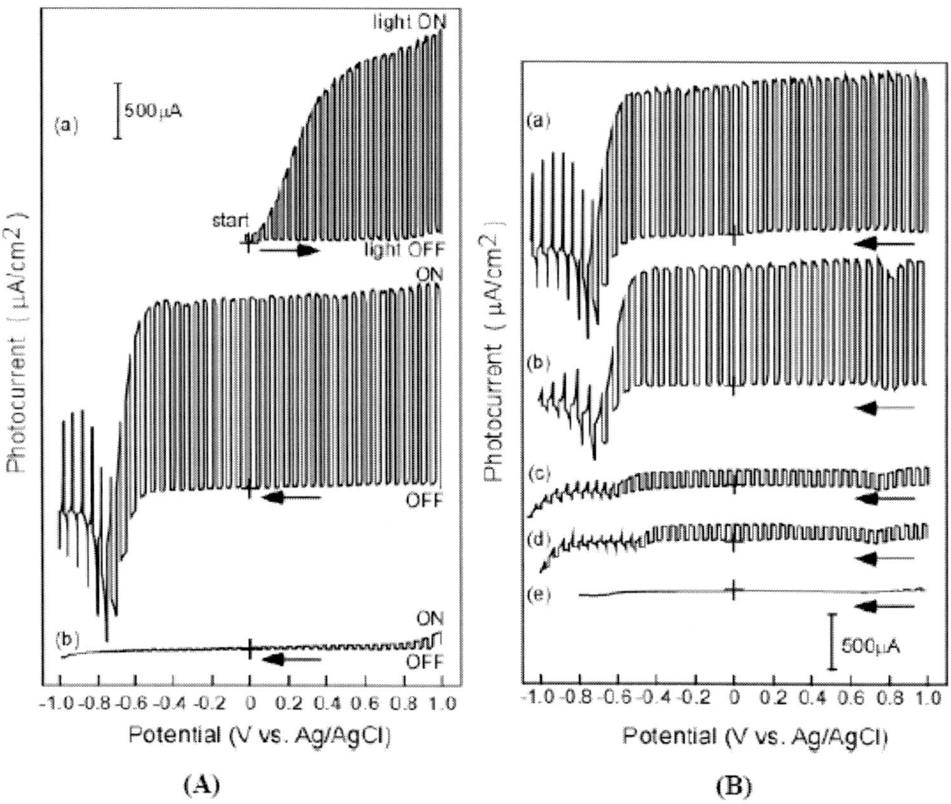

Figure 25. (A) Voltammograms of $[Nb_6O_{17}]^{4-}$ electrode (n=1) in the first sweep (a) in 0.1M CH_3OH + 0.5M K_2SO_4; (b) in 0.5M K_2SO_4. (B) Voltammograms of the LBL electrodes (n=1) in 0.1M CH_3OH + 0.5M K_2SO_4; (a) $[Nb_6O_{17}]^{4-}$ (b) $[Ca_2Nb_3O_{10}]^{-}$ (c) $[Ti_4O_9]^{2-}$ (d) $[Ti_{(2-x/4)}\square_{x/4}O_4]^{-x}$; (e) $[MnO_2]^{-0.45}$ [Reproduced with permission from *J. Phys. Chem. B* 2006, 110, 4645-4650. Copyright 2006 Am. Chem. Soc.].

Layer-by-Layer assembly method for film preparation is an efficient method to test the effect of the number of layers on the photoelectrochemical properties of films. Interestingly, a number of layers hardly influenced the intensity of photocurrent and same amount of photocurrent was obtained even with a monolayer adsorbed on a substrate [124]. We know that alcohol molecules promote the photocatalytic activity. Thus, adsorption of alcohol molecules on the surface of film should be the controlling mechanism in the photocatalytic activity rather than the intercalative feature of layered oxide films. Furthermore, hydrophilic property of the surface controls the adsorption of alcohols onto the surface of an oxide sheet. It was also demonstrated that alcohol molecules with lower molecular weight gave higher photocurrent. Sakai et al reported that increase in diffusion length, that is a higher number of layers, causes higher recombination rates of electron hole pair [125]. Moreover, hydrophilicity is higher at lower number of layers on the surface of the film and is promoted by band gap excitation of semiconductor sheets. Consequently, adsorption of lower molecular weight alcohol molecules on the surface of a monolayer, which is photochemically driven by hydrophilicity under illumination, creates the most appropriate conditions for the highest activity.

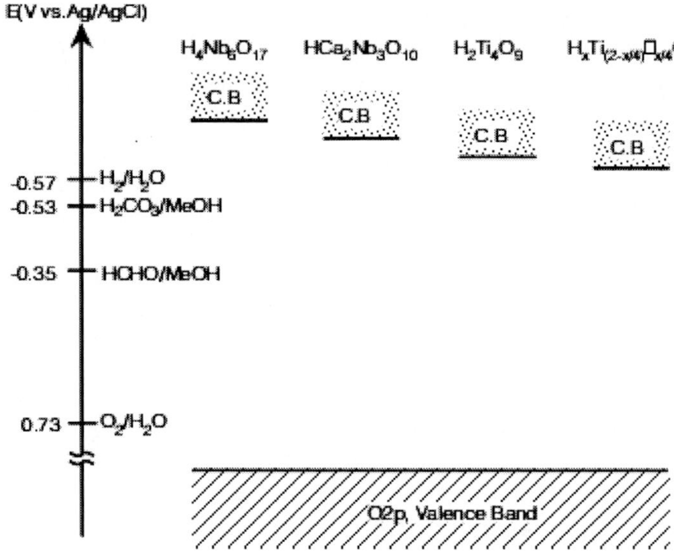

Figure 26. Band structures of semiconductor photocatalyst and energy levels of electrolyte species. Conduction level bands were approximated to the flat band potential as these photocatalysts act as an n-type semiconductor. Redox potentials are at pH=6.0 vs Ag/AgCl standard electrode.

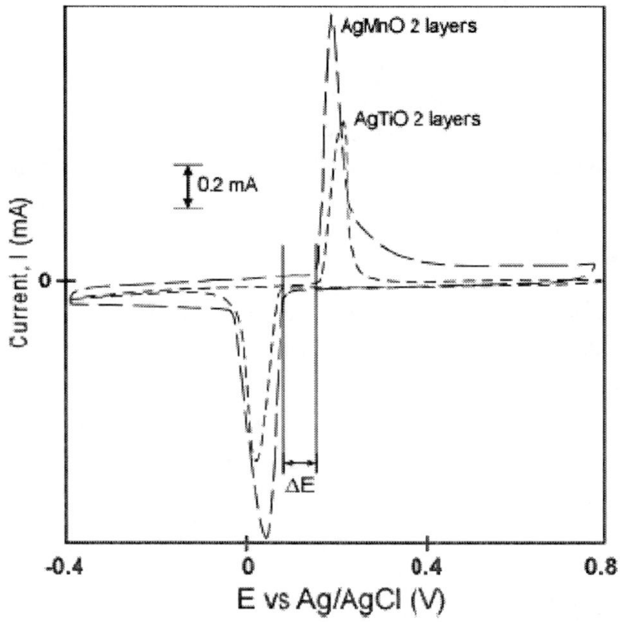

Figure 27. Cyclic Voltammograms of Ag^+ intercalated $H_xTi_{(2-x/4)}\square_{x/4}O_4$ and $H_{0.45}MnO_2$ layered oxides. Films were prepared by LBL method and consist of two intercalation layers. Cyclic Voltammograms were performed in 0.1 M K_2SO_4, pH=6.

Layered oxides were also tested for their potential capacity as a host to electrochemically active species. It has been already revealed that the interlayer is electrochemically active in terms of electron transfer and intercalative feature [52,53,66,118]. In this respect, we have investigated possible electrochemical reactions of various species in the interlayer. Most of the metal complexes did not give any electrochemical response in the interlayer. Among the tested complexes are amine complexes of transition metal cations. NH_3 ligands suppress the electron transfer between host layer and transition metal in these systems. On the other hand, $Ru(bpy)_3^{+}$ shows redox reaction of Ru^{2+}/Ru^{3+} couple in the interlayer of $Cs_xTi_{(2-x/4)}\square_{x/4}O_4$ film. [52]. Redox reactions are accompanied with the intercalation/deintercalation of electrolyte species to keep the charge balance in the interlayer.

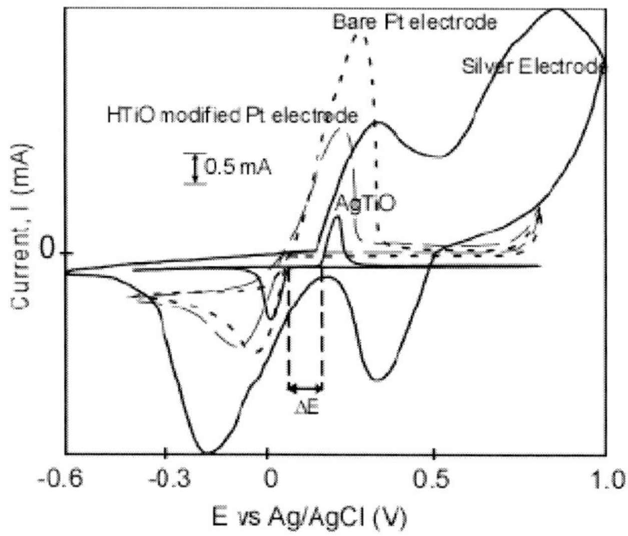

Figure 28. Comparison of the cyclic voltammograms of various modified electrodes. Electrolyte for Bare and $H_xTi_{(2-x/4)}\square_{x/4}O_4$ modified Pt electrodes: 0.01 M $Ag(NH_3)_2$ + 0.1 M K_2SO_4, pH=12. Electrolyte for Silver and Ag^+ intercalated $H_xTi_{(2-x/4)}\square_{x/4}O_4$ modified electrodes: 0.1 M K_2SO_4, pH=6.0. Electrode area: 1 cm^2, Scan rate: 20 mV/s.

One of the most distinct examples to the interlayer electrochemistry of layered oxide is very unique electrochemical reaction of silver species intercalated into host layers. Layered oxides acted as a very good host for silver species and supplied an outstanding environment for a unique electrochemical behavior. Electrochemical behavior of silver species has been studied in various host materials but the reaction taking place in those host layers is generally complex and hard to be resolved. For example, zeolites have been studied as a host to silver species and silver species showed complex and irreversible chemical reactions in zeolitic channels depending on the ion exchange site and site-to-site conversion [126-128]. Therefore, cyclic voltammograms of silver exchanged zeolites or layered materials give several peaks on their cyclic voltammograms. Tested layered oxides in our study, however, gave clearly resolvable redox peaks with high current as seen in Figure 27. Redox peak appearing in the range of 0.05-0.2 V vs Ag/AgCl reference electrode was assigned to Ag^+/Ag redox couple. The most distinctive feature of the redox behavior is the potential difference of around 0.1 V between onset potentials of anodic and cathodic currents. Typical cyclic voltammograms of silver electrode or layered oxide modified electrode in Ag^+ containing electrolyte did not show such characteristic as seen in Figure 28. The difference was assigned to the existence of an energy barrier in host layer. Another feature of the film is that higher current was obtained from the films with a higher number of layers and thus a larger amount of Ag^+ species in the interlayer[1]. As seen in Figure 29, electric charge, which is calculated from the area under redox peaks on cyclic voltammograms, and the amount of silver in the interlayer linearly increase with the number of layers. Among tested layered oxides, manganese oxide films intercalated with Ag^+ gave the highest current, where we obtained the lowest current from niobate films. Scan rate and pH controlled experiments revealed that the reaction taking place in the interlayer is a diffusion controlled process. Reduction and oxidation peak currents increased linearly with increasing scan rate. In addition, dependence of peak currents on pH and the concentration of electrolyte species like K^+ bring us to a conclusion that electrochemical reaction in the interlayer is controlled by the diffusion of H^+ or K^+ (for K_2SO_4 electrolyte) into the interlayer.

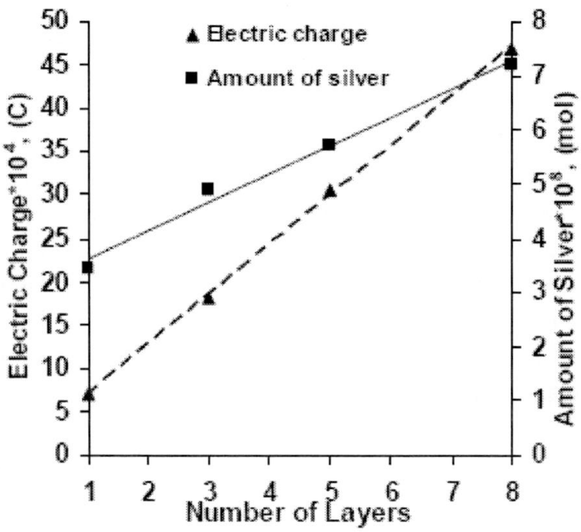

Figure 29. Electric charge (C) and the amount of Ag (mol) in the interlayer as a function of the number of layers deposited on a substrate.

7. Optical Properties

Semiconductive nature of nanosize sheets composing layered oxides offers spectacular physical and chemical properties and these properties make them available in various practical applications, such as photochemical splitting of water, photocatalytic reactions, and photoluminescence. These applications depend particularly on the optical properties of the layered oxides. Thus, understanding and evaluating these properties help synthesis and development of more advanced materials suitable for certain applications with higher compatibility and yield. Intercalative route puts a new dimension in modifying layered oxides for the sake of a desired application. Readily semiconductive host oxide layers can absorb light in UV light range and moreover, the absorption range can be extended to visible light by

[1] Films were prepared by Layer-by-Layer assembly method by adsorbing single sheets from an exfoliation solution and Ag^+ species from an aqueous solution alternatively on a Pt substrate.

introducing photoactive molecules sensitive to visible light into the interlayer. Synthesis of layered oxide on the course of more efficient applications requires better understanding and analysis of host-guest or guest-guest relations. Revealing the mechanism in the interlayer in terms of energy transfer and interaction between guest and host lead to the synthesis of new intercalation materials in more rational path. In order to do so, optical spectroscopy is a powerful technique and layered oxides have been widely examined with UV-vis and photoluminescence spectroscopic techniques.

Layered oxides absorb light below 400 nm in their bulk form. Protonation of interlayer makes slight change on the absorption spectra but introduction of photoactive molecules, nanosize semiconductor particles or exfoliation into single sheets bring about considerable spectral changes. While bulk $H_xTi_{(2-x/4)}\square_{x/4}O_4$ results a broad absorption peak with an absorption edge around 400 nm, its exfoliated form gives a well resolved sharp peak blue-shifted to 266 nm as given in Figure 30. The blue shift corresponds to increase in the band gap energy relative to the bulk form and was assigned to the size-quantization effect resulted by exciton confinement in nanothick (0.75 nm) sheets [79,94]. Thickness of one sheet is in a comparable range with a size of an exciton. The peak at 266 nm is the typical peak for exfoliated sheets of $H_xTi_{(2-x/4)}\square_{x/4}O_4$ in a colloidal solution and optical density at this wavelength was estimated as 0.065 for a single titania sheet. Size quantization effect was also observed for other exfoliated layered oxides such as $H_{0.45}MnO_2$ [81]. Plain UV-vis absorption curve of bulk MnO_2 was altered to absorption curve with a broad peak centered around 380 nm. Absorption peak for exfoliated sheets has taken as a reference point for monitoring adsorption of layers on a substrate by Layer-by-Layer assembly method. Adsorption of each layer on a substrate causes a linear increase in absorbance.

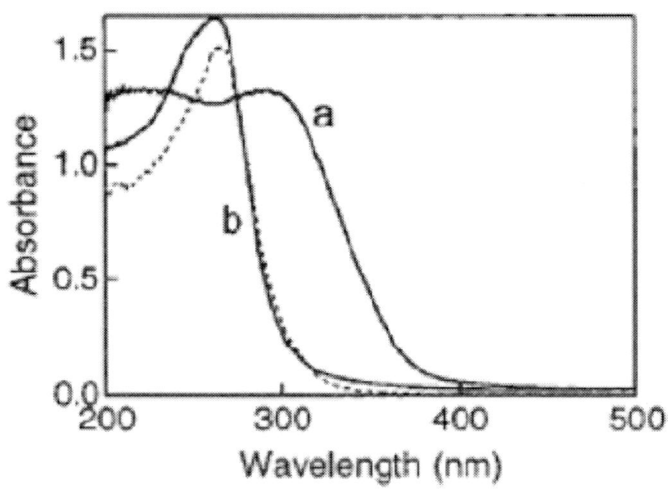

Figure 30. UV-vis spectra of (a) bulk $H_xTi_{(2-x/4)}\square_{x/4}O_4$ and (b) a film of exfoliated $H_xTi_{(2-x/4)}\square_{x/4}O_4$ sheets on a Si substrate. Plot with dotted lines is spectrum of colloidal suspension of nanosize sheets [Reprinted with permission from *J. Phys. Chem. B.*; 1997; 101(49); 10159-10161. Copyright 1997 American Chemical Society.].

Introducing photoactive molecules into the interlayer of layered oxides brings about ability of absorption in the visible light range. Thus, layered oxides can be sensitized to absorb in the visible light range for photocatalytic and photoelectrochemical purposes. $K_4Nb_6O_{17}$ is one of the most extensively studied layered oxides because of its unique two-layer layered structure and thus high photocatalytic activity. Sensitization is carried out by introducing various photoactive molecules, such as dyes [58-63]. Intercalation of these molecules into the interlayer carries the absorption range into the visible light region as seen in Figure 31. Under illumination, these molecules are excited to their Metal-to-Ligand Charge Transfer (MLCT) state, which can be easily located on the UV-vis absorbance spectra.

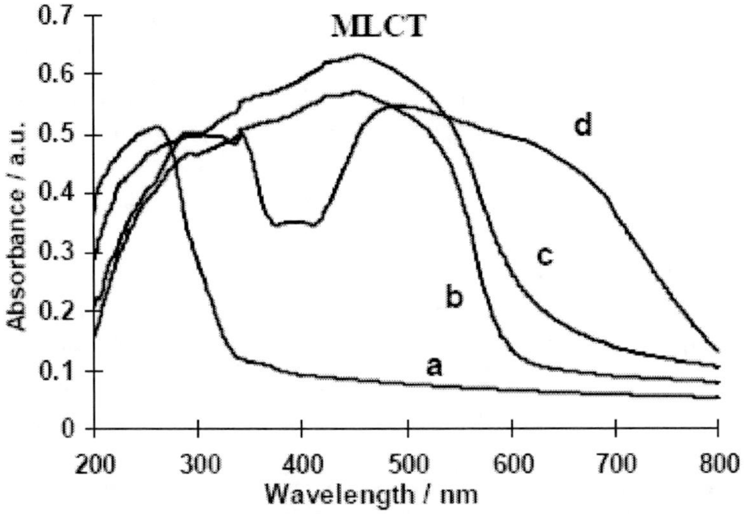

Figure 31. UV-vis absorbance spectra of the films deposited from (a) exfoliated niobate nanosheets, and $Ru(bpy)_3^{2+}$ intercalated niobate layered oxide (b) before and (c) after heat-treatment at 300 ^{0}C, and (d) Methylene Blue intercalated niobate layered oxide. [Reproduced from J. Solid State Chem. 2006, 179, 33-40].

We have observed in our study that sensitized layered oxides are not stable enough in their as-prepared state but require heat-treatment at a certain temperature in order to obtain stability [52,53]. Although sensitized layered oxides can absorb light in visible light range, a low amount of photoactivity is obtained from as-prepared samples. Photoactive molecules are surrounded with water molecules in the interlayer which results in weak interaction between the host and guest. Applying heat-treatment or dehumidification of interlayer brings about more stable molecules in the interlayer and higher state of interaction between the host and guest, which results in better photoelectrochemical activity. The result was confirmed by UV-vis spectra of heat-treated samples (Figure 30). Heat-treatment at a certain temperature causes red-shift in the UV-vis spectra of sensitized layered oxides, which could be assigned to a better interaction state between the host and guest. The general mechanism of charge transfer is as follows for the sensitized layered oxides: Interlayer photoactive guest molecules are excited to their MLCT state under visible light illumination and these molecules feed host layer with electrons at this state.

A detailed analysis of $Ru(bpy)_3^{2+}$ intercalated $K_4Nb_6O_{17}$ with femtosecond transient absorption spectroscopy by Furube et al showed that a very fast energy transfer occurs to the host layer after the excitation of interlayer $Ru(bpy)_3^{2+}$ under illumination [50]. Taking into account that a portion of excited electrons will relax back to ground state with non-radiative electron transfer and there will be back-electron transfer from the host layer to guest molecules, the maximum quantum yield that can be obtained from the system was estimated to be 25%. Similarly, photoluminescent behavior of intercalated compounds gives valuable information about the interaction between host and guest. According to the report by Nakato et al, interlayer guest density and cointercalation of alkylamine molecules in the interlayer affect the concentration quenching as a result of energy transfer between guest molecules [49]. In addition, host-guest interaction is regulated by interlayer space. Quenching occurs by the guest as result of distortion on interlayer molecules at shorter interlayer distances. Thus, the general mechanism for energy transfer in intercalated layered oxides can be investigated in two parts: One is energy transfer taking place between guests in the interlayer and the other one is quenching as a result of host-guest interaction, which also is the primary origin of photocatalytic activity in visible light. Factors affecting the quenching by the guest itself and host layer are associated with the concentration and structural differences of guest and the level of distortion under the stress by the host layer, respectively [49-51,54].

Photoluminescent Layered Oxides

Photoluminescence is one of the principal optical properties of layered oxides. It is also a useful technique to reveal primary mechanisms taking place in photochemical applications. Under band gap excitation, a semiconductive layered oxide goes essentially into three possible processes in a host-guest system, as shown in Figure 32 [129]. First one is self luminescence of host layer (Figure 32a). Second and third processes are luminescence from in-matrix or interlayer guest molecules or ions, respectively, as a result of charge transfer from the host layer to the guest (Figure 32b and c). Recombination of the electrons and holes produced at host occurs on the guest yielding luminescence. Thus, electrons and holes act as charge carriers leading to photocatalytic reactions in the interlayer. For some certain systems with appropriate band structures, the conduction and valence bands of oxide host layer should satisfy the redox reactions of interlayer guest molecules to give luminescence.

Semiconductive layered oxides are known to show luminescence under illumination that takes place as a result of excitation in the host layer (Figure 32) [129-132]. Luminescence is generally observed at temperatures as low as 77 ^{o}C. For the luminescence from layer itself, the emission center is MO_6 units in the layer. Alternatively, doping a lanthanide cation or introducing a lanthanide molecule into the interlayer gives more complex systems producing luminescence both at the guest lanthanide molecules and in the host layer.

An important parameter for luminescence of layered oxides containing lanthanide cations is the structure of metal-oxygen bonding. Angle between M-O-M bonds and length of M-O bond have quite an important role on the delocalization of excited states in the host layer. In this manner, shorter M-O bonds have higher contribution to the luminescence. In addition, the closer the bond angle of O-M-O to 180^0 the more the excitation energy is delocalized [130-134]. For lanthanum doped layered oxides, emission centers are mainly doped lanthanum units in the layer and providing that M-O bonds are shorter and M-O-M is closer to 180^0, excitation energy produced in the interlayer under illumination easily migrates to the doped lanthanide cations and is trapped to produce luminescence. In some cases like lanthanide doped $KLaNb_2O_7$, emission was observed only from the lanthanide cations or both lanthanide cations and host layer depending on the doped lanthanide cation [129,132,135]. Redox levels of doped cations determine the quenching level of host emission. Tendency of easier changing in the oxidation level of doped cation brings about complete quenching of host emission. For example, Eu^{3+} and Sm^{3+} can go into their divalent state easier than other lanthanide trivalent cations can, and as a result charge transfer to Eu^{3+} and Sm^{3+} is more feasible. Otherwise, host emission is not complete as for Er^{3+} and Dy^{3+} doped $KLaNb_2O_7$ and both host emission and lanthanide emission are observed [130-132].

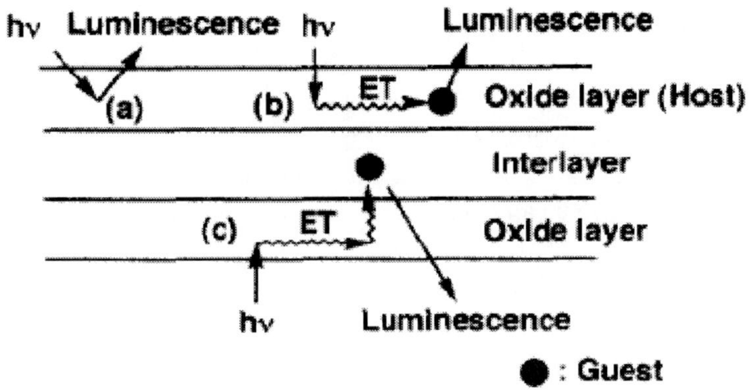

Figure 32. Photoluminescence processes induced by host excitation in host-guest systems based on layered oxides[2].

It has been revealed that interlayer molecules have effect on the luminescence properties of layered oxides. Protonation of interlayer or exchanging with other molecules may result in the alterations in excitation and emission spectrums. Replacing K^+ of $KTiNbO_5$ with H^+, for example, results in red shift on the emission spectrum [129], which was assigned to the alteration in the bond order of the short Nb-O bonds open to the interlayer. Interlayer H^+ acts as trapping sites for the electrons produced in the host layer and higher concentrations of H^+ in the interlayer causes more drastic reduction in the host layer emission intensity. When lanthanide cations were placed in the interlayer, luminescence is observed due to the movement and trapping of energy of the host layer by means of electron-hole pair to the interlayer lanthanide cations. Satisfaction of redox levels of interlayer lanthanide cations by conduction and valence band levels of host layer is an inevitable condition in order to observe lanthanide cation emission, because cations should acquire other oxidation levels by receiving energy from excited host layer [137].

We have studied several layered oxides such as titanate and niobates in order to reveal the luminescent properties of lanthanide cation intercalated layered oxides [96]. The intercalation of lanthanide cations was carried out with Layer-by-Layer methods. We have studied the photoluminescence properties of $H_xTi_{(2-x/4)}\square_{x/4}O_4$, $H_2Ti_4O_9$ and $H_4Nb_6O_{17}$ intercalated with various lanthanide cations. According to current results, Eu^{3+} intercalated $H_xTi_{(2-x/4)}\square_{x/4}O_4$ and Tb^{3+} intercalated $H_2Nb_6O_{17}$ have given the most intense red and green emissions, respectively. Emission could be seen quite clearly with naked eye. As demonstrated in Figure 33, emission was observed from the lanthanide cations only. Linear improvement in emission intensity with the increasing number of layer was observed for the

[2] Reprinted from Microporous and Mesoporous Materials 21 Akihiko Kudo Eichi Kaneko, Photoluminescence properties of ion exchangeable layered oxides 615-620., Copyright (1998), with permission from Elsevier

films prepared by Layer-by-Layer method. For Eu^{3+} intercalated $H_xTi_{(2-x/4)}\square_{x/4}O_4$ films, peaks at 593.2 and 614.4 nm were assigned to the $^5D_0\rightarrow{}^7F_1$ and $^5D_0\rightarrow{}^7F_2$ transitions, respectively. The most intense peak was observed at 614.4 nm, which represents red emission. It is known that the local symmetry of the Eu^{3+} ion in the host matrix determines the relative dominance of the $^5D_0\rightarrow{}^7F_1$ or $^5D_0\rightarrow{}^7F_2$ transition over one another. Relatively strong parity forbidden $^5D_0\rightarrow{}^7F_2$ electric dipole transition peak over $^5D_0\rightarrow{}^7F_1$ magnetic dipole transition peak shows that Eu^{3+} cations in the interlayer occupy sites without inversion symmetry. In the case of Tb^{3+} intercalated $H_4Nb_6O_{17}$ film, a strong green emission was observed and these emission peaks from 480 to 650 nm were associated with the $^5D_4$-^7F_J (J=3-6) transition among which the green $^5D_4$-$^7F_5$ emission at around 545 nm was the strongest. Excitation spectra of the intercalated films gave a broad peak below 400 nm as a result of the excitation in the host layer as seen in Figure 33 and according to the excitation spectra, observed emission is mainly contributed by the band gap excitation in the host layer. Thus, electrons and holes produced by the band gap excitation migrate in the host layer and easily move into the interlayer to be trapped by the lanthanide cations.

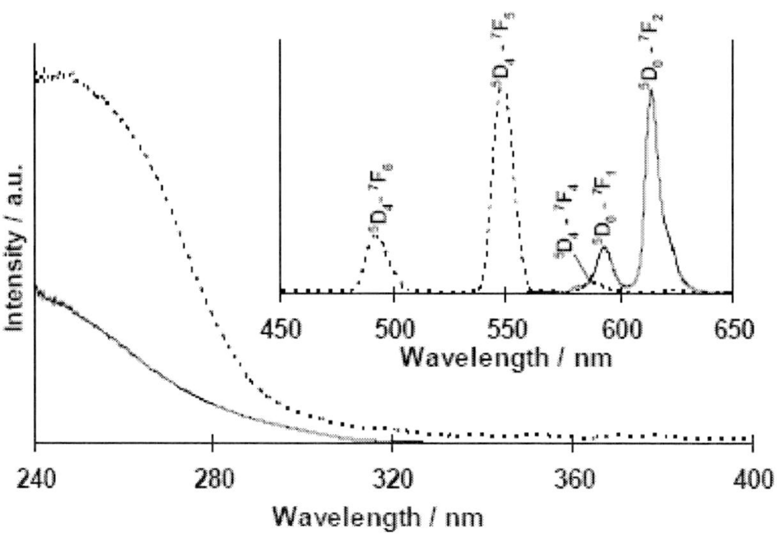

Figure 33 Excitation and emission spectra of Eu^{3+} intercalated $H_xTi_{(2-x/4)}\square_{x/4}O_4$ (solid line) and Tb^{3+} intercalated $H_4Nb_6O_{17}$ (dashed line) films composed of 6 layers by Layer-by-Layer assembly method. Emission wavelengths, λ_{em}, are 614 and 545 nm for Eu^{3+} and Tb^{+}, respectively. Excitation wavelength, λ_{ex}, is 254 nm for both films.

It is well known that the energy transfer between the host layer and the guest molecules occurs when the defect state level of host is higher than the excitation levels of the guest lanthanide cation [137,138]. In addition, redox levels of lanthanide cations are also important because band gap energy levels of host layer should satisfy redox potentials of the interlayer guest cation. Among studied systems, Tb^{3+} has higher luminescence excited state than Eu^{3+} has and moreover, defect state level of $H_xTi_{(2-x/4)}\square_{x/4}O_4$ is the lowest among studied layered oxides [137,138]. Energy diagram giving the possible energy transfer between the host and the guest was drawn in Figure 34. Energy diagram shows that $H_xTi_{(2-x/4)}\square_{x/4}O_4$ provides the most feasible conditions of energy transfer between Eu^{3+} and host layer. In addition,

$H_4Nb_6O_{17}$, which has larger band gap energy, provides sufficiently lower defect state levels for Tb^{3+} to cause an energy transfer between the conduction band of $H_4Nb_6O_{17}$ and $^5D_4$ level of Tb^{3+}. Although the conduction band level of $H_4Nb_6O_{17}$ is high enough for a reasonable energy transfer to Eu^{3+}, luminescence of Eu^{3+} in the interlayer of $H_4Nb_6O_{17}$ is weak, probably because of transfer of the part of the energy to the host layer luminescence, as we observed weak host layer luminescence for this intercalation compound. Taking advantage of these two systems producing both red and green luminescence, we have attempted to fabricate a heterogeneous system containing both intercalation compounds as given in Figure 35. The film was deposited by Layer-by-Layer Assembly method and contains both Eu^{3+} intercalated $H_xTi_{(2-x/4)}\square_{x/4}O_4$ and Tb^{3+} intercalated $H_4Nb_6O_{17}$. Emission spectra in Figure 35 shows that this system is able to produce both red and green luminescence and emission of Eu^{3+} and Tb^{3+} cations are independent from each other and also shows that emission increases with the increasing number of units composed of both intercalated layered oxide layers.

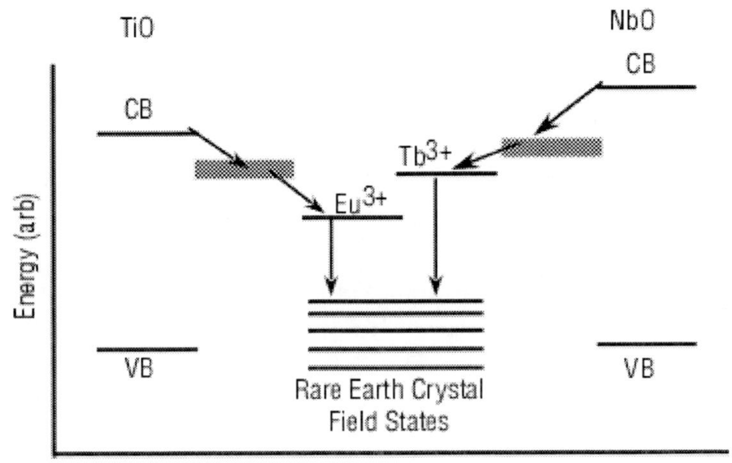

Figure 34. Suggested energy transfer mechanism between the energy levels of host and guest. Shaded boxes represent the possible defect state energy levels of host layers.

Luminescence of these host-guest systems generally occurs at low temperatures, i.e. 77 K, and room temperature luminescence is quenched by the interlayer H^+ or water molecules. We have, however, observed in our research that the existence of interlayer water is inevitable to obtain room temperature luminescence. Dehumidified or heat treated films gave lower luminescence than as-prepared or humidified films did as demonstrated in Figure 36. According to the structural analysis of the films with methods like XRD, TG/DTA and FTIR, lanthanide cations in the interlayer are coordinated with 7-10 water molecules to form their aqua complexes [96]. Higher intensity of emission peaks in the case of water coordinated lanthanide cations suggests that water molecules coordinating lanthanide cations result in better environmental symmetry and higher covalency around lanthanide cations. This determination is strongly supported by higher ratio of the intensities of $^5D_0 \rightarrow ^7F_2$ peak to $^5D_0 \rightarrow ^7F_1$ peak for Eu^{3+} for the humidified film [96]. When lanthanide cations are coordinated with water molecules in the interlayer, the crystal field strength applied on lanthanide cations is stronger as a result of their confined state. Consequently, excitation of host layer produces an electron-hole pair and this electron-hole pair drifts in the host layer and moves simultaneously through the coordinating water molecules to interlayer lanthanide cations to yield emission rather than giving radiationless quenching via energy transfer to OH vibration on water molecules. The suggested mechanism was given in Figure 36.

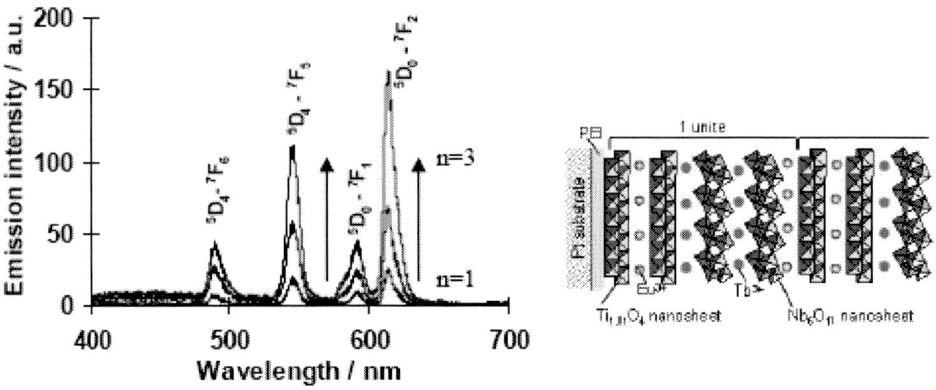

Figure 35. Emission spectra (λ_{ex}, is 254 nm) of heterogeneous film with the structure given in on the right figure. Film is composed of Eu^{3+} and Tb^{3+} cations intercalated between $H_xTi_{(2-x/4)}\square_{x/4}O_4$ and $H_4Nb_6O_{17}$ nanosize sheets, respectively. The direction of the arrow on the left figure shows the increase in the number of units (denoted by n on the figure) in the film.

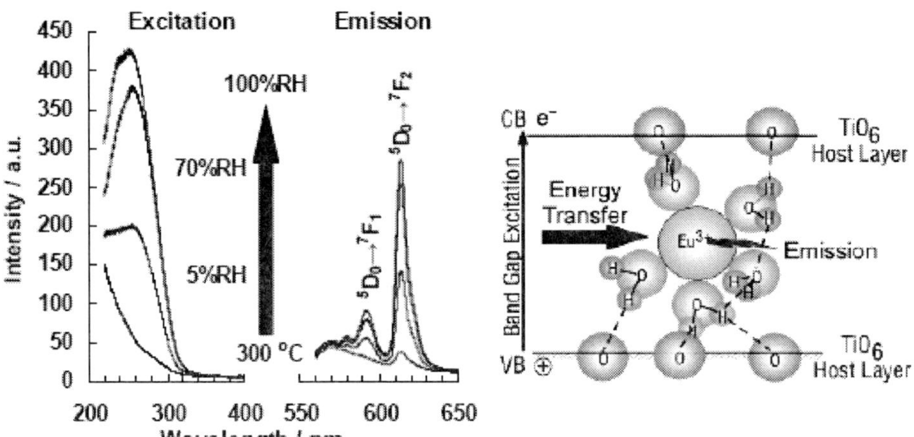

Figure 36 Heat treatment and humidity effect on the emission spectra (λ_{ex}=280 nm) of Eu^{3+} intercalated $H_xTi_{(2-x/4)}\square_{x/4}O_4$ film by Layer-by-Layer assembly. Right image demonstrates the possible energy transfer from excited host layer to the interlayer guest cation. [Reproduced with permission from *J. Phys. Chem. B* 2005, 109, 12748-12754. Copyright 2005 Am. Chem. Soc.].

Contribution of water to luminescence was further tested with various techniques. For example, Eu^{3+} intercalated $H_xTi_{(2-x/4)}\square_{x/4}O_4$ films were illuminated with UV light at various wavelengths. At wavelengths less than the band gap energy of host layer, no change was observed in the luminescence properties of films. On the other hand, intensity of luminescence decreased significantly when the wavelength of UV light exceeded band gap energy of host layer. Excitation in band gap produces electron hole pair to decompose interlayer water molecules [66] and disturbs the coordination symmetry around interlayer lanthanide cations. In addition, luminescence analysis of films was carried out in H_2O and D_2O[3].

In this case, films produced more intense emission in D_2O which coordinates interlayer lanthanide cations instead of H_2O. Higher emission in D_2O can be assigned to the reason that the phonon energy of D_2O is lower than that of H_2O, so tunnelling of electrons through D_2O molecules is more feasible. In conclusion, water molecules coordinating interlayer lanthanide cations play an important role in the luminescence behavior of lanthanide-layered oxide systems. The state of water molecules determines the electron-hole transfer from the excited host layer to the guest lanthanide cation.

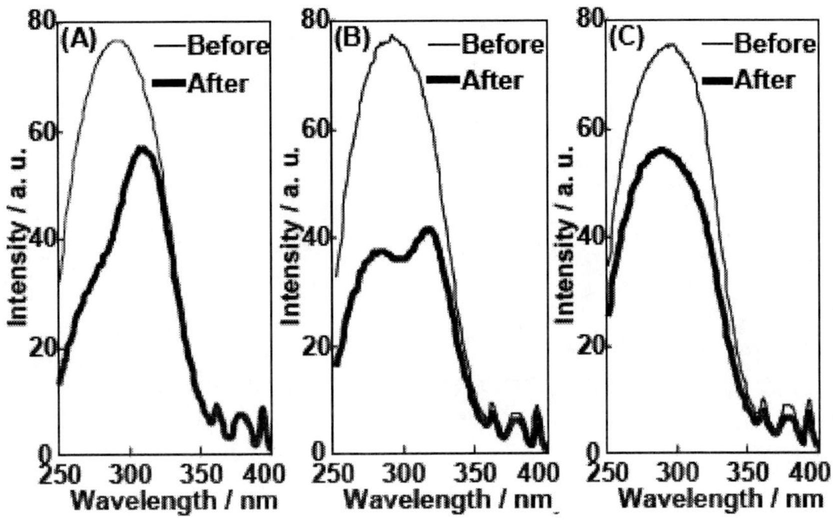

Figure 37. The excitation spectra of Eu^{3+} intercalated $H_xTi_{(2-x/4)}\square_{x/4}O_4$ films before and after irradiation with UV light at (A) 270, (B) 300 and (C) 325 nm. The excitation spectra were obtained by monitoring at 614 nm at room temperature.

Illuminating the film at various wavelengths in the UV region gave interesting results. Figure 37 shows alteration in the excitation spectra of the Eu^{3+} intercalated $H_xTi_{(2-x/4)}\square_{x/4}O_4$ films under different illumination conditions. Film was exposed to UV light at a certain wavelength for 100 sec and its excitation spectrum was measured. The interesting point is that the excitation intensity at corresponding wavelengths of UV light (270, 300 and 325 nm, respectively) was decreased when compared with those before illumination. These spectral changes are similar to spectral hole burning observed in Eu^{3+} doped materials. We have assigned this spectral change to the formation of electron-hole pairs at different energy levels depending on the excitation wavelength of UV light. These electron-hole pairs at the respective energy level alter the state of surrounding water and undergo a relevant energy transfer to the interlayer Eu^{3+} cations. This specific energy transfer between the host and the

[3] Films were placed in a quartz cell filled with H_2O or D_2O and luminescence analysis was carried out.

guest accordingly results in the spectral change on the excitation spectra at corresponding wavelength.

Another feature that we observed from lanthanide intercalated layered oxide films is the decay in emission under excitation over a long period of time. Interestingly, the decay in emission was recovered after a period of a couple of hours as given in Figure 38. Eu^{3+} intercalated $H_xTi_{(2-x/4)}\square_{x/4}O_4$ film was monitored at λ_{ex}=300 nm for 30 sec to observe the decay. Re-excitation of the film after a while resulted in the improvement of the intensity to a certain extent depending on the time gap between excitation periods. Eventually, the film gained its initial emission intensity after a time gap of 3 hours.

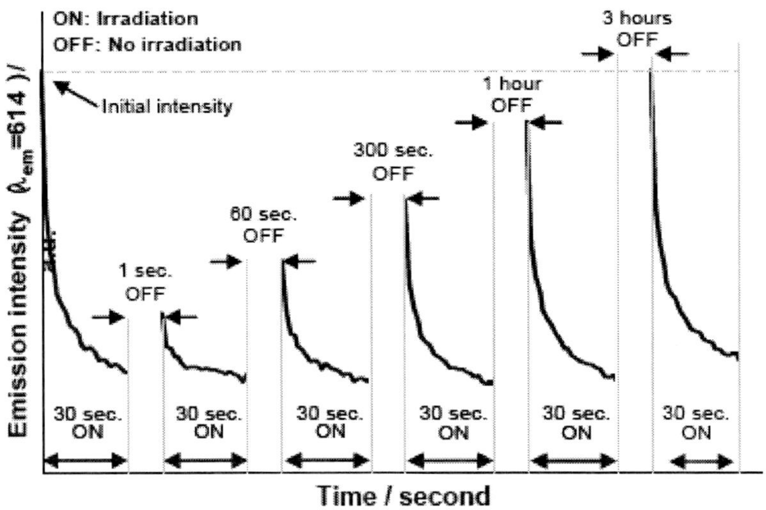

Figure 38. Recovery times for decayed emission of Eu^{3+} intercalated $H_xTi_{(2-x/4)}\square_{x/4}O_4$ film.

8. Photocatalytic Properties and Applications

Photocatalytic applications and water splitting under illumination are the fields that semiconductive layered oxides have been studied and utilized most. One of the most advantageous characteristics of semiconductive layered oxides compared to other semiconductives used in same field is the interlayer space that allows one to modify for the purpose of application. In addition, exfoliated two dimensional sheets can be easily put into any form that gives flexibility in the form of usage. Taking advantage of the semiconductive characteristics together with the two dimensional layered structure, the absorbed light energy is utilized to decompose water or organic molecules in the interlayer or catalyze photoreactions. The research on the use of visible light energy in energy conversion or storage has been extensively carried out. Availability of the band gap in semiconductive layered oxides makes use of the light energy particularly in the process of splitting water. Illuminating the system consisting of a layered oxide yields electron-hole pairs in the host layers, only if the energy of light is higher than the band gap energy of layered oxide. Electrons excited from the conduction band to the valence band drift to the surface or interlayer to take place in a photocatalytic reaction. On the other hand, one of the main shortcomings of the semiconductive layered materials is their inability to absorb visible light as a consequence of possessing relatively large band gap. Band gap is formed by the potential difference between the Md (M: Nb, Ta or Ti) and O2p orbital forming the layered oxides, and is generally around 3 eV or higher. The level of conduction and valence bands of the layered oxides determines the ability of oxidation or reduction of water into O_2 and H_2, respectively. The conduction and valence band of a semiconductor photocatalyst should be higher than the redox potential of H^+/H_2O and lower than the redox potential of O_2/H_2O for water splitting, respectively as seen in Figure 39. Higher activity, however, can be obtained by only providing that there is an enough gap between the energy levels of the bands and the redox potential. For example, higher activity of $Sr_2Ta_2O_7$ than that of $Sr_2Nb_2O_7$ was attributed to the higher conduction and deeper valence band levels of former than that of latter [133] although both have similar layered perovskite structures.

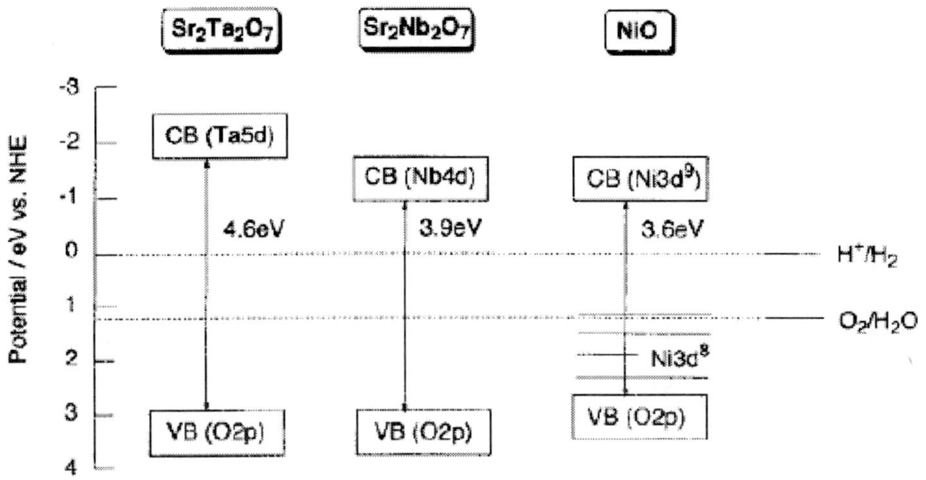

Figure 39. Band structures of $Sr_2M_2O_7$ (M =Nb and Ta) photocatalysts and NiO cocatalyst [Reprinted with permission from *J. Phys. Chem. B* 2000, 104, 571-575. Copyright 2000 American Chemical Society.].

Narrow enough band gap is necessary for efficient visible light absorption. Because as the band gap increases, the semiconductor covers only short range of the visible light spectrum and, in addition, the number of absorbed photon decreases, which results in lower efficiency. Another parameter affecting the photocatalytic activity of layered oxides is the effect of delocalization of electron-hole pair after excitation. It was reported that the closer the bond angle of O-M-O to 180°, the more the excitation energy is localized. The more mobile the electron-hole pair is, the more probable the electrons and holes can reach the reaction sites [129,133]. In addition, M-O bond length is another determining factor as energy transfer is more favorable through short bonds [131].

Doping or ion exchanging of transition metals into the layered oxides also enhances the photocatalytic activities. Ni or Pt loaded $K_4Nb_6O_{17}$ or similar layered oxides possessing two distinctive interlayer exhibited a significant increase in the photocatalytic activity by the contribution of ultrafine metal particles existing in interlayer I depending on the concentration [17,123,139-141] (Figure 40). It was proposed that the reduction of water to H_2 was promoted by the ultrafine metal particles in the interlayer I. In the meantime, interlayer II contributed to the splitting of water by oxidizing water to O_2 [17]. Excited electrons in host layer under band gap illumination are utilized in H_2 production on metal and holes take place in O_2 evolution as seen in Figure 40. Electron transfer is more likely to occur from higher conduction band levels to the co-catalyst as in NiO loaded $Sr_2Ta_2O_7$ (Figure 39). It was also reported that pretreatment by reduction and oxidation at certain temperatures is inevitable to achieve optimum photocatalytic activity. Pretreatment has been applied in various systems to activate or to increase the activity of catalyst. The effect of the amount and size of the interlayer molecules was also investigated and ultrafine particles as fine as 5Å at low concentration in the interlayer were found to be responsible for the high photocatalytic activity [17,142]. Increase in the amount of size of particles might have a negative effect on the photocatalytic activity as it might bring about higher number of defect sites, which act as recombination sites for electron-hole pair [143]. For layered oxides with only one type of interlayer, a quite different mechanism was proposed [144]. In this case, Ni particles as large as 100-200 nm at the external surface of particles acted as reaction sites for H_2 production via H_2O reduction (Figure 40, right image).

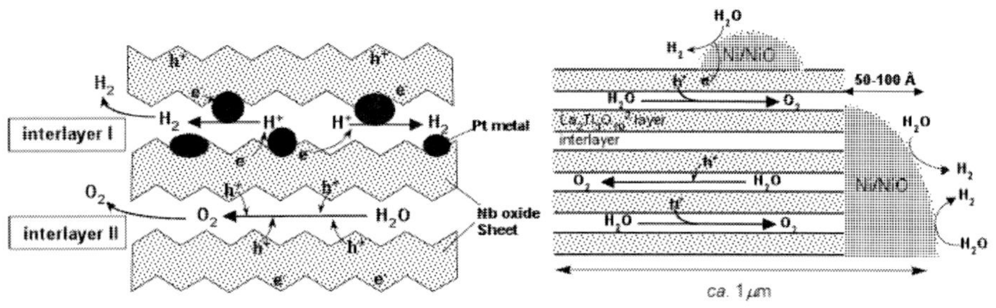

Figure 40. Mechanisms of water splitting in metal loaded layered oxides. [Left Image, Reprinted from J. Photochem. Photobiol. A 114 A. K. Sayama, K. Yase H. Arakawa, K. Asakura, A. Tanaka, K. Domen, T. Onishi, 125-135, Copyright (1998), with permission from Elsevier. Right Image, Reprinted with permission from *Chem. Mater.* 1997, 9, 1063-1064. Copyright 1997 Am. Chem. Soc.].

Oxidation of H_2O to produce O_2 took place in the interlayer of layered oxides. Moreover, addition of alkali hydroxides like KOH or NaOH has promoted the photocatalytic activity [142]. It has been also found that doping host layer with lanthanum cations results an enhancement in the photocatalytic activity. In this case, 4f level of lanthanum cations creates a path to photocatalytic activity, which was found to be dependent on the doped lanthanum. The degree of the hybridization of Ln4f orbital with O2p and Ta5d orbitals is closely related with the photocatalytic activity for $RbLnTa_2O_7$ [145].

Utilization of the visible light can also be accomplished in the presence of photoactive molecules in system. Photosensitization of a layered oxide with a dye or photoactive compound makes it able to use visible light efficiently to some extent. Mallouk et al exhibited photodecomposition of acidic KI solution by the surface sensitized niobates and titanates in the presence of a Ru complex under visible light [146]. Excitation of the complex to its MLCT state supplies electrons to the layered oxide to be consumed in H^+ evolution on Pt sites in the interlayer. Furube et al has described the charge transfer mechanism between $Ru(bpy)_3^{2+}$ and $K_4Nb_6O_{17}$ in detail with spectroscopic measurements [50]. Introduction of photoactive materials such as poly(pyridyl) complexes or dyes results in the absorption in visible light range. We have demonstrated that $H_4Nb_6O_{17}$ intercalated with photoactive molecules such as $Ru(bpy)_3^{2+}$ was able to produce H_2 from water in the presence of methanol as a sacrificial agent [53]. System was active enough to split water under visible light illumination. As given in Figure 41, the maximum amount of H_2 evolution was reached at 0.28 μmol after 1.5 h. Photocatalytic H_2 evolution was ceased, however, after a while as a result of deintercalation of interlayer $Ru(bpy)_3^{2+}$ into the solution. Photocatalytic reaction takes place in the interlayer of $H_4Nb_6O_{17}$ as explained in section 6 and figure 23. Host layer gains electrons from the excited $Ru(bpy)_3^{2+}$ molecules in the interlayer and these electrons react with interlayer water molecules or H_3O^+ ions to produce H_2.

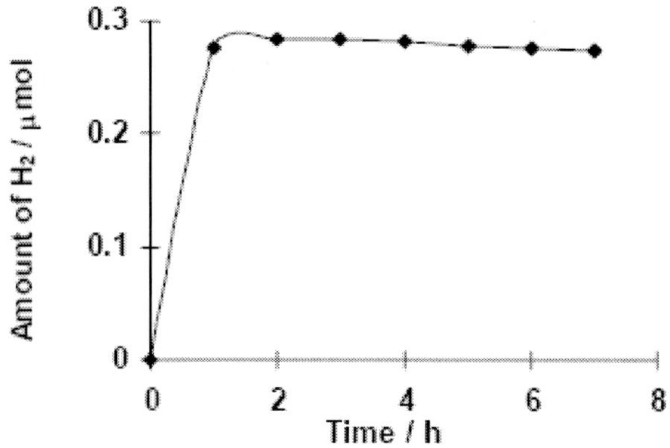

Figure 41. Photocatalytic H_2 gas evolution on $Ru(bpy)_3^{2+}$ intercalated niobate layered oxide from methanol–water solution under illumination in the visible light range.

For some systems, addition of sacrificial agents such as methanol is recommended process to improve the photocatalytic activity [87-147]. Quantum efficiency increases up to 50% for sacrificial agent added system from <5% for system without any sacrificial agent [17]. Alcohols as sacrificial reagents strongly promote the H_2 evolution by its preferential photooxidation, as they act as hole scavenger. Alcohols with smaller molecular size such as CH_3OH and C_2H_5OH promote the photocatalytic H_2 evolution more strongly than those with larger molecular size [122,131,148,149]. The size selectivity for the photooxidation of alcohol is explained by the intercalation mechanism. Alcohols with smaller molecular size are preferentially oxidized compared with those with larger molecular size because of easier intercalation of smaller alcohol molecules into the interlayer. Thus, the intercalation mechanism is one of the key points in the photocatalytic behavior of the layered oxide materials. However, when we investigated the photocatalytic mechanism of H_2O splitting in the presence of alcohols as a sacrificial agent, we have found that the interlayer space is not photoelectrochemically essential for the thin films of various layered oxides such as $K_4Nb_6O_{17}$ or $KCa_2Nb_3O_{10}$ [124]. Instead, adsorption of alcohol controlled by hydrophilicity of the surface of a monolayer film has determined the photocatalytic activity of layered oxide film as described in section 6.

It has been revealed that the introduction of the fine semiconductor particles such as, TiO_2 [74,150,151], Fe_2O_3 [75,152], CdS [77,78], ZnS [76] or their combinations into the interlayer resulted in noticeable increase in the photocatalytic activity. The band gap energies of the pillar semiconductor-layered oxide composites are generally lower than that of the starting materials but higher than that of the pillar semiconductors. In spite of this, fine semiconductor pillared layered oxides are capable of visible light absorption. The increase in the photocatalytic activity is related to the retardation of electron-hole recombination by the pillaring semiconductor and the host-guest interaction allowing quick charge transfer in between. The quick charge transfer from the guest to the host layer depresses the electron-hole recombination. Another advantage of the pillared layered oxides is having a large surface area, which increases the number of reaction sites. One of the main requirements for the photocatalytic activity of layered oxides is the interlayer space as much as the semiconductive properties, because most of the photocatalytic reactions take place in the interlayer space of layered oxides. Thus, large interlayer space, that is higher surface area, brings about an advantage of more reaction sites. To do so, pillaring of interlayer space with semiconductor nanosize particles is a well defined solution. In such a system, size of pillars also affects the photocatalytic activity. Fine enough pillaring semiconductors result in smaller band gap and a shorter diffusion path for the photoinduced electron-hole pair towards reaction sites [76,78]. Pillared systems also possess higher thermal and chemical stability than parent layered oxides do.

9. Conclusion

In this chapter, we have given the concept of semiconductive layered oxides supported by our own data and a concise collection of research carried out up to date. We have especially tried to cover titanate, niobate and tantalates in this chapter because these are mostly studied semiconductive layered oxides and main interest of our group. Since the breakthrough of semiconductive layered oxides and discovery of their potential in various applications, plenty of research has been carried out. In the earlier stage of the research, their ion exchange capacities have been studied and ion exchange and intercalation behavior of the layered oxides have been well resolved. Exchangeable interlayer cations let layered oxides to be modified on the course of a desired application. The negatively charged nature of layers makes it possible to introduce a wide range of cations or positively charged molecules into the interlayer. There are, however, some restrictions. For example, high layer charge density is a shortcoming for some layered oxides to go in an intercalation reaction with large molecules. Another difficulty is interlayer distance that is not wide enough to allow large molecules to penetrate into the interlayer. These problems could be solved after realizing the solid acidic properties of layered oxides.

Taking advantage of acid-base reactions that these materials can easily undergo under appropriate conditions, a variety of rational paths to desired intercalation compounds have been developed. For instance, layered distance is initially opened wide enough by short chain alkylamines and then desired molecule is intercalated into the interlayer in next step. If the concentration of alkylamine molecules or alternatively, sterically bulky organic bases exceeds a certain level, layers are separated into single nanosize sheets, which is called exfoliation. Colloidal sheets as a result of exfoliation provide another approach for intercalation. These colloidal sheets serve as building blocks for intercalation compounds and nanosize assemblies. Alteration in the ionic strength or addition of any cationic molecule into colloidal solution results in reassembling of sheets with counterions. We were able to intercalate various metal complex molecules into the interlayer of titanates and niobates using this approach. In addition, Layer-by-Layer assembly of nanosize sheets and counterions on a substrate is another way for fabrication of intercalated films of layered oxides.

These processes can be carried out with a soft-chemical process in mild conditions without destroying the main framework of the host layer. Soft chemical approach is a rational way for synthesis of various types of layered oxides. Starting from the parent compound, one can fabricate a desired product in a low temperature process by following the steps which are planned in advance. Soft chemistry approach can be more feasible for synthesis of a layered oxide phase rather than conventional solid state reactions.

Apart from ion exchange and intercalation characteristics, layered oxides have prominent physicochemical properties, which can be utilized in various applications. Semiconductive feature of layered oxides is beneficial for photocatalytic activities, especially in water splitting processes. Layered oxides alone, however, are not productive enough because of their large band gap energies and deficiencies in photochemical reactions. Thus, heterogeneous systems were developed involving layered oxides and co-catalysts or photoactive materials. In spite of the development in the photocatalytic applications of layered oxides, there is yet to discover and improve, especially photoelectrochemical behavior of layered oxides in aqueous solutions and capability of using visible light in the presence of photoactive molecules.

The research on semiconductive layered oxides is continuously expanding to a variety of fields. These types of layered oxides have great potential in the field of photocatalytic applications, photoluminescence, host-guest systems, and battery applications. We will be continuing to focus on possible electrochemical and photoluminescence applications of these materials in the future.

REFERENCES

[1] Rouxel, J.; Tournoux, M. *Solid State Ionics* 1996, 84, 141-149.

[2] Schaak, R.E.; Mallouk, T.E. *Chem. Mater*. 2002, 14, 1455-1471.

[3] Chu, C. W.; Hor, P. H.; Meng, R. L.; Gao, L.; Huang, Z. J.; Wang,Y. Q. *Phys. Rev. Lett.* 1987, 58, 405-407.

[4] Hor, P. H.; Meng, R. L.; Wang, Y. Q.; Gao, L.; Huang, Z. J.;Bechtold, J.; Forster, K.; Chu, C. W. *Phys. Rev. Lett*. 1987, 58, 1891-1894.

[5] Takada, K.; Sakurai, H.; Muromachi, E.T.; Izumi, F.; Dilanian, R.A.; Sasaki T. *Nature* 2003, 422, 53-55.

[6] Bruce, P.G. *Chem. Comm*. 1997, 1817-1824.

[7] Hervieu, M.; Raveau, B. *Rev. Chim. Min*. 1981, 18, 642-649.

[8] Grey, I. E.; Li, C.; Madsen, I. C.; Watts, J. A. *J. Solid State Chem*.1987, 66, 7-19.

[9] Sasaki, T.; Watanabe, M.; Hashizume, H.; Yamada, H.; Nakazawa, H. *J. Am. Chem.* Soc. 1996, 118, 8329-8335.

[10] Sasaki, T.; Ebina, Y.; Kitami, Y.; Watanabe, M.; Oikawa, T. *J. Phys. Chem*. B 2001, 105, 6116-6121.

[11] Fukuda, K.; Nakai, I.; Oishi, C.; Nomura, M.; Harada, M.; Ebina, Y.; Sasaki, T. *J. Phys. Chem.* B. 2004, 10, 13088-13092.
[12] Sasaki, T.; Watanabe, M.; Komatsu, Y.; Fujiki, Y. *Inorg. Chem.*1985, 24, 2265-2271.
[13] Grey, I.E.; Madsen, I.C.; Watts, J.A.; Bursill, L.A.; Kwiatkowska, J. *J. Solid State Chem.* 1985, 58, 350-356.
[14] Bursill, L.A.; Smith, D.J.; Kwiatkowska, J. *J. Solid State Chem.* 1987, 69, 360-368.
[15] Gasperin, M.; Le Bihan, M.T. *J. Solid State Chem.* 1980, 33, 83-39.
[16] Gasperin, M.; Le Bihan, M.T. *J. Solid State Chem.* 1982, 43, 346-353.
[17] Kudo, A.; Sayama, K.; Tanaka, A.; Asakura, K.; Domen, K.; Maruya, K.; Onishi, T. J. *Catal.* 1989, 120, 337-352.
[18] Abe, R.; Kondo, J.K.; Hara, M.; Domen, K. *Supramolecular Science* 1998, 5, 229-233.
[19] Sayama, K.; Yase, K.; Arakawa, H.; Asakura, K.; Tanaka, A.; Domen, K.; Onishi, T. J. *Photochem. Photobiol. A. Chem.* 1998, 114, 125-135.
[20] Fukuoka, H.; Isami, T.; Yamanaka, S. *J. Solid State Chem.* 2000, 151, 40-45.
[21] Fair, G.; Shemkunas, M.; Petuskey, W.T.; Sambasivan, S. *J. Eur. Ceram. Soc.* 1999, 19, 2437-2447.
[22] Jacobson, A.J.; Johnson, J.W.; Lewandowski, J.T. *Inorg. Chem.* 1985, 24, 4-3729.
[23] Uma, S.; Raju, A.R.; Gopalakrishnan, J. *J. Mater. Chem.* 1993, 3, 709-713.
[24] Kendall, K.R.; Navas, C.; Thomas, J.K.; zur Loye, H.-C. *Chem. Mater.* 1996, 8, 642-649.
[25] Sasaki, T.; Izumi, F.; Watanabe, M. *Chem. Mater.* 1996, 8, 777-782.
[26] Tong, Z.; Shichi, T.; Oshika, K.; Takagi, K. *Chem. Lett.* 2002, 31, 876-877.
[27] Airoldi, C.; Nunes, L.M.; de Farias, R.F. *Mater. Res. Bull.* 2000, 35, 2081-2090.
[28] Liu Z.; Yang, X.; Ooi, K. *J. Coll. Int. Sci.*2003, 265, 115-120.
[29] Sasaki, T.; Komatsu, Y.; Fujiki, Y. *Inorg. Chem.* 1989, 28, 2276-2279.
[30] Sasaki, T.; Watanabe, M.; Michiue, Y.; Komatsu, Y.; Izumi, F.; Takenouchi, S. *Chem. Mater.* 1995, 7, 1001-1007.
[31] Takagaki, A.; Sugisawa, M.; Lu, D.; Kondo, J. N.; Hara, M.; Domen, K.; Hayashi, S. *J. Am. Chem. Soc.* 2003, 125, 5479-5485.
[32] Takagaki, A.; Yoshida, T.; Lu, D.; Kondo, J. N.; Hara, M.; Domen, K.; Hayashi, S. *J. Phys. Chem.* B. 2004, 108, 11549-11555.
[33] Takagaki, A.; Lu, D.; Kondo, J. N.; Hara, M.; Hayashi, S.; Domen, K. *Chem. Mater.* 2005, 17, 2487-2489.
[34] Schaak, R. E.; Mallouk, T. E. *Chem. Mater.* 2000, 12, 3427-3434.
[35] England, W.A.; Birkett, J.E.; Goodenough, J.B.; Wiseman, P.J. *J. Solid State Chem.* 1983, 49, 300-308.
[36] Sasaki, T.; Komatsu, Y.; Fujiki, Y. *Inorg. Chem.* 1998, 28, 2776-2779.
[37] Sasaki, T.; Komatsu, Y.; Fujiki, *Chem. Mater.* 1992, 4, 894-899.
[38] Abe, R.; Hara, M.; Kondo, J.N.; Domen, K. *Chem. Mater.* 1998, 10, 1647-1651.
[39] Abe, R.; Ikeda, S.; Kondo, J.N.; Hara, M.; Domen, K. *Thin Solid Films* 1999, 343-344, 156-159.
[40] Sumida, T.; Takahara, Y.; Abe, R.; Hara, M.; Kondo, J.N.; Domen K.; Kakihana, M.; Yoshimura, M. *Phys. Chem. Chem. Phys.* 2001, 3, 640-644.
[41] Schaak, R. E.; Mallouk, T. E. *J. Am. Chem. Soc.* 2000, 122, 2798-2803.
[42] Sugimoto, W.; Shirata, M.; Kuroda, K.; Sugahara, Y. *Chem. Mater.* 2002, 14, 2946-2952.

[43] Zhong, Z.; Ding, W.; Hou, W.; Chen, Y.; Chen, X.; Zhu, Y.; Min, N. *Chem. Mater.* 2001, 13, 538-542.
[44] Gopalakrishnan, J.; Bhat, V.; Raveau, B. *Mater. Res. Bull.* 1987, 22, 413-417.
[45] Uma, S.; Gopalakrishnan, J. *Mater. Sci. Eng.* B 1995, 34, 175-179.
[46] Bhuvanesh, N.S.P.; Gopalakrishnan, J. *Mater. Sci. Eng.* B 1998, 53, 267-271.
[47] Takahasi, S.; Nakato, T.; Hayashi, S.; Sugahara, Y.; Kuroda, K. *Inorg. Chem.* 1995, 34, 5065-5069.
[48] Nakata, T.; Sakamoto, D.; Kuroda, K.; Kato, C. *Bull. Chem. Soc. Jpn.* 1992, 65, 322-328.
[49] Nakato, T.; Kusunoki, K.; Yoshizawa, K.; Kuroda, K.; Kaneko, M. *J. Phys. Chem.* 1995, 99, 17896-17905.
[50] Furube, A.; Shiozawa, T.; Ishikawa, A.; Wada, A.; Domen, K.; Hirose, C. *J. Phys. Chem.* B 2002, 106, 3065-3072.
[51] Yao, K.; Nishimura, S.; Imai, Y.; Wang, H.; Ma, T,; Abe, E.; Tateyama, H.; Yamagishi, A. *Langmuir* 2003, 19, 321-325.
[52] Unal, U.; Matsumoto, Y.; Tanaka, N.; Kimura, Y.; Tamoto, N. *J. Phys. Chem.* B 2003, 107, 12680-12689.
[53] Unal, U.; Matsumoto, Y.; Tamoto, N.; Koinuma, M.; Machida, M.; Izawa, K. J. Solid State Chem. 2006, 179, 33-40.
[54] Yao, K.; Nishimura, S.; Ma, T,; Okamoto, K.; Inoue, K.; Abe, E.; Tateyama, H.; Tateyama, H.; Yamagishi, A. *J. Electroanal. Chem.* 2001, 510, 144-148.
[55] Nakato, T.; Miyashita, H.; Yakabe, S. *Chem. Lett.* 2003, 32, 72-73.
[56] Bizeto, M.A.; de Faria, L.A.; Constantino, V.R.L. *J. Mater. Sci. Lett.* 1999, 18, 643-646.
[57] Yamaguchi, Y.; Yui, T.; Takagi, S.; Shimada, T.; Inoue, H. *Chem. Lett.* 2001, 30, 644-645.
[58] Miyata, H.; Sugahara, Y.; Kuroda, K.; Kato, C. *J. Chem. Soc.* Faraday Trans. 1 1988, 84, 2677-2682
[59] Nakato, T.; Miyata, H.; Kuroda, K.; Kato, C. *React. Solid.* 1988, 6, 231-238.
[60] Nakato, T.; Kuroda, K.; Kato, C. *Catal. Today* 1998, 16, 471-478.
[61] Nakato, T.; Sugahara, Y.; Kuroda, K.; Kato, C. *Mat. Res. Soc. Symp. Proc.* 1991, 233, 169-173.
[62] Shinozaki, R.; Nakato, T. *Langmuir* 2004, 20, 7583-7588.
[63] Miyamoto, N.; Kuroda, K.; Ogawa, M. *J. Am. Chem. Soc.* 2001, 123, 6949-6950.
[64] Kaito, R.; Kuroda, K.; Ogawa, M. *J. Phys. Chem.* B. 2003, 107, 4043-4047.
[65] Yakabe, S.; Nakato, T. *J. Mater. Sci.* 2003, 38, 3809-3812.
[66] Matsumoto, Y.; Funatsu, A.; Matsuo, D.; Unal, U.; Ozawa, K. *J. Phys. Chem.* B 2001, 105, 10893-10899.
[67] Nakano, S.; Sasaki, T.; Takemura, K.; Watanabe, M. *Chem. Mater.* 1998, 10, 2044-2046.
[68] Choy, J.H.; Lee, H.C.; Jung, H.; Hwang, S.J. *J. Mater. Chem.*, 2001, 11, 2232-2234.
[69] Lee, H.C.; Jung, H.; Oh, J.M.; Choy, J.H. *Bull. Korean Chem. Soc.* 2002, 23, 477-480.
[70] Choy, J.H.; Lee, H.C.; Jung, H.; Kim, H.; Boo, H. *Chem. Mater.* 2002, 14, 2486-2491.
[71] Andersen, M.W.; Klinowski, J. *Inorg. Chem.* 1990, 29, 3260-3263.
[72] Cheng, S.; Wang, T.C. *Inorg. Chem.* 1989, 28, 1283-1289.
[73] Kooli, F.; Sasaki, T.; Watanabe, M. *Micropor. Mesopor. Mater.* 1999, 28, 495-503.

[74] Kooli, F.; Sasaki, T.; Watanabe, M.; Martin, C.; Rives, V. *Langmuir* 1999, 15, 1090-1095.
[75] Yanagisawa, M. *Solid State Ion* 2001, 141-142, 575-581.
[76] Fujishiro, Y.; Uchida, S.; Sato, T. *Int. J. Inorg. Mat.* 1999, 1, 67-72.
[77] Tawkaew, S.; Fujishiro, Y.; Kin, S.; Sato, T. *Coll. Surf.* A 2001, 179, 139-144.
[78] Shangguan, W.; Yoshida, A. *Sol.Energy Mater. Sol. Cells* 2001, 69, 189-194.
[79] Sasaki, T. *Supramolecular Science*, 1998, 5, 367-371.
[80] Sasaki, T.; Watanabe, M. *J. Am. Chem. Soc.* 1998, 120, 4682-4689.
[81] Omomo, Y.; Sasaki, T.; Wang, L.; Watanabe, M. *J. Am. Chem. Soc.* 2003, 125, 3568-3575.
[82] Xu, F.F.; Ebina, Y.; Bando, Y.; Sasaki, T. *J. Phys. Chem.* B 2003, 107, 9638-9645.
[83] Liu, Z.; Ooi, K.; Kanoh, H.; Tang, W.; Tomida, T. *Langmuir,* 2000, 16, 4154-4164.
[84] Kaschak, D. M.; Johnson, S.A., Hooks, D.E.; Kim, H.N.; Ward, M.D.; Mallouk, T.E. *J. Am. Chem. Soc.* 1998, 120, 10887-10894.
[85] Saupe, G.B.; Waraksa, C.C.; Kim, H.N.; Han, Y.J.; Kaschak, D.M.; Skinner, D.M.; Mallouk, T.E. *Chem. Mater.* 2000, 12, 1556-1562.
[86] Kim, H.-N.; Keller, S. W.; Mallouk, T. E.; Schmitt, J.; Decher, G *Chem. Mater.* 1997, 9, 1414-1421.
[87] Abe, R.; Shinohara, K.; Tanaka, A.; Hara, M.; Kondo, J. N.; Domen, K. *Chem. Mater.* 1997, 9, 2179-2184.
[88] Du, G.H.; Peng, L.-M.; Chen, Q.; Zhang, S.; Zhou, W.Z. *Appl. Phys. Lett.* 2003, 83, 1638-1640.
[89] Bizeto, M.A.; Constantino, V.R.L. *Micropor. Mesopor. Mater.* 2005, 83, 212-218.
[90] Bizeto, M.A., Constantino, V.R.L. *Mater. Res. Bull.* 2004, 39, 1811-1820.
[91] Han, Y.S.; Park, I.; Choy, J.H. *J. Mater. Chem.* 2001, 11, 1277-1282.
[92] Bruzaud, S.; Levesque, G. *Chem. Mater.* 2002, 14, 2421-2426.
[93] Tahara, S.; Takeda, Y.; Sugahara, Y. *Chem. Mater.* 2005, 17, 6198-6204.
[94] Sasaki, T.; Watanabe, M. *J. Phys. Chem.* B. 1997, 101, 10159-10161.
[95] Han, Y.S.; Park, I.; Choy, J.H. *J. Mater. Chem.* 2001, 11, 1277-1282.
[96] Matsumoto, Y.; Unal, U.; Kimura, Y.; Ohashi, S.; Izawa, K. *J. Phys. Chem.* B. 2005, 109,12748-12754.
[97] Abe, R.; Shinohara, K.; Tanaka, A.; Hara, M.; Kondo, J.N.; Domen, K. *Chem. Mater.* 1998, 10, 329-333.
[98] Yamaki, T.; Shinohara, R.; Asaki, K. *Thin Solid Films*, 2001, 393, 154-160.
[99] Muramatsu, M.; Akatsuka, K.; Ebina, Y.; Wang, K.; Sasaki, T.; Ishida, T.; Miyake, K.; Haga, M. *Langmuir*, 2005, 21, 6590-6595.
[100] Decher, G. *Science,* 1997, 277, 1232-1237.
[101] Fendler, J.H. *Chem. Mater.* 1996, 8, 1616-1624.
[102] Keller, S.W.; Kim, H.N.; Mallouk, T.E. *J. Am. Chem. Soc.* 1994, 116, 8817-8818.
[103] Fang, M.; Kaschak, D.M.; Sutorik, A.C.; Mallouk, T.E. *J. Am. Chem. Soc.* 1997, 119, 12184-12191.
[104] Kerimo, J.; Adams, D.M.; Barbara, P.F.; Kaschak, D.M.; Mallouk, T.E. *J. Phys. Chem.* B 1998, 102, 9451-9460.
[105] Kovtyukhova, N. I.; Olliver, P.J.; Martin, B.R.; Mallouk, T.E.; Chizhik, S.A.; Buzaneva, E.V.; Gorchinsky, A.D. *Chem. Mater.* 1999, 11, 771-778.

[106] Fang, M.; Kim, C.H.; Saupe, G.B.; Kim, H.N.; Waraksa, C.C.; Miwa, T.; Fujishima, A.; Mallouk T.E. *Chem. Mater*. 1999, 11, 1526-1532.
[107] Schaak, R.E.; Mallouk, T.E. *Chem. Mater*. 2000, 12, 2513-2516.
[108] Sasaki, T.; Ebina, Y.; Tanaka, T.; Harada, M.; Watanabe, M.; Decher, G. *Chem. Mater*. 2001, 13, 4661-4667.
[109] Sasaki, T.; Ebina, Y.; Fukuda, K.; Tanaka, T.; Harada, M.; Watanabe, M. *Chem. Mater*. 2002, 14, 3524-3530.
[110] Harada, M.; Sasaki, T.; Ebina, Y.; Watanabe, M. J. *Photochem. Photobiol*. A 2002, 148, 273-276.
[111] Wang, Z.S.; Ebina, Y.; Takada, K.; Watanabe, M.; Sasaki, T. *Langmuir* 2003, 19, 9534-9537.
[112] Wang, L. Z.; Omomo, Y.; Sakai, N.; Fukuda, K.; Nakai, I.; Ebina, Y.; Takada, K.; Watanabe, M.; Sasaki, T. *Chem. Mater*. 2003, 15, 2873-2878.
[113] Wang, Z.S.; Sasaki, T.; Muramatsu, M.; Ebina, Y.; Tanaka, T.; Wang, L. Z.; Watanabe, M. *Chem. Mater*. 2003, 15, 807-812.
[114] Ma, R.; Sasaki, T.; Bando, Y. *J. Am. Chem. Soc*. 2004, 126, 10382-10388.
[115] Wang, L. Z.; Ebina, Y.; Takada, K.; Sasaki, T. *J. Phys. Chem*. B 2004, 108, 4283-4288.
[116] Bruce, P. G. *Chem. Commun*. 1997, 1817-1824.
[117] Sakai, N.; Ebina, Y.; Takada, K.; Sasaki, T. *J. Am. Chem. Soc*. 2004, 126, 5851-5858.
[118] Koinuma, M.; Seki, H.; Matsumoto, Y. J. *Electroanal. Chem*. 2002, 531, 81-85.
[119] Sakai, N.; Ebina, Y.; Takada, K.; Sasaki, T. *J. Phys. Chem*. B. 2005, 109, 9651-9655.
[120] Domen, K.; Yoshimura, J.; Sekine, T.; Tanaka, A.; Onishi, T. *Catal. Lett*. 1990, 4, 339-344.
[121] Ebina, Y.; Sasaki, T.; Harada, M.; Watanabe, M. *Chem. Mater*. 2002, 14, 4390-4395.
[122] Yin S.; Wu, J.; Aki, M.; Sato, T. Int. J. *Inorg. Mat*. 2000, 2, 325-331.
[123] Takata, T.; Tanaka, A.; Hara, M.; Kondo, J.N.; Domen, K. *Catal. Today* 1998, 44, 17-26.
[124] Izawa, K.; Yamada, T.; Unal, U.; Ida, S.; Altuntasoglu, O.; Koinuma, M.; Matsumoto, Y. *J. Phys. Chem*. B 2006, 110, 4645-4650.
[125] Sakai, N.; Fukuda, K.; Shibata, T.; Ebina, Y.; Takada, K.; Sasaki, T. *J. Phys. Chem*. B 2006, 110, 6198-6203.
[126] Baker, M.D.; Senaratne, C. *Phys. Chem. Chem. Phys*. 1999, 1, 1673-1677.
[127] Baker, M.D.; Zhang, J. *J. Phys. Chem*. 1990, 94, 8703-8708.
[128] Li, J.; Pfanner, K.; Calzaferri, G. *J. Phys. Chem*. 1995, 99, 2119-2126.
[129] Kudo, A.; Kaneko, E. *Micropor. Mesopor. Mater*. 1998, 21, 615-620.
[130] Kudo, A.; Sakata, T. *J. Phys. Chem*. 1996, 100, 17323-17326.
[131] Kudo, A.; Kondo, T. *J. Mater. Chem*. 1997, 7, 777-780.
[132] Kudo, A. *Chem. Mater*. 1997, 9, 664-669.
[133] Kudo, A.; Kato, H.; Nakagawa, S. *J. Phys. Chem*. B. 2000, 104, 571-575.
[134] Kudo, A.; Shibata, T.; Kato, H. *Chem. Lett*. 1999, 9, 959-960.
[135] Kudo, A.; Sakata, T. *J. Phys. Chem*., 1995, 99, 15963-15967
[136] Kudo, A.; Kaneko, E. *Chem. Comm*. 1997, 349-350
[137] Xin H.; Ma, R.; Wang, L.; Ebina, Y.; Takada, K.; Sasaki, T. *J. Appl. Phys. Lett*., 2004, 85, 4187-4189.
[138] Frindell, K.L; Bartl, M. H.; Robinson, M. R.; Bazan, G.C.; Popitsch, A.; Stucky, G. D. *J. Solid State Chem*. 2003, 172, 81-88.

[139] Kudo, A.; Tanaka, A. ; Domen, K.; Maruya, K.; Aika, K.; Onishi, T. *Journal of Catalysis* 1988, 111, 67-76.

[140] Sayama, K.; Arakawa, H.; Domen, K. *Catalysis Today*, 1996, 28, 175-182.

[141] Sayama, K.; Tanaka, A.; Domen, K.; Maruya, K.; Onishi T. *J. Phys. Chem*. 1991; 95m, 1345-1348.

[142] Sayama, K.; Tanaka, A.; Domen, K.; Maruya, K.; Onishi T. *Catal. Lett*. 1990, 4, 217-222.

[143] Ikeda, S.; Tanaka, A.; Shinohara, K.; Hara, M.; Kondo, J.N.; Maruya, K.; Domen, K. *Micropor. Mater*. 1997, 9, 253-258.

[144] Takata, T.; Furumi, Y.; Shinohara, K.; Tanaka, A.; Hara, M.; Kondo, J. N.; Domen, K. *Chem. Mater*. 1997, 9, 1063-1064.

[145] Machida, M.; Yabunaka, J.; Kijima, T.; Matsushima, S.; Arai, M. *Int. J. Inorg. Mat*. 2001, 3, 545-550.

[146] Kim, Y. I.; Salim, S.; Huq, M.J.; Mallouk, T.E. *J. Am. Chem. Soc*. 1991, 113, 9561-9563.

[147] Ebina, Y.; Tanaka, A.; Kondo, J. N.; Domen, K. *Chem. Mater*. 1996, 8, 2534-2538.

[148] Domen, K.; Ebina, Y.; Ikeda, S.; Tanaka, A.; Kondo, J. N.; Maruya, K. *Catal. Today* 1996, 28, 167-174.

[149] Ohtani, B.; Ikeda, S.; Nakayama, H.; Nishimoto, S. Phys. Chem.Chem. Phys. 2000, 2, 5308-5313

[150] Uchida, S.; Yamamoto, Y.; Fujishiro, Y.; Watanabe, A.; Ito, O.; Sato, T. *J. Chem. Soc*. Faraday Trans. 1997, 93, 3229-3234.

[151] Yanagisawa, M.; Uchida, S.; Sato, T. *Int. J. Inorg. Mat*. 2000, 2, 339-346.

[152] Sato, T.; Yamamoto, Y.; Fujishiro, Y.; Uchida, S. *J. Chem. Soc*. Faraday Trans. 1996, 92, 5089-5092.

In: Progress in Solid State Chemistry Research
Editor: Ronald W. Buckley, pp. 55-116
ISBN 1-60021-313-8

Chapter 2

STRUCTURAL, OPTICAL AND PHOTOELECTRICAL PROPERTIES OF LOW-DIMENSIONAL SEMICONDUCTORS DEPOSITED IN THIN FILM FORM BY CHEMICAL AND SONOCHEMICAL METHODS

Biljana Pejova*
Institute of Chemistry, Faculty of Natural Sciences and Mathematics,
Sts. Cyril and Methodius University, PO Box 162, 1001 Skopje, Macedonia

ABSTRACT

Quantum dots of several metal chalcogenide semiconducting compounds were deposited in thin film form by chemical and sonochemical methods. The studied systems included ZnSe, SnSe, Bi_2S_3, In_2S_3, $AgBiS_2$ and $CuInS_2$. All of the developed methods were based on controlled colloidal precipitation reactions in the presence or absence of 20 kHz ultrasonic irradiation, applied using a direct immersion ultrasonic horn. The predominant crystal growth mechanism during the deposition was studied by coherent light scattering experiments. The synthesized materials in thin film form as well as bulk precipitates were identified by X-ray diffraction method. Besides for identification, the XRD data were used to calculate the average crystal size of the synthesized materials within the Debye-Scherrer approach. On the basis of XRD patterns, a refinement of the unit cell parameter values was also performed employing multiple regression analysis techniques and further conclusions concerning the existence and magnitude of lattice strain in the samples were derived. The dislocation densities were calculated as well. Optical absorption measurements in the UV-VIS-NIR spectral region were used to calculate the band gap energies of the studied systems within the framework of Fermi's golden rule for interband electronic transitions employing the parabolic approximation for the dispersion relation. Blue-shifted band gap values with respect to the corresponding bulk specimen, which exhibit red shifts upon thermal annealing of the thin film samples, were observed for all synthesized materials, further confirming their low-dimensional character. The size quantization effects were rationalized within the

* E-mail: biljana@iunona.pmf.ukim.edu.mk

framework of the models of Brus, Nosaka and the hyperbolic band model. In the case of semiconductors in which valence band is split due to spin-orbit coupling, the evolution of splitting energy upon average crystal size increase was followed and discussed. On the basis of temperature dependence of thin films' resistivity, thermal band gap energies and impurity levels ionization energies were calculated. The changes in intercrystalline barriers to electronic transport properties in the case of synthesized nanocrystalline semiconductor films upon thermal treatment were studied as well. In the case of systems exhibiting photoconduction properties, the spectral dependence of stationary nonequilibrium conductivity was measured by the constant field method. These photoconductivity spectral response data were used to calculate the parameters relevant to the band structures of the studied semiconducting films. Photoconductivity relaxation dynamics upon interaction of the films with light impulses was studied by the oscilloscopic method. On the basis of data concerning the photoconductivity time-decay, conclusions about the kinetics of nonequilibrium charge carriers' recombination processes were derived and also their average lifetimes were calculated. Further conclusions on the photoconduction dynamics were derived on the basis of measured lux-ampere characteristics of the synthesized nanocrystalline thin films.

1. Introduction

The physics and chemistry of low-dimensional systems, particularly semiconductors, is a rapidly expanding field of contemporary science [1-3]. According to many authors, the rate of development of the subject in question has been even alarming in certain periods [1,2]. This is due to reasons which may be roughly grouped into two categories: the purely applicative interest for these systems and the significance of the dimensionality reducing for fundamental solid state sciences in general. The potential application of semiconductors in the microelectronics era is self-evident. In order to provide a very brief illustration of the previous statement, we are going to recapitulate only the most general fields in which such systems have been extensively used. These include solar cells, light-emitting diodes, electrochromic windows, batteries, optoelectronic devices, photoelectrochemical devices, thermoelectric coolers, electrical switching devices, solar selective coatings etc [4-9]. A particularly interesting and rapidly emerging aspect of contemporary science, closely related to low-dimensional structures, is the possibility offered for supramolecular chemistry by phenomena arising from the mechanism of self-assembly of such structures [10-16].

Besides the purely applicative aspects, as mentioned before, low-dimensional systems have played an important role in understanding the transition from isolated molecules or clusters to the corresponding solids and liquids [3,17]. Since the present paper is devoted to low-dimensional semiconductors, a brief introductory overview is going to be devoted to this particular class of materials. A number of new, sometimes unexpected phenomena arise when semiconductors are synthesized in form of low-dimensional structures [17,18]. The most prominent one, perhaps, is the change in color upon particle size reduction. However, this easily detectable manifestation of size quantization effects is related to the variation of probably the most important parameter which characterizes these materials - the band gap energy. Band gap energy variation with particle size is a quantum phenomenon. The quest for a solid theoretical background for understanding of the experimentally observed trends, regarding the discussed physical quantity, has led to important insights into effect which arises from the spatial confinement of charge carriers' motions in semiconductors. On the other hand, the effect itself has led to a possibility to engineer the band gap energy of a particular semiconductor only by controlling the dimensionality and the particle size. This simple possibility has further encouraged the shift of the contemporary technology to nano-range of physical dimensions of the system. Even before the rise of nanotechnology era, it has been known that semiconductor properties are particularly sensitive to the synthetic route employed for their synthesis. The additional possibility to control the particle size by controlling the synthetic conditions has further encouraged the quest for novel approaches to semiconductor synthesis. Perhaps the most notable alternative to conventional chemical synthetic methods for low-dimensional semiconductors is the sonochemical approach [19-24] which employs ultrasonic irradiation of the reaction systems. Such unconventional way of introducing the energy required for chemical transformation into the reaction system enables further possibilities for material properties control.

Upon particle size decrease, especially in the nanometer spatial region, a prominent increase of the surface-to-volume ratio occurs. Such increase of the fraction of surface atoms induces a large number of unsaturated ("dangling") bonds. This effect causes certain structural consequences, like inducement of lattice strain which is of residual *i.e.* inherent character [25]. Certain physical properties of the semiconducting material in question are further affected by the strain. The large exposed surface area makes the nanoparticulate materials particularly interesting in the field of heterogeneous catalysis. Although a number of potential applications of nanomaterials will certainly include individual nanoparticles (such as, *e.g.* in medicine [26]), other fields of interest would require self-assembled structures.

As a summary of the introductory section we conclude that contemporary nanoscience and nanotechnology involve two most important aspects: being able to synthesize a nanostructure with required size and shape and being able to understand the properties of both individual nanoparticles and their aggregates. The present review mostly contains a recapitulation of the author's modest contribution to this rapidly expanding field of contemporary science [27-43]. The second chapter is devoted to conventional and sonochemical synthetic methods developed by the author to synthesize quantum dots of various semiconductors (in thin film form as well as bulk precipitates), together with the basic overview of the physics and chemistry of the synthetic routes. Third chapter is related to the structural characteristics of the synthesized low-dimensional semiconductors. Fourth chapter discusses the optical properties of the synthesized semiconducting quantum dots with a special emphasis on size quantization effects, possibilities for band gap engineering, evolution of semiconductor band structure from an isolated QD to bulk solid etc. The fifth and sixth chapters are devoted to the transport phenomena of charge carriers in the synthesized quantum dots in thin film form under equilibrium and non-equilibrium conditions respectively.

2. Controllable Chemical and Sonochemical Synthetic Routes to Variable-Sized Nanostructured Semiconductors in Thin Film Form

2.1. Thermodynamical Aspects of Chemical Synthetic Routes to Nanocrystalline Semiconductors

The chemical methods for synthesis of low-dimensional semiconductor structures in thin film form developed by our group were based on controlled precipitation reactions in aqueous solution. In order to be able to understand the fundamental basis behind the possibilities offered by the chemical approaches to "control" the synthesis of semiconductor nanocrystals, in the sense of their size and shape, two aspects have to be taken into account: thermodynamical and kinetical. When a solid semiconducting compound is synthesized by "wet" chemical route (*i.e.* starting from an initially homogeneous system), the key phenomenon governing the physics of the process is formation of a new phase, consisting of nucleation and subsequent crystal growth. Three types of nucleation processes are relevant to the present topic: homogeneous, heterogeneous and secondary nucleation [25,44,45]. As the self-explanatory names imply, homogeneous nucleation process occurs even in the absence of phase boundaries within the reaction system. The main driving force behind this process, from a thermodynamical viewpoint, is the energetic instability of a supersaturated system. Formation of a spherical solid phase nucleus with radius r within the initially homogeneous liquid phase is accompanied by a change in Gibbs energy given by:

$$\Delta G(r) = \frac{4}{3} r^3 \pi \, \Delta G_V + 4 r^2 \pi \gamma \tag{1}$$

where γ is the surface Gibbs free energy per unit surface area, while ΔG_V is the Gibbs free energy change per unit volume of stable condensed phase. This quantity is given by:

$$\Delta G_V = -\frac{k_B T}{V} \ln(S) \tag{2}$$

where S is the relative supersaturation of the solution (*i.e.* the supersaturation ratio), T is the thermodynamic temperature and k_B is Boltzmann constant. In the case of positive supersaturation ratio, the function $\Delta G(r)$ increases for small values of r (for which the second term is dominant), achieves a maximum and subsequently exhibits a decreasing trend for large r values. The overall shape of $\Delta G(r)$ implies that nucleus formation is an activated process. The maximum of the curve corresponds to the activation energy for nucleation. The value of r^* at which $\Delta G(r)$ achieves a maximum is the critical radius of the nuclei. This quantity may be found from the condition for the extremum, *i.e.*:

$$\frac{\partial(\Delta G(r^*))}{\partial r} = 0 = 4 (r^*)^2 \pi \, \Delta G_V + 8 \pi r^* \gamma \tag{3}$$

$$r^* = -\frac{2\gamma}{\Delta G_V} \tag{4}$$

For a particular value of S, nuclei with radius larger than r^* will have a pronounced tendency to grow (as their further growth will be accompanied by a decrease in the Gibbs free energy), while nuclei characterized by $r < r^*$ will tend to dissolve. Obviously, it follows that a higher supersaturation ratio implies a smaller critical size of the nuclei. The rate of homogeneous nucleation J_{hom} is directly related to $\Delta G(r^*)$ and is given by:

$$J_{\mathrm{hom}} = \left(\frac{\mathrm{d}N}{\mathrm{d}t}\right)\frac{1}{V} = A_{\mathrm{hom}} \exp\left(-\frac{\Delta G(r^*)}{k_{\mathrm{B}}T}\right) \tag{5}$$

where A_{hom} is a constant of the order of magnitude of $10^{25}\ \mathrm{s}^{-1}\ \mathrm{m}^{-3}$.

When phase boundaries are present in the reaction solution, the homogeneous nucleation model needs to be generalized in order to include the possibility for heterogeneous nucleation processes. The rate of heterogeneous nucleation J_{het} is related to J_{hom} in the following way:

$$J_{\mathrm{het}} = A_{\mathrm{het}} \exp\left(-\frac{\Delta G_{\mathrm{het}}(r^*)}{k_{\mathrm{B}}T}\right) \tag{6}$$

where:

$$\Delta G_{\mathrm{het}}(r^*) = f(\theta)\Delta G(r^*) = \frac{(2+\cos\theta)(1-\cos\theta)^2}{4}\Delta G(r^*) \tag{7}$$

In the last equation θ is the contact angle between the nucleus and the surface on which the nucleation process occurs. Small contact angles lead to values of the function $f(\theta)$ significantly smaller than 1 and such conditions favor the heterogeneous nucleation. This situation is in practice encountered in case when the adhesion forces between the surface and the newly formed solid phase are rather pronounced.

It is, certainly, far beyond the scope of this author's review to cover various aspects of nucleation theory that may be employed to effectively design chemical methods for synthesis of nanocrystalline solid phases. The previously outlined arguments were recapitulated in order to follow the basic reasoning behind the experimental design in the present case. In fact, the most easily controllable quantity in (1) by a classical chemical approach is the supersaturation ratio. Consider a semiconducting compound P which is in dynamical equilibrium with its soluble part ($\mathrm{P_{(aq)}}$). The dissociation of this compound leads to formation of ionic species $\mathrm{R_i}$ according to:

$$\mathrm{P(s)}\ ^{\mathrm{a}}\ \mathrm{P(aq)}\ ^{\mathrm{a}}\ \sum_{\mathrm{i}} \phi_{\mathrm{i}} \mathrm{R_i\,(aq)} \tag{8}$$

where ϕ_i are the corresponding stoichiometric coefficients. The supersaturation ratio for the case considered is given by:

$$S = \frac{\prod_i a_i^{\phi_i}}{\left(\prod_i a_i^{\phi_i}\right)_e} = \frac{\prod_i a_i^{\phi_i}}{K_s} \tag{9}$$

where a_i are the activities of the relevant species present in the system. From the previous arguments, it follows that S may be most easily tuned to a desired value by controlling the activities of the present ionic species (*i.e.* their concentrations). In order to synthesize semiconductor nanocrystals predominantly in thin film form, the supersaturation ratio has to be kept within a certain interval throughout the deposition process. Too large values of S would lead to predominant deposition of bulk precipitate, while too small values of this quantity would lead to larger crystal size of the deposited material.

Tuning of S values in the studied reaction systems was achieved by controlling the concentrations of relevant cationic and anionic species in the deposition solution. Cation concentrations were controlled by a careful choice of a suitable complexing agent, while controllably low anionic concentrations were achieved by indirect introduction of these species in the reaction system. This was done by introducing the anions via suitable precursors. Chemical aspects of the deposition processes are covered in the following chapter. Further reduction of the crystal size of synthesized materials was achieved by a continuous irradiation of the reaction systems by ultrasound (*i.e.* employing sonochemical synthetic routes). This is described in more details in chapter 2.3.

Subsequent to nucleation, the crystal growth process occurs in the deposition solution. From a fundamental viewpoint, two possible mechanisms may be involved in the overall crystal growth process. The first one is *ion-by-ion* mechanism, consisting of successive incorporation of metal and chalcogenide ions in the semiconductor crystal lattice:

$$(M_xA_y)_n + xM^{z+} + yA^{2-} \text{ a } (M_xA_y)_{n+1} \quad (10$$

$$(M_xA_y)_{n+1} + xM^{z+} + yA^{2-} \text{ a } (M_xA_y)_{n+2} \quad (11)$$

The second, so-called *cluster* or *colloidal* mechanism, involves adsorption and coagulation of colloidal particles, *i.e.*:

$$(M_xA_y)_l + (M_xA_y)_k \text{ a } (M_xA_y)_{l+k} \quad (12)$$

where $l,k > n$, n being the number of formula units forming a stable nucleus.

The crystal growth mechanism for the investigated films was studied by coherent light scattering experiments [46]. We have demonstrated that in the case of all considered reaction systems, crystal growth occurs predominantly by the cluster mechanism. This was also in line with the structural characteristics of the synthesized films and bulk precipitates obtained from the same reaction system which are discussed further in this review.

2.2. Chemical Aspects of the Developed Synthetic Routes to Nanocrystalline Semiconductors

Although all of the chemical methods developed in our previous works were based on the basic logic of the chemical bath deposition approach, substantial effort was made to synthesize the mentioned semiconducting compounds from reaction systems containing as fewer components as possible. This is especially convenient if high purity of the synthesized material is desired in both chemical and crystallographic sense (*i.e.* the phase purity) or if the material in question should be obtained in the form as close as possible to an intrinsic semiconductor. The previous statement may be illustrated through the example of zinc selenide thin films. While most of the known literature methods for synthesis of ZnSe in thin film form usually lead to mixture of cubic and hexagonal phases of this semiconductor [47-49], often containing significant quantity of $Zn(OH)_2$ or ZnO impurities, the route developed by our group [39] allows synthesis of only the cubic phase of ZnSe with a pronounced nanocrystalline character. Besides these aspects, the developed synthetic methods contain certain novelties in the chemistry of the approach, as compared to the known literature methods, especially in the case of ZnSe, SnSe, In_2S_3 and $AgBiS_2$. Also, for most of the mentioned materials, sonochemical routes for synthesis of the corresponding nanocrystalline thin films were optimized for the first time by our group.

2.2.1. Metal Selenides

For deposition of metal selenides thin films, we have used sodium selenosulfate as selenide ions precursor in the reaction system. Since this substance is commercially unavailable, we have prepared it in the form of water solution by adding gray selenium to a hot water solution of sodium sulfite and stirring the formed heterogeneous mixture for 1 hour at 90 °C. After that, we have removed the excess (unreacted) selenium by filtration. As the synthesized solution of sodium selenosulfate is relatively unstable (especially upon exposure to light), it had to be prepared prior to thin film deposition process and it also had to be kept in a dark container. The stability of selenosulfate solution is improved by the sulfite ions, which are present in excess in the system, due to the following reasons. Selenosulfate anionic species generate H_2Se by spontaneous hydrolysis:

$$SeSO_3^{2-}(aq) + H_2O(l) \rightleftharpoons H_2Se(aq) + SO_4^{2-}(aq) \tag{13}$$

H_2Se further dissociates in two steps, generating free Se^{2-} ionic species in the solution, present in very small concentrations, due to the very low value of the second ionization constant [50,51]:

$$H_2Se(aq) + H_2O(l) \rightleftharpoons HSe^{-}(aq) + H_3O^{+}(aq) \qquad K_1^{\circ} \approx 10^{-4} \tag{14}$$

$$HSe^{-}(aq) + H_2O(l) \rightleftharpoons Se^{2-}(aq) + H_3O^{+}(aq) \quad K_2^{\circ} \approx 10^{-14} \tag{15}$$

The sequence of previous three equilibria is, in fact, the basis of controllable generation of selenide ions in the reaction system by chemical means. However, selenide ions may be relatively easily oxidized to elemental selenium by the oxygen dissolved in the reaction solution [50,52]:

$$Se^{2-}(aq) + H_2O(l) + 1/2O_2(g) \rightarrow Se(s) + 2OH^-(aq) \tag{16}$$

Occurrence of this process is prevented in the selenosulfate solution by the sulfite ions, due to the reducing properties of these species, which are enhanced in alkaline medium. This follows from analysis of the Nernst equation for the corresponding redox system [50]:

$$SO_4^{2-}(aq) + H_2O(l) + 2e \rightleftharpoons SO_3^{2-}(aq) + 2OH^-(aq) \quad E^\circ_{SO_4^{2-}/SO_3^{2-}} = -0.90\ V \tag{17}$$

$$E_{SO_4^{2-}/SO_3^{2-}} = E^\circ_{SO_4^{2-}/SO_3^{2-}} - \frac{0.059V}{2}\log\frac{a(SO_3^{2-})}{a(SO_4^{2-})} + 0.059V(14 - pH) \tag{18}$$

Also, the oxidation properties of the dissolved oxygen become less pronounced in alkaline medium, as implied by the Nernst equation for the corresponding redox pair:

$$O_2(g) + 4H^+(aq) + 4e \rightleftharpoons 2H_2O(aq) \quad E^\circ_{O_2/H_2O} = +1.23\ V \tag{19}$$

$$E_{O_2/H_2O} = E^\circ_{O_2/H_2O} + \frac{0.059V}{4}\log a(O_2) - 0.059V\,pH \tag{20}$$

It therefore follows that, from the aspects of stability of selenide ions generator, alkaline medium is very suitable for synthesis of metal selenide thin films.

All nanocrystalline semiconducting films were deposited onto standard microscope glasses. Prior to the deposition process, the used substrates were treated in a diluted solution of tin(II) chloride and than annealed at 200 °C for about twenty minutes. During this substrate treatment, small and stochastically distributed crystals of tin(II) oxide are created onto the glass surface, which, afterwards, initiate heterogeneous nucleation and improve film adhesion [27-43].

Controllably low concentration of metal cations in the reaction systems was achieved by using suitable complexing agents. Since the cationic systems are characterized by substantially different chemical behavior in reaction solutions, the exact conditions for chemical deposition of semiconducting nanocrystals in thin film form were optimized using various approaches, depending on the particular cation in question.

In the case of ZnSe thin films, controllably low concentration of Zn^{2+} ions in the reaction system was maintained by its complexation with hydroxide ions. After a detailed careful analysis of the existing equilibria in the system, it was concluded that if pH is maintained at rather high value (pH = 14), the Zn^{2+} ions under these conditions will be almost completely present in the form of $[Zn(OH)_4]^{2-}$ complex species (in equilibrium with the Zn^{2+} ions). Precipitation of ZnSe, induced by generation of Se^{2-} ions, may thus be carried out including only the equilibrium:

$$[Zn(OH)_4]^{2-}(aq) \rightleftharpoons Zn^{2+}(aq) + 4OH^-(aq) \tag{21}$$

In other words, under such conditions, one avoids the presence of $Zn(OH)_2$ in the reaction system and therefore its incorporation in the synthesized ZnSe. This aim is essentially impossible to achieve at lower alkaline pH values, even in the presence of some other complexing agent (as in the case of other methods for synthesis of this semiconductor proposed in the literature [47-49]). This is due to the fact that precipitation of $Zn(OH)_2$ occurs in a very wide range of pH values starting from about 7.5 if the initial concentration of Zn^{2+} ions in the reaction system is 0.01 mol/dm^3. Alkaline pH is also required to provide conditions under which the oxidation of selenosulfate (or, more precisely, selenide or hydrogenselenide) species to elemental selenium is avoided, as discussed before. The formation of zinc selenide takes place in accordance with the overall stoichiometric equation:

$$[Zn(OH)_4]^{2-}(aq) + SeSO_3^{2-}(aq) \quad ZnSe(s) + SO_4^{2-}(aq) + 2OH^-(aq) + H_2O(l) \tag{22}$$

In order to initiate the process of precipitation of ZnSe, the presence of hydrazine in the reaction system is necessary. In the absence of this agent, the system remains homogeneous and the reaction does not occur at all. We attribute the role of hydrazine to its reduction properties [50,52] which are pronounced in alkaline medium:

$$\overset{0}{N}_2(g) + 4H_2O(l) + 4e \rightleftharpoons \overset{-2}{N}_2H_4(aq) + 4OH^-(aq) \; E^{\circ}_{N_2/N_2H_4} = -1.16 \text{ V} \tag{23}$$

$$Se\overset{+6}{S}O_3^{\;2-}(aq) + 2e \rightarrow Se^{2-}(aq) + \overset{+4}{S}O_3^{\;2-}(aq) \tag{24}$$

Due to the rather large number of chemical equilibria which are involved in the overall deposition process (*i.e.* the large number of relevant experimental variables affecting the thin film properties), it is essentially impossible to perform a full factorial experimental design of the optimal experimental conditions for synthesis of nanocrystalline ZnSe thin films. Besides the chemical complexity of the considered system, there are also several possible target functions with respect to which the optimization could be performed, which further complicate the applicability of the full factorial approach. Similar reasoning is valid for other semiconductor systems studied by our group and therefore the optimization of experimental conditions was carried out by a classical approach.

The chemical synthesis of SnSe by hydrothermal or related approach (either as bulk precipitate or in thin film form) appears to be a rather difficult task, in line with the rather low number of attempts to achieve this goal which have been published so far [53]. This is mostly due to the pronounced tendency of Sn^{2+} ions towards hydrolysis:

$$Sn^{2+}(aq) + H_2O(l) \rightleftharpoons Sn(OH)^+(aq) + H^+(aq) \; K^{\circ} = 1 \cdot 10^{-2} \tag{25}$$

Obviously, the previous equilibrium is displaced to left in acidic medium which prevents the hydrolysis of Sn^{2+} species. Having in mind this, we have dissolved the used tin(II) salt in acidic EDTA solution with c(EDTA) = 0.1 mol/dm^3 (to prevent the occurrence of hydrolysis process (25)). The pH value of the starting EDTA solution was 5. Since EDTA is weak four-base acid, four anionic species originating from this complexing agent may be present in the solution, depending on its pH value. By analysis of the distribution diagram for various $H_nY^{(4-n)-}$ species as a function of pH [54], it may be concluded that at pH = 5, the H_2Y^{2-} species are predominant in the solution. Having in mind the known stoichiometry of the reaction between Sn^{2+} ions and EDTA [54], the complex ion formation upon tin(II) salt dissolution may be thus represented by the following equation:

$$Sn^{2+}(aq) + H_2Y^{2-}(aq) \quad [SnY_4]^{2-}(aq) + 2H^{+}(aq) \tag{26}$$

In order to be able to precipitate SnSe from this reaction system, however, its pH value has to be raised significantly. We have achieved this goal by adding an appropriate amount of ammonia buffer to the deposition solution, keeping the pH value at 9 during the thin film deposition process.

The exact experimental conditions during the deposition process were optimized with respect to the photoelectrical performances of the deposited tin(II) selenide films. This was done since our main goal was to develop a controllable chemical route for synthesis of *nanocrystalline photoconductive* SnSe thin films. With this respect, the optimal temperature of the reaction system was found to be 60 °C. The obtained films were homogeneous and with a constant quality.

2.2.2. Metal Sulfides

In the case of deposition of metal sulfide thin films, either sodium thiosulfate or thioacetamide were used as sulfide ions generators. The first precursor was used for synthesis of Bi_2S_3 and $AgBiS_2$, while the second one was employed to synthesize nanocrystalline thin films of In_2S_3 and $CuInS_2$. Thioacetamide undergoes a hydrolytic process in aqueous medium:

$$CH_3CSNH_2(aq) + H_2O(l) \rightleftharpoons CH_3CONH_2(aq) + H_2S(aq) \tag{27}$$

and the formed H_2S further dissociates in two steps, finally generating sulfide anions in controllably low concentration. Thiosulfate anion, on the other hand, also undergoes a slow decomposition which may be represented by the equation [54]:

$$S_2O_3^{2-}(aq) + H_2O(l) \rightleftharpoons H_2S(aq) + SO_4^{2-}(aq) \tag{28}$$

However, at the same time, the $S_2O_3^{2-}$ species are in dynamical equilibrium with the sulfite ions:

$$SO_3^{2-}(aq) + S(s) \rightleftharpoons S_2O_3^{2-}(aq) \tag{29}$$

The last equilibrium is due to the fact that in acidic media, thiosulfate ions are converted to thiosulfuric acid, which is highly unstable and undergoes a disproportion according to the equation:

$$S_2O_3^{2-}(aq) + 2H^+(aq) \rightleftharpoons H_2S_2O_3(aq) \rightarrow S(s) + SO_2(g) + H_2O(l) \qquad (30)$$

It follows from the previously outlined deduction that the equilibrium (29), which involves thiosulfate and sulfite ions, is displaced to left in acidic media. On the other hand, thiosulfate ions possess certain reduction properties [51,52,54]:

$$S_4O_6^{2-}(aq) + 2e \rightleftharpoons 2S_2O_3^{2-}(aq) \quad E^\circ_{S_4O_6^{2-}/S_2O_3^{2-}} = 0.08\ V \qquad (31)$$

and may thus reduce the generated elemental S, whose oxidation properties increase upon decrease of the pH value. This may be concluded by analysis of the Nernst equation for the redox equilibrium:

$$S(s) + 2H^+(aq) + 2e \rightleftharpoons H_2S(aq) \qquad E^\circ_{S/H_2S} = 0.142\ V \qquad (32)$$

$$E_{S/H_2S} = E^\circ_{S/H_2S} - \frac{0{,}059\ V}{2}\log a(H_2S) - 0{,}059\ V\,pH \qquad (33)$$

In the case of Bi_2S_3 and $AgBiS_2$ thin film synthesis, thiosulfate ions, besides as sulfide-generating agents, also act as complexing agents, forming stable complexes with metal cations. For In_2S_3 and $CuInS_2$, on the other hand, sodium thiosulfate was also used as complexing agent. However, synthesis of high-quality thin films of the last two semiconductors was not possible in the absence of thioacetamide as sulfide ions generator. Much more technical details about the optimal deposition conditions may be found in our papers devoted to the mentioned systems [42,43].

2.3. The Sonochemical Synthesis

Besides the conventional chemical approaches, for synthesis of some of the studied thin film materials, we have used sonochemical routes as well. In the case of sonochemical synthetic approach, the deposition solutions were continuously irradiated by a high-intensity ultrasonic probe (100 W cm^{-2}; frequency 20 kHz) using a direct-immersion ultrasonic horn.

We have employed sonochemistry in our studies due to the fact that our primary aim was to develop synthetic methods for deposition of low-dimensional semiconductors in thin film form. It has been recognized that ultrasound irradiation may influence significantly the properties of the synthesized materials, mostly due to the specific way in which the energy is supplied to the reaction system. In fact, when a reaction system is irradiated by ultrasonic waves with sufficient intensity, extreme conditions are imposed in particular microareas of the reaction system. The chemical effects of the ultrasound are not a consequence of a direct interaction between ultrasonic waves and the atoms or molecules of the reaction systems. They arise due to the phenomenon of acoustic cavitation induced by the propagation of high-intensity ultrasound in liquid media [19-24,55-60]. Consider a one-dimensional acoustic wave spreading parallel to the x-axis. Due to the longitudinal character of the acoustic waves, an additional acoustic pressure is exhibited in the medium which simply superimposes to the existing hydrostatic pressure. Acoustic pressure is a periodic function of time, which may be expressed for a one-dimensional case mentioned before with [57,59]:

$$\widetilde{P} = A\omega\rho v\cos\left[\omega\left(t - \frac{x}{v}\right)\right] \tag{34}$$

In the previous equation, A is the amplitude of the ultrasonic wave, ρ is the density of the medium in which the wave is propagating, while v is the speed of sound propagation. From the previous equation it follows that the amplitude of hydrostatic pressure is given by the term $A\omega\rho v$. If this quantity is large enough, the local inhomogeneties formed in the liquid medium may grow into cavities (bubbles) with macroscopic dimensions filled with a vapor phase (this phenomenon is referred to as *acoustic cavitation*) [19-24,55-60]. These cavities are, however, instable and undergo an implosive collapse within very short time intervals. As a consequence of this implosive collapse the energy of ultrasonic wave is concentrated, *i.e.* localized in small volumes of the system, usually called hot points. These hot points are characterized by extremely high temperatures of about 5000 K, pressures of about 1000 atm and heating/cooling rates of the order of 10^{10} K s^{-1} [19-24,55-60]. In other words, acoustic cavitation enables existence of extreme physical and chemical conditions in localized areas of reaction system, whose overall temperature is, on the other hand, relatively low. From a hydrodynamic viewpoint, the extremely high temperatures of the hot spots in reaction systems owe to the fact that the implosion rate is significantly higher than the rate of heat transfer in the medium.

In sonochemical reaction systems two types of cavitation processes could be distinguished: *homogeneous* and *heterogeneous*. In the case of homogeneous cavitation, implosion collapse of the bubbles is spherically symmetric. This leads to formation of two sonochemical centers. The first one, which is of primary importance to sonochemical effects, is the vapor phase within the cavity formed in the course of cavitation process. The second sonochemical center corresponds to the liquid phase located in the immediate vicinity of the imploding microcavity. It has been shown that the temperatures within the two sonochemical centers are about 5000 K and 2000 K respectively [19-24,55-60]. When the reaction system is heterogeneous, *i.e.* contains phase boundaries (solid-liquid in the most important examples) the implosive collapse is not spherically symmetrical and leads to a variety of heterogeneous sonochemical effects. As the reaction systems encountered in our works are heterogeneous during the main part of the deposition processes, the heterogeneous sonochemical effects will be illustrated exactly for these cases. As mentioned before, the crystal growth phase of the overall deposition process of metal chalcogenide thin films occurs via cluster (*i.e.* colloidal) mechanism. A sequence of implosive collapses of bubbles formed by acoustic cavitation processes in the vicinity of phase boundaries leads to formation of shockwaves and microjects [55,56]. If the frequency of used ultrasonic irradiation is 20 kHz, the formed shockwaves and microjets will cause high-speed movement of the formed semiconductor nanoclusters which are smaller than 200 μm. On the other hand, in the case of semiconductor nanoclusters larger than 200 μm, microjets cause erosion of the exposed micro-solid phase surfaces. In this process the overall potential energy of the microcavity is transformed into the kinetic energy of the microjet. The high kinetic energy of nanoclusters enables high-energy collisions between them. This may result in reducing or increasing the average crystal size of the deposited semiconductor nanoparticles. The second effect occurs when the energy liberated during high-energy collisions is large enough that can cause local melting of the material nanoparticles. In the case of all investigated systems for which sonochemical deposition routes were optimized in our previous works, the ultrasonic irradiation of reaction systems resulted in decrease of the average crystal size of the deposited semiconductor nanoparticles.

The equations of the effective ionic reactions, on which the syntheses of the considered semiconducting nanomaterials in thin film form are based, in the presence of ultrasonic irradiation of the system may be written in the following form:

$$2In^{3+}(aq) + 3S^{2-}(aq) \xrightarrow{)))} In_2S_3(s) \quad (35)$$

$$Ag^{+}(aq) + Bi^{3+}(aq) + 2S^{2-}(aq) \xrightarrow{)))} AgBiS_2(s) \quad (36)$$

$$Cu^{+}(aq) + In^{3+}(aq) + 2S^{2-}(aq) \xrightarrow{)))} CuInS_2(s) \quad (37)$$

In (35)-(37) the symbol ")))" denotes that the reaction system has been ultrasonically irradiated.

The fact that ultrasonic irradiation influences the average crystal size of nanoparticles deposited as continual thin films onto glass substrates certainly implies that all of the optical and photoelectrical properties of the synthesized nanomaterials will be influenced by the ultrasonic treatment. These aspects will be covered in more details in the following chapters. In this context, it is important to mention that besides the differences in optoelectrical properties of the synthesized films, sonochemical effects also influence the overall film quality, the kinetics of the deposition processes and also the reaction yield. These macroscopic effects may be attributed to the more efficient mass transport induced by acoustic streaming, leading to more efficient turbulent mixing in particular microareas of the reaction systems. Formation of microjets and shockwaves, on the other hand, lead to activation of the reaction centers due to local increase of temperature. The mentioned processes induced by heterogeneous sonochemical effects also cause more efficient cleaning of the exposed surfaces relevant to thin film deposition processes, leading to improvement of the adhesion properties.

3. Structural Characteristics

3.1. Estimation of the Average Crystal Size of Nanocrystalline Semiconductor Thin Films and the Corresponding Bulk Precipitates

The Debye-Scherrer method of X-ray diffraction from polycrystalline samples was used to identify the synthesized materials as both bulk precipitates and in thin film form and also to derive conclusions about their structural characteristics. *X*-ray diffraction patterns were recorded on a Philips PW 1710 diffractometer, using monochromatic Cu-K_α radiation. On the basis of intrinsic broadening of the diffraction maxima, which arises from the small crystal size of the studied polycrystalline samples, using the approach of Debye-Scherrer, the average crystal size was determined for each of the studied systems. This was done for the materials deposited in thin film form and as bulk precipitates obtained from the same reaction system. The basic working equation, which was used for that purpose, was the Debye-Scherrer one [61,62]:

$$\langle d \rangle = \frac{4}{3} \frac{0.9\ \lambda}{\sqrt{\beta_m^2 - \beta_s^2}\ \cos\theta} \tag{38}$$

where $\langle d \rangle$ is the average diameter of the crystals (in spherical approximation), λ is the wavelength of used X-ray radiation and θ is the angle which corresponds to diffraction maximum. β_m is the raw value of the full width at half maximum intensity (FWHMI) of the peak (which contains both instrumental and intrinsic broadening factors), while β_s is a standard FWHMI value referring to a correspondent (closely appearing) peak in the case of macrocrystalline and strain-free Al_2O_3 sample. To determine precisely all of the parameters in eq. (38), the diffraction peaks were interpolated with linear combinations of Gaussian and Cauchy (Lorentzian) functions by the least-squares based methodology. Each peak was thus interpolated by a function of the form:

$$I(2\theta) = \gamma \frac{2}{H\pi}\left[1 + 4(2\theta)^2\right]^{-1} + (1-\gamma)\frac{\sqrt{4\ln 2}}{H\sqrt{\pi}}\exp\left[-4\ln 2(2\theta)^2\right] \tag{39}$$

where the parameters H and γ were determined by least-squares techniques. Note that the function of the form (39) is frequently called a "pseudo-Voigt" profile in the context of Rietveld analysis.

It was, however, found that neglect of the instrumental broadening in (38) leads to relative errors in the calculated average crystal radii of less than few percents (even less than 1 % in most cases). The experimentally obtained $<r>$ values calculated according to (38) are summarized in Table 1. A typical XRD pattern of a nanocrystalline thin film and the corresponding bulk precipitate is shown in Fig. 1 for a specimen of ZnSe (as deposited, as well as thermally treated at several temperatures). The average crystal radii of as-deposited thin films and the corresponding bulk precipitates synthesized from the same reaction systems were found to be identical in the case of each of the studied systems. This is a strong further support to the predominance of the cluster mechanism of crystal growth in all cases. As can be also seen from Table 1, all of the synthesized materials are nanocrystalline which is confirmed by other investigations carried out in our previous works as well [27-43].

As discussed briefly in the previous chapter, the influence of ultrasonic irradiation of the reaction systems on the average crystal size of the synthesized materials was rationalized within the concepts of heterogeneous sonochemical effects [19-24,55-60], having in mind the predominance of the cluster mechanism of crystal growth. These effects are a direct consequence of the implosive collapse of acoustically generated bubbles near phase boundaries, leading to periodical shockwaves and microjets. The formed colloidal nanoparticles in the reaction system under these conditions achieve relatively high velocities and are thus driven into frequent mutual collisions. These collisions could lead to either decrease or increase of the average crystal size. The second effect takes place when the locally generated energy during interparticle collision is large enough to cause local melting.

Table 1. Average crystal radii of synthesized materials in thin film form by the conventional chemical and sonochemical routes and their evolution upon post-deposition thermal annealing treatment

Material			<*r*>/nm
ZnSe	as-deposited		1.4
	thermally treated at 200 °C		1.8
	thermally treated at 250 °C		2.0
SnSe	as-deposited		14.8
	thermally treated at 150 °C		23.3
Bi_2S_3	as-deposited		9.0
	thermally treated at 250 °C		24.2
In_2S_3	as-deposited	chemical	2.5
		sonochemical	2.0
	thermally treated at 200 °C		4.1
$AgBiS_2$	as-deposited	chemical	8.4
		sonochemical	4.2
	thermally treated at 250 °C		27.1
$CuInS_2$	as-deposited	chemical	2.1
		sonochemical	1.7

As obvious from Table 1, in the case of all presently studied systems, ultrasonic irradiation of the reaction systems leads to effective *decrease* of average crystal size. Sonification of the reaction systems significantly affects the deposition kinetics if all other factors are unchanged. It also leads to a significant increase of the terminal thin film thickness. These experimental findings may be attributed to the more effective transport of reactants which occurs due to turbulent mixing in particular microareas of the reactor (taking place as a consequence of acoustic streaming) and also to more effective subsequent cleaning and activation of the substrate surface occurring as a consequence of ultrasonic irradiation of the systems.

Post-deposition thermal annealing of the thin films and bulk precipitates leads to increase of the average crystal radius of the nanoparticles (as can be seen from Table 1), which is reflected in their optical and photoelectrical properties as well, as discussed further in the text.

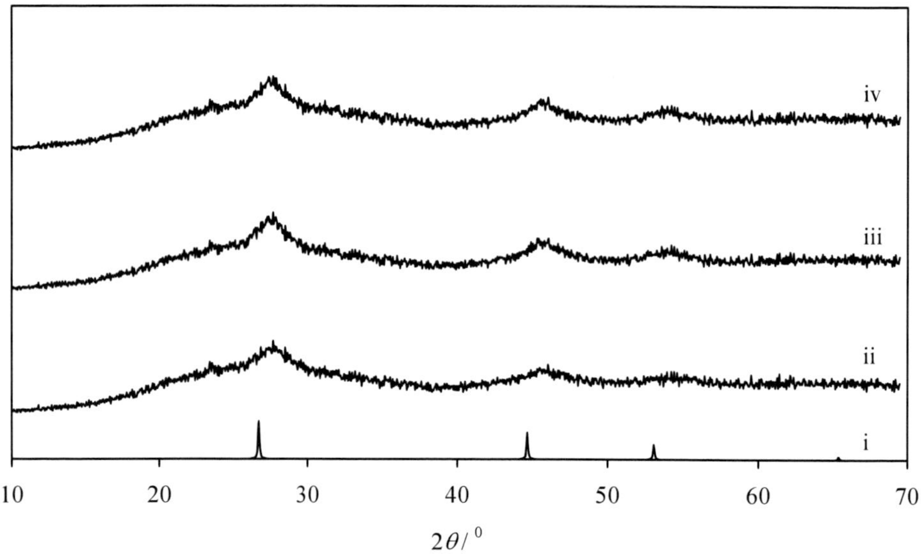

a)

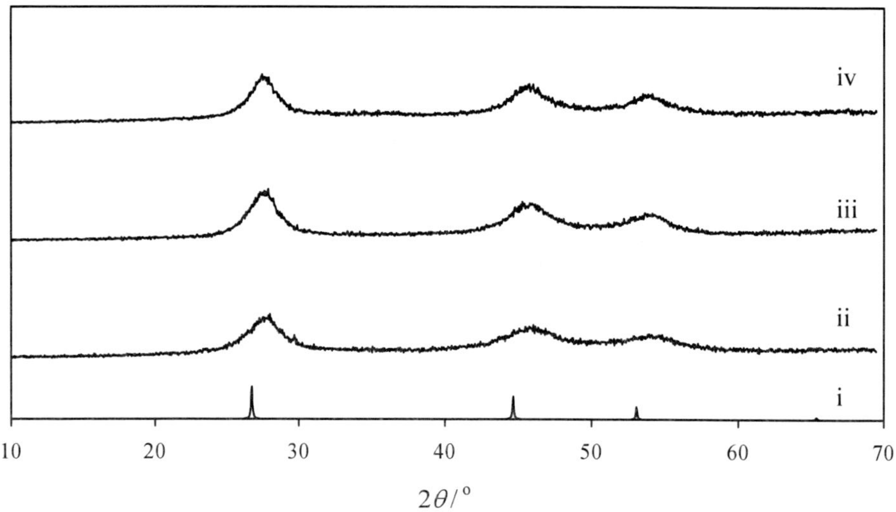

b)

Figure 1. XRD pattern of as-deposited and thermally treated nanocrystalline ZnSe thin film (a) and the corresponding bulk precipitate (b): i) - standard; ii) - as deposited sample; iii) sample annealed at 200 °C; iv) sample annealed at 250 °C.

3.2. Refinement of the Unit Cell Parameters of Low-Dimensional Semiconductor Structures

XRD analysis was also used to perform a refinement of the unit cell parameters in the case of synthesized materials. The refinement procedures were carried out on the basis of equations which arise from combination of the Bragg's law and the expressions which connect the interplanar distances d_{hkl} with the unit cell parameters, depending on the crystalline system [63]. The Bragg's law is usually written in the following form:

$$2d_{hkl}\sin\theta_{hkl} = n\lambda \tag{40}$$

where d_{hkl} is the interplanar distance between planes characterized by Miller indices h, k and l, λ is the wavelength of the incident X-ray radiation, while n is the diffraction (reflection) order. It is, however, convenient to write the Bragg's law in a different form:

$$2\frac{d_{hkl}}{n}\sin\theta_{hkl} = \lambda \tag{41}$$

Having in mind that:

$$\frac{d_{hkl}}{n} = d_{nh,nk,nl} \tag{42}$$

it follows that by suitable reindexing of the planes Bragg's law may always be written in the form:

$$2d_{nh,nk,nl}\sin\theta_{nh,nk,nl} = \lambda \tag{43}$$

i.e.:

$$2d_{h'k'l'}\sin\theta_{h'k'l'} = \lambda \tag{44}$$

In other words, higher-order reflections from planes with Miller indices h, k and l may be regarded as first-order reflections from planes with indices nh, nk and nl (h', k' and l'). In a most general case of triclinic crystallographic singony, on the other hand, the equation relating the interplanar distances to the unit-cell parameters has the form:

$$\begin{aligned}\frac{1}{d_{hkl}^2} &= (1-\cos^2\alpha-\cos^2\beta-\cos^2\gamma+2\cos\alpha\cos\beta\cos\gamma)^{-1}\cdot\\ &\cdot\left[\frac{h^2}{a^2}\sin^2\alpha+\frac{k^2}{b^2}\sin^2\beta+\frac{l^2}{c^2}\sin^2\gamma+\frac{2kl}{bc}(\cos\beta\cos\gamma-\cos\alpha)+\right.\\ &\left.+\frac{2lh}{ac}(\cos\gamma\cos\alpha-\cos\beta)+\frac{2hk}{ab}(\cos\alpha\cos\beta-\cos\gamma)\right]\end{aligned} \tag{45}$$

where a, b and c denote the modules of basis vectors defining the unit cell, and α, β and γ are the angles between these vectors. The refinement of unit cell parameters of the synthesized low-dimensional semiconductors was done using a combination of equations (44) and (45).

We are going to illustrate this procedure for a particular case of a semiconductor crystallizing in orthorhombic singony.

In the case of orthorhombic crystallographic system, the following equation holds:

$$\sin^2\theta_{hkl} = \frac{\lambda^2}{4a^2}h^2 + \frac{\lambda^2}{4b^2}k^2 + \frac{\lambda^2}{4c^2}l^2 \tag{46}$$

or, in a more compact form:

$$\sin^2\theta_{hkl} = Ah^2 + Bk^2 + Cl^2 \tag{47}$$

Obviously, then, $\sin^2\theta_{hkl}$ should depend linearly on h^2, k^2 and l^2 as three independent variables. Using our experimental XRD data for as-deposited and annealed films belonging to orthorhombic crystal system, we performed multiple regression analyses for both cases to determine the values of constants A, B and C in (47). Subsequently, the unit cell parameters were calculated. The results from these correlation analyses are summarized in Tables 2-5. Multiple regression analyses were performed with the Statistica program package [64].

Table 2. The calculated unit cell parameters in the case of synthesized nanocrystals of the cubic modifications of ZnSe, In_2S_3 and $AgBiS_2$ in thin film form

thin film		R^2	A	a/Å
ZnSe	as-deposited	0.9998	0.01898 ± 0.00009	5.5885
	thermally treated at 200 °C	0.9997	0.01888 ± 0.00010	5.6033
	thermally treated at 250 °C	0.9998	0.01887 ± 0.00009	5.6054
In_2S_3	as-deposited	0.9997	0.02074 ± 0.00017	5.3467
	thermally treated at 200 °C	0.9994	0.02047 ± 0.00013	5.3819
$AgBiS_2$	as-deposited	1	0.01841 ± 0.00001	5.6750
	thermally treated at 250 °C	1	0.01850 ± 0.00001	5.6612

Table 3. The calculated unit cell parameters in the case of synthesized nanocrystals of the tetragonal modification of $CuInS_2$ in thin film form

$CuInS_2$	R^2	A	C	a/Å	c/Å
as-deposited	1	0.02123 ± 0.00008	0.00515 ± 0.00005	5.2850	10.7297

Table 4. The calculated unit cell parameters in the case of synthesized nanocrystals of the orthorhombic modification of Bi_2S_3 in thin film form

Bi_2S_3	as-deposited	thermally treated at 250 °C for 10 min
R^2	0.9997	0.9997
A	0.005106 ± 0.000042	0.005092 ± 0.000043
B	0.004544 ± 0.000088	0.004538 ± 0.000091
C	0.035090 ± 0.000907	0.035133 ± 0.000944
a / Å	10.7758	10.7906
b / Å	11.4228	11.4303
c / Å	4.1110	4.1080

Table 5. The calculated unit cell parameters in the case of synthesized nanocrystals of the orthorhombic modification of SnSe in thin film form

SnSe	as-deposited	thermally treated at 150 °C
R^2	0.999999	0.999999
A	0.004549 ± 0.000001	0.004548 ± 0.000001
B	0.033803 ± 0.000006	0.033792 ± 0.000006
C	0.029842 ± 0.000004	0.029832 ± 0.000004
a / Å	11.4209	11.4220
b / Å	4.1897	4.1904
c / Å	4.4591	4.4598

3.3. The Lattice Strain and Dislocation Density In Low-Dimensional Semiconductor Structures Deposited as Thin Films

On the basis of XRD data, further conclusions about the existence and magnitude of lattice strain in the synthesized thin films were derived. This was done employing two approaches as explained in details in our papers [39-43]. For the purpose of this review, we briefly recapitulate few basic aspects concerning the issue of quantitative specification of the state of deformation in a solid specimen. As for all other physical properties of crystals, the deformation with respect to the equilibrium state has to be rigorously explained via tensor formalism [25,65]. For the purpose of clarity of the presentation, we will first focus on one-dimensional case.

Consider an extendible string placed parallel to the x-axis of a laboratory coordinate system (Fig. 2).

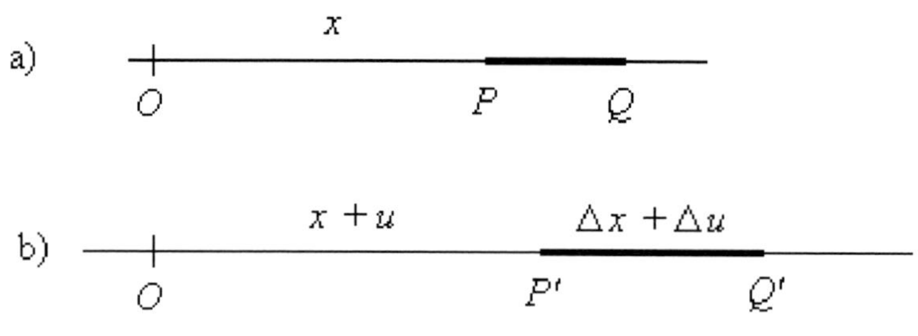

Figure 2. Elastic deformation of an extendible string: a) unstretched; b) stretched.

Let us fix the coordinate system origin at point O. Upon stretching, let the point P move to the position P' and denote the OP and OP' distances by x and $x + u$ respectively. The displacement u may vary either linearly or non-linearly with distance x. In the first case, the stretching of the string is homogeneous, while the nonlinearity of the function $u(x)$ implies an inhomogeneous stretch. If Q is an arbitrarily chosen point on the string (placed close to P), upon stretching deformation it will move to Q'. Denoting the initial distance between P and Q by Δx, it follows that $P'Q' = \Delta x + \Delta u$. However, when studying the strain in crystalline materials, one is actually not concerned about the absolute values of the displacements of points, but rather about their relative displacements. The average strain of the sequence PQ is defined as:

$$\langle \varepsilon \rangle = \frac{P'Q' - PQ}{PQ} = \frac{\Delta u}{\Delta x} \tag{48}$$

The actual strain at the point P is a natural generalization of (48), *i.e.*:

$$\varepsilon = \lim_{Q \to P} \frac{P'Q' - PQ}{PQ} = \lim_{\Delta x \to 0} \frac{\Delta u}{\Delta x} = \frac{du}{dx} \tag{49}$$

In the case of a homogeneous strain, the previous equation may be integrated to give:

$$u = u_0 + \varepsilon x \tag{50}$$

where u_0 denotes the displacement of the point at the origin. Generalization to a two-dimensional case (considering deformation of an extendible plane sheet) implies that in order to specify the strain at an arbitrary point within the plane (x_1, x_2) four quantities of the type (49) are required. Within the language of tensor calculus, this physical property is a second-rank tensor. Its components are defined as:

$$\varepsilon_{ij} = \frac{\partial u_i}{\partial x_j}; \ (i, j = 1, 2) \tag{51}$$

Subsequently:

$$\Delta u_i = \sum_j \frac{\partial u_i}{\partial x_j} \Delta x_j = \sum_j \varepsilon_{ij} \Delta x_j \tag{52}$$

Finally, for a complete specification of the strain in a three-dimensional body (such as a crystal), nine components of the form (49) are required, *i.e.* one deals with a third-rank tensor:

$$\varepsilon_{ij} = \frac{\partial u_i}{\partial x_j};\ (i, j = 1, 2, 3) \tag{53}$$

In the case of a homogeneous distortion, all components of the strain tensor are constants and we may write:

$$u_i = (u_0)_i + \sum_j \varepsilon_{ij}\, x_j \tag{54}$$

In the case of semiconductor nanocrystals deposited in thin film form, studied by our group, the lattice strain is of residual, *i.e.* internal character, arising from a large fraction of surface atoms and dangling bonds [25,66]. As can be seen from the Table 2 given in the previous chapter, the lattice constant a in the case of as-deposited ZnSe quantum dots in thin film form is significantly smaller than the bulk literature value and it converges to the bulk value upon thermal annealing. Thus, the average $\Delta a = a_{\text{bulk}} - a_{\text{nanocrystal}}$ value is 0.0785 Å in the case of as-deposited ZnSe quantum dot thin films obtained from sulfate reaction solution, while it reduces to 0.0616 Å upon thermal annealing at 250 °C. Thus, in the case of ZnSe thin films, the intrinsic lattice strain is of the order of 1 % with a decreasing trend upon annealing and the average $a_{\text{nanoparticle}}$ value converging to the a_{bulk} one. This convergence seems to be quite faster in the case of thin films obtained from the acetate reaction solution, as discussed in more details in Ref. [39]. If we adopt a simple definition of the isotropic average lattice strain as [25]:

$$\langle \varepsilon \rangle = \frac{|\Delta a|}{a_{\text{bulk}}} \tag{55}$$

where $\Delta a = a_{\text{bulk}} - a_{\text{nanocrystal}}$ is the difference between the bulk and nanocrystal lattice constant, then according to the previously outlined data in Table 2, the lattice strain is relaxed upon thermal annealing. A somewhat more involved treatment of the lattice strain, based on the Bragg's law may be derived as follows.

In the case of a polycrystalline aggregate in which the crystals are bent, stretched or compressed etc., due to such lattice strains the reflections from various sets of lattice planes will be shifted with respect to the "unstrained" values. Let us consider an ensemble of quantum dots with large fraction of surface atoms and dangling bonds, leading to "uni-directional" strain, *i.e.* to shrinkage of the nanocrystalline volume in comparison to the case when the dangling bonds are absent. If we consider a reflection from a set of planes with the interplanar spacing d_{hkl} (in the case of an unstrained crystal), the Bragg's law for the unstrained (bulk) crystal, for the first order reflections ($n = 1$) may be written as:

$$2d_{hkl} \sin \theta_{hkl} = \lambda \tag{56}$$

If the strain in the nanocrystal is denoted by ε then the interplanar spacing between the previously considered set of planes will be d_{hkl} $(1 + \varepsilon)$ and the corresponding Bragg angle θ_{hkl} + $\Delta\theta_{hkl}$, leading to the following form of the Bragg's law [67]:

$$2d_{hkl}(1+\varepsilon)\sin(\theta_{hkl} + \Delta\theta_{hkl}) = \lambda \tag{57}$$

After a few elementary trigonometric transformations, it could be easily shown that the previous expression gives:

$$\Delta\theta_{hkl} = -\varepsilon \tan(\theta_{hkl}) \tag{58}$$

which allows the average lattice strain to be calculated straightforwardly from the experimental data. The ε values calculated according to this method are in excellent agreement with those obtained on the basis of eq. (55), which further confirms the correctness of the outlined approach.

A somewhat larger value for ε was found in the case of $CuInS_2$ films, of the order of 4 %, with the same trend upon annealing. For all other synthesized semiconductors the lattice strain was found to be negligible (*i.e.* less than 1 %).

The dislocation density δ (defined as the overall dislocation lines length per unit crystal volume, or, equivalently, as the number of dislocations that intersect a unit area of a random section) was calculated using the expression [68]:

$$\delta = \frac{n}{D^2} \tag{59}$$

where n is a constant, with a value which is usually close to 1 [69]. The results are summarized in Table 6.

As can be seen, along with the lattice strain, upon thermally-induced average crystal size increase, the dislocation density decreases in the case of all studied systems. The values for this quantity were further used as quantitative measures of the degree of structural disorder in the synthesized nanocrystalline materials in analyzing the sub-band-gap absorption tails and other phenomena related to their optical properties.

Table 6. The dislocation densities of synthesized materials in thin film form by the conventional chemical and sonochemical routes and their evolution upon post-deposition thermal annealing treatment

thin film			δ / nm^{-2}
ZnSe	as-deposited		$10 \cdot 10^{-2}$
	thermally treated at 200 °C for 1 hour		$8 \cdot 10^{-2}$
	thermally treated at 250 °C for 1 hour		$6 \cdot 10^{-2}$
SnSe	as-deposited		$10 \cdot 10^{-4}$
	thermally treated at 150 °C for 1 hour		$5 \cdot 10^{-4}$
Bi_2S_3	as-deposited		$30 \cdot 10^{-4}$
	thermally treated at 250 °C for 1 hour		$4 \cdot 10^{-4}$
In_2S_3	as-deposited	chemical	$4 \cdot 10^{-2}$
		sonochemical	$6 \cdot 10^{-2}$
	thermally treated at 200 °C for 1 hour		$1 \cdot 10^{-2}$
$AgBiS_2$	as-deposited	chemical	$30 \cdot 10^{-4}$
		sonochemical	$100 \cdot 10^{-4}$
	thermally treated at 250 °C for 1 hour		$3 \cdot 10^{-4}$
$CuInS_2$	as-deposited	chemical	$6 \cdot 10^{-2}$
		sonochemical	$9 \cdot 10^{-2}$

4. Optical Properties of Semiconductor Quantum Dots Deposited in Thin Film Form Studied by Absorption Electronic Spectroscopy in the UV-VIS-NIR Spectral Range

4.1. Band Structure of Semiconductors and Absorption of Electromagnetic Radiation

The optical (electronic) properties exhibited by semiconducting compounds are crucial in a large number of materials applications. These properties are, on the other hand, determined by the band structures of the solid materials. According to the contemporary solid-state physics, the band structure of semiconducting materials is a function of a three-dimensional wave vector ($\vec{k}$) within the Brillouin zone which depends on the crystal structure and corresponds to the Wigner-Zeits-type primitive unit cell of the reciprocal lattice [17,70]. The simplest method for probing the band structure of a semiconducting material is to measure its absorption spectrum. In general, the optical absorption by semiconductors may be roughly divided into two parts: absorption below the band gap (*i.e.* the gap separating the highest occupied valence band and the lowest unoccupied conduction band) and absorption above the band gap. The phenomena of sub-band gap absorption may be related to free-carrier absorption in strongly degenerate semiconductors, localized tail states in amorphous and strongly disordered systems, existence of impurity levels in the forbidden energy gap, as well as Urbach and excitonic absorption [17,70-73]. The last mechanism is particularly relevant at very low temperatures. Although we have been involved in studies of these phenomena, in the present review we are going to cover only the aspects related to absorption of electromagnetic radiation above the fundamental gap.

When a semiconductor thin film is studied by absorption/transmission spectroscopy, usually the spectral dependence of transmission coefficient (transmittance) is measured, *i.e.* the function $T(\hbar\omega)$. The spectral dependence of absorption coefficient $\alpha(\hbar\omega)$ is subsequently calculated by:

$$\alpha(\hbar\omega) = \frac{1}{d}\ln\frac{I_0(\hbar\omega)}{I(\hbar\omega)} \tag{60}$$

where d is the film thickness and $I_0(\hbar\omega)$ and $I(\hbar\omega)$ are spectral dependencies of the incident and the transmitted light intensities correspondingly. In our investigations, we have determined thin film thickness by either gravimetric or interferometric methods [74]. In the case of the latest approach, the following equation for determination of d was employed:

$$d = \frac{1}{2}\frac{\lambda_1\lambda_2}{n(\lambda_1)\lambda_2 - n(\lambda_2)\lambda_1} \tag{61}$$

where λ_1 and λ_2 are wavelengths of first and second interferometric maximum, while $n(\lambda_1)$ and $n(\lambda_2)$ are refractive indexes at λ_1 and λ_2 correspondingly.

In order to probe the band structure of semiconductor thin films on the basis of experimentally measured spectral dependencies of α, the relation between this dependence and the characteristic features of the semiconductor band structure provided by the band theory of solid state has to be used [71,75-81]. According to this theory, the function $\alpha(\omega)$ is given by:

$$\alpha(\omega) = \frac{2\pi}{V}\frac{e^2|p_{CV}|^2}{m_0^2\varepsilon_0 c_0 n\omega}\sigma_{CV}(\omega) \tag{62}$$

where V is the volume of the semiconductor, p_{CV} is the transition dipole moment matrix element, σ_{CV} is the joint (combined) density of states while all other symbols have their usual meanings. The joint density of states is given by:

$$\sigma_{CV}(\omega) = \frac{V}{4\pi^3}\int\delta[\hbar\omega_C(k) - \hbar\omega_C(k) - \hbar\omega]d^3k = V\frac{k^2}{\pi^2}\left(\frac{dE_C(k)}{dk} - \frac{dE_V(k)}{dk}\right)^{-1} \tag{63}$$

In the case of spherical parabolic bands, the dispersion relations may be expressed by:

$$E_V(k) = \frac{\hbar^2k^2}{2m_V^*} \tag{64}$$

and

$$E_C(k) = \frac{\hbar^2 k^2}{2m_C^*} + E_g \tag{65}$$

for the valence and conduction band correspondingly. In the last equations, m_V^* and m_C^* are hole and electron effective masses correspondingly. Due to the supposed spherical symmetry of the bands, the wave vector is written as a scalar quantity. Substituting the last two expressions in (63) for a joint density of states the following relation is obtained:

$$\sigma_{CV}(\omega) = V\frac{(2\mu^*)^{3/2}}{2\pi^2\hbar^3}\sqrt{\hbar\omega - E_g} \tag{66}$$

Using this final expression for the joint density of states, one is led to the following working expression for calculation of the band gap energy of semiconductors on the basis of their optical absorption spectra:

$$\alpha(\omega)\hbar\omega = A\sqrt{\hbar\omega - E_g} \tag{67}$$

or, in a more familiar form:

$$(\alpha(\omega)\hbar\omega)^2 = const.(\hbar\omega - E_g) \tag{68}$$

Note that both of the constant multipliers in the last two equations contain the transition dipole moment matrix element p_{CV}. It follows from the last equation that in case when p_{CV} has a non-zero value (*i.e.* in the case of dipole-allowed band to band transitions) the band gap energy may be found by plotting the semiconductor absorption function $(\alpha(\omega)\hbar\omega)^2 = f(\hbar\omega)$ and subsequently extrapolating the linear part to $\alpha(\omega)\hbar\omega = 0$. Equation (68) has been derived on the basis of Fermi's golden rule for transitions between continually distributed states (energy bands), employing parabolic approximation for the dispersion relation. Also, the electrostatic interaction between photogenerated charge carriers has not been taken into account in derivation of (68). It is therefore expected that a linear dependence of $(\alpha(\omega)\hbar\omega)^2$ on $\hbar\omega$ would hold in the vicinity of $\hbar\omega = E_g$. This is so since the parabolic approximation for *E*(*k*) is valid near the critical points of the joint density of states. We have therefore determined band gap energies in the case of the studied systems by a careful analysis of the experimental data for $\alpha(\hbar\omega)$ on the basis of combined interpolation-extrapolation procedures, as will be discussed below. It should also be noted that eq. (68) is valid for direct gap semiconductors (for which the valence band maximum and the conduction band minimum correspond to the same points in *k*-space, *i.e.* the band to band transition is not accompanied by a change in electronic momentum).

In cases when the transition dipole moment matrix element is zero, the corresponding transition is dipole forbidden. Expanding the p_{CV} in McLaurin series in vicinity of the relevant k-point, it may be shown that the following relation holds for such a case:

$$\left(\alpha_{forb}(\omega)\hbar\omega\right)^{2/3} = const.\left(\hbar\omega - E_g\right) \tag{69}$$

When a semiconductor has an indirect band structure, *i.e.* when the valence band maximum and the conduction band minimum correspond to different points in momentum space, the band to band transition is accompanied by a change in electronic momentum. In accordance with the energy and momentum conservation laws, for such transition to be possible, a lattice phonon has to be involved, which in fact supplies the necessary momentum and has a negligible energy in comparison to E_g. Such transitions are often called phonon-assisted. Denoting the phonon energy by $\hbar\Omega$ and employing the Bose-Einstein statistics to describe the filling of phonon states, the following equation for the spectral dependence of absorption coefficient may be derived for dipole-allowed indirect band to band electronic transitions [71,75-81]:

$$\alpha_{ind}(\hbar\omega)\hbar\omega = const\left[\frac{\left(\hbar\omega - E_{\mathrm{g,ind}} - \hbar\Omega\right)^2}{1-\exp\left(-\dfrac{\hbar\Omega}{k_B T}\right)} + \frac{\left(\hbar\omega - E_{\mathrm{g,ind}} + \hbar\Omega\right)^2}{\exp\left(-\dfrac{\hbar\Omega}{k_B T}\right)-1}\right] \tag{70}$$

In the last equation, the constant arises from the Fermi's golden rule for fundamental electronic transitions within the framework of parabolic approximation for the dispersion relation, the sign + or – before $\hbar\Omega$ refers to emission or absorption of phonons in the course of light absorption correspondingly. Since usually $\hbar\Omega \ll E_g$, it could be easily shown that the previous relation gets a much simpler form:

$$(\alpha_{\mathrm{ind}}(\omega)\hbar\omega)^{1/2} = const(\hbar\omega - E_{\mathrm{g,ind}}) \tag{71}$$

For a dipole-forbidden indirect electronic transition, on the other hand, the corresponding semiconductor absorption function has the form:

$$(\alpha_{\mathrm{ind}}(\omega)\hbar\omega)^{1/3} = const(\hbar\omega - E_{\mathrm{g,ind}}) \tag{72}$$

On the basis of the previously outlined arguments, by constructing, plotting and analyzing the semiconductor absorption functions of the form $\left(\alpha(\omega)\hbar\omega\right)^n = f(\hbar\omega)$ for all physically acceptable values of n in the case of the semiconducting thin films synthesized by our group, we have determined both the types of electronic transitions and the corresponding band gap energies. This was done by the following procedure. Linear least-squares fits of the $\left(\alpha(\omega)\hbar\omega\right)^n$ *vs.* $\hbar\omega$ dependencies (for all physically acceptable values of n) were carefully carried out in the relevant energy ranges, with successive inclusion or elimination of a number of neighboring points in the correlation ranges and parallel monitoring of the R^2 value. In this way, the sets of relevant points belonging to the linear $(\alpha(\omega)\hbar\omega)^n$ *vs.* $\hbar\omega$ dependency corresponding to (or, more precisely, originating from) a given band-to-band transition were determined. Subsequent extrapolations to $\alpha(\omega)\hbar\omega = 0$ were used to obtain the corresponding transition energies, on the basis of previously derived correlation equations. Also, the order of magnitude of the absorption coefficient α in the region above the band gap served as an indication for the type of interband transition.

4.2. Optical Properties of ZnSe Quantum Dots Deposited in Thin Film Form

In order to provide a good theoretical basis for interpretation of optical properties of three-dimensional quantum dots of ZnSe, we will begin with a brief description of the band structure of the corresponding bulk material. The most important features of bulk ZnSe band structure (near $k = 0$) are given in Fig. 3. The conduction band of this semiconducting compound arises predominantly due to the overlap of metal atom *s* orbitals, whereas the valence band is mostly constituted of the selenide anion *p* ones [82-84]. The conduction band is thus non-degenerate and nearly isotropic while the valence band is threefold orbitally degenerate. Accounting for the spin-orbit interaction, at $k = 0$ the valence band is split into an upper (in an energetic sense) Γ_8 component and a lower Γ_7 component. Including spin, the conduction band Γ_6 is twofold degenerate, the valence band Γ_8 component is fourfold degenerate, while the corresponding Γ_7 component is twofold degenerate. This language will be very useful in interpreting the electronic absorption spectra of the non-quantized ZnSe thin films. However, in order to have a theoretical basis for assignment of the electronic transitions in the case of quantized films, the effects of size-quantization (*i.e.* three-dimensional quantum confinement in the present case) on the degenerate valence band have to be known. According to the model proposed by Chestnoy, Hull and Brus [83], based on previous works by Baldereschi and Lipari [84], as a result of the three-dimensional confinement of the electronic and hole motion, the band structure of small ZnSe clusters gets a discrete character – Fig. 4. As can be seen from Fig. 4, this analysis predicts existence of two low-lying hole states which originate from valence band. They are split by a value Δ, which is expected to be similar to the Δ_{SO} (the spin-orbit splitting, usually denoted by Δ_0 as well) in the case of bulk material.

The $(\alpha h\nu)^2$ *vs.* $h\nu$ dependence of as-deposited ZnSe thin film is presented in Fig. 5. With extrapolation of linear dependence of $(\alpha h\nu)^2$ *vs.* $h\nu$ we have obtained a value of 3.10 eV for the band gap energy of as-deposited ZnSe thin film. This value refers to room temperature. The blue shift of room temperature band gap energy of 0.52 eV in the case of our chemically deposited zinc selenide thin films, compared with the corresponding value of bulk material (2.58 eV [85]), is due to the confinement effects *i.e.* the ZnSe crystals behave as quantum dots (three-dimensionally confined structures).

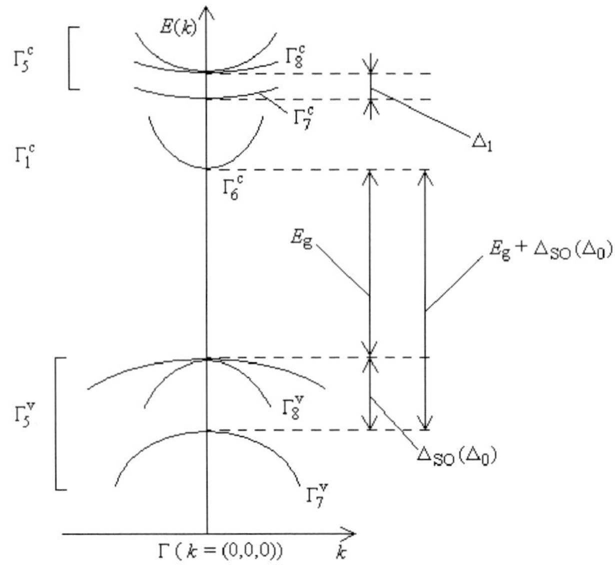

Figure 3. The most important features of bulk ZnSe band structure (near $k = 0$).

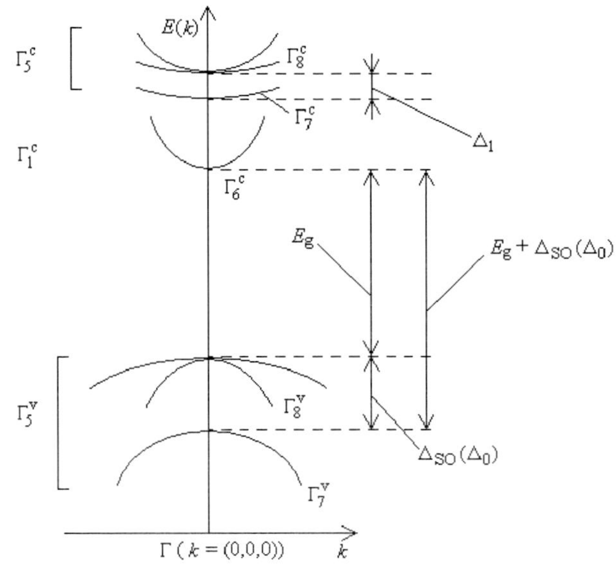

Figure 4. The discrete energy level structure of small ZnSe clusters, according to [83].

The zinc selenide quantum dots existence is in agreement with the average crystal size estimation on the basis of X-ray line broadening, as well as with the estimated value of the Bohr excitonic radius for bulk ZnSe given in our paper [39] and elaborated further in this review. Besides the transition appearing at 3.10 eV, as can be seen from Fig. 5, an additional electronic transition is also observed at about 3.50 eV. The splitting between these two transitions is thus nearly 0.40 eV. Accounting for the discrete energy level structure of small ZnSe clusters given in Fig. 4, the nature of the two transitions observed in the case of ZnSe quantum dots in thin film form could be easily explained. We attribute these two gap energies to the transitions from the low-lying hole states (split by the spin-orbit interaction) to the 1*S* electronic state arising from the bulk conduction band (denoted by arrows in Fig. 4). The "next" allowed $1P \rightarrow 1P$ transitions are expected to be considerably further blue-shifted. It may be thus concluded that both of the energetically lowest-lying electronic transitions in the case of ZnSe quantum dots in thin film form are blue-shifted with respect to the expected transitions in the case of bulk material. The splitting between them (which is essentially due to the spin-orbit interaction), however, remains essentially unchanged.

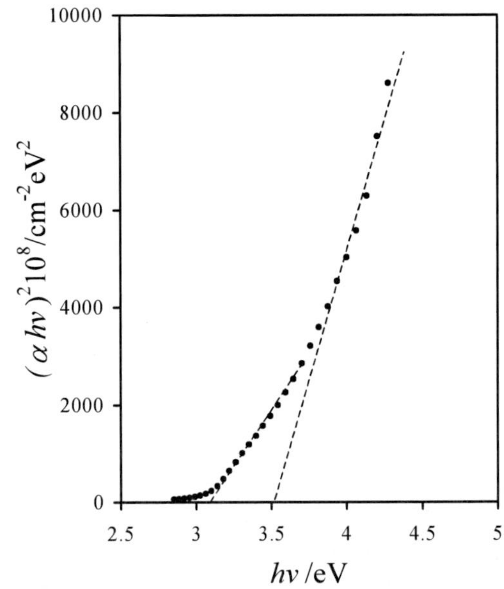

Figure 5. The $(\alpha h\nu)^2$ *vs.* $h\nu$ dependence of as-deposited ZnSe thin film together with the extrapolations of linear parts to $\alpha h\nu = 0$.

The reasons behind the unchanged spin-orbit splitting energy of the valence band components, due to particle size reduction, have been proposed in a detailed and rigorous analysis given in the original paper by Chestnoy and co-workers [83]. In the case of three-dimensionally confined ZnSe nanostructures the bulk ZnSe valence band components Γ_7 and Γ_8 are split into discrete levels, which may be characterized with various overall angular momentum, as depicted in Fig. 4. On the other hand, the low-lying $1S$ hole states arising from valence band Γ_7 and Γ_8 components are relatively pure (*i.e.* do not mix with other, higher angular momentum states) in the case of small nanocrystallites. As shown by the analyses of Chestnoy and Baldereschi [83,84], both of these states would be expected to shift together to higher (in absolute value) energies via the *isotropic* hole mass (not the light or heavy hole mass). This, on the other hand, means that although there is overall band gap blue shift (manifested as a shift in optical transition energies), the splitting between the two transitions remains unchanged.

The next issue that we address is the influence of post-deposition thermal treatment of the films on the optical absorption properties. The post-deposition annealing of the films leads to a continual increase of the average crystal size, which is accompanied by a continual red shift of the absorption onset [39]. Films annealed at 250 °C are characterized by an absorption onset at ≈ 475 nm (≈ 2.60 eV), essentially equal to the bulk value. A plot of $(\alpha h\nu)^2$ *vs.* $h\nu$ in the case of a film annealed at 250 °C with thickness of 75 nm is shown in Fig. 6.

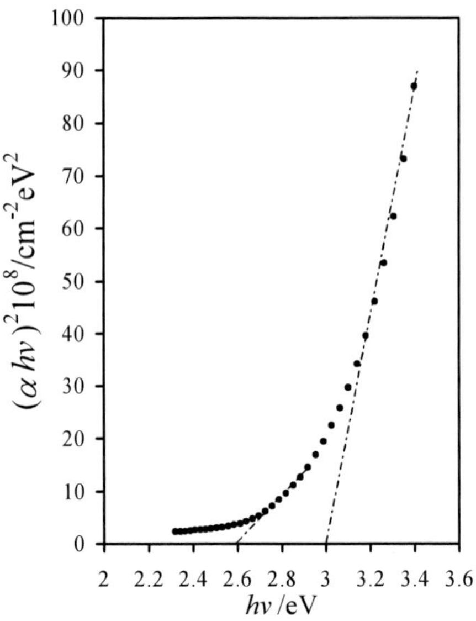

Figure 6. The $(\alpha h\nu)^2$ *vs.* $h\nu$ dependence of ZnSe thin film annealed at 250 °C together with the extrapolations of linear parts to $\alpha h\nu = 0$.

The type of electronic transitions remains unchanged upon thermal annealing, as well as the order of magnitude of α above the absorption onset. Besides the energetically lowest-lying electronic transition at 2.60 eV, an additional one was detected at 3.00 eV. The first value corresponds to the band gap value of bulk ZnSe, i.e. using the notation from Refs. [83,84], to direct transitions from the highest valence band to the lowest conduction band at the Γ point ($\Gamma_{15}^{v} \rightarrow \Gamma_{1}^{c}$ in single-group notation). Since the spin-orbit interaction splits the Γ_{15}^{v} valence band into Γ_{8}^{v} and Γ_{7}^{v} components (with splitting energy Δ_0), the corresponding optical transitions at $k = 0$ (or near $k = 0$) relevant to the present study are: the $\Gamma_{8}^{v} \rightarrow \Gamma_{6}^{c}$, occurring at energy E_g (or also denoted as E_0), and the $\Gamma_{7}^{v} \rightarrow \Gamma_{6}^{c}$ transition, occurring at energy $E_g + \Delta_0$ ($E_0 + \Delta_0$).

Another important aspect that we paid attention to in our studies was the evolution of the magnitude of α upon annealing treatment of the films. Comparisons between the $(\alpha h\nu)^2$ *vs.* $h\nu$ dependencies for the cases of ZnSe thin films annealed at various temperatures (*i.e.* with continually increasing average crystal sizes) showed a systematically decreasing trend of the absorption coefficient α (at a given photon energy) upon increase of the average crystal size [39]. The size dependence of the exciton oscillator strengths in the case of nanocrystalline semiconductors is still only partially solved problem [86]. We have employed the following arguments to explain the experimentally observed situation in our case. The oscillator strength f of the exciton is given by the expression [86]:

$$f = \frac{2m}{\hbar^2}\Delta E|\mu|^2|U(0)|^2 \quad (73)$$

where m is the electron mass, ΔE the energy of the transition, μ is the transition dipole moment, while $|U(0)|^2$ represents the probability of finding the electron and hole at the same site (a factor of overlap). The absorption coefficient is determined by the ratio of the overall (total) oscillator strength per cluster – $f_{cluster}$ and the cluster volume V (*i.e.* the ratio $f_{cluster}/V$). In the case of clusters with radius larger than the Bohr exciton radius ($R >> r_B$), the overlap factor $|U(0)|^2$ is size-independent, the macroscopic transition dipole moment determining the oscillator strength. In cases when $R < r_B$, $|U(0)|^2$ increases upon cluster volume decrease. The $f_{cluster}$ value is thus only weakly dependent on the cluster size in this regime. The absorption coefficient (*i.e.* the ratio $f_{cluster}/V$) is thus expected to increase upon decrease of the cluster size. This is exactly observed in our experiments, although the absorption bands corresponding to excitonic transitions are still wiped away by the continuum absorption.

Another interesting aspect of the optical properties of nanocrystalline ZnSe thin films is related to the fact that ZnSe quantum dots are closely packed when deposited in thin film form. As shown in several recent studies, certain interdot electronic coupling may occur in the case of closely-packed quantum dot solids [87-92]. According to the recently accumulated data [87-92], such coupling would be expected to occur only between very closely packed and very small quantum dots, mainly by tunneling mechanism, in which the charge carrier leakage outside the nanoparticle is significant. The main manifestations of such interdot coupling in the optical properties of the solid materials are broadening of the excitonic peaks (as compared to diluted ensembles of quantum dots). In the case of presently studied ZnSe quantum dots in thin film form, the non-observance of clearly defined excitonic absorption peaks, besides to other relevant reasons, may be attributed to exactly such additional broadening occurring as a result of interdot electronic couplings. The observance of a structureless absorption spectrum in the case of our chemically deposited ZnSe quantum dot thin films is thus a consequence of the formation of collective electronic states in an ensemble of closely-packed quantum dots. These states, which are delocalized within a finite number of nanocrystals, are formed due to the interdot electronic coupling. Post-deposition annealing, however, similarly as in the case of CdSe quantum dots, significantly reduces the possibility of interdot electronic coupling due to the crystal size increase as a result of sintering [87-92].

4.3. Optical Properties of Nanocrystalline SnSe Thin Films. Evidence for Size-Quantization Of The Indirect Band Gap Energy

By analysis of the corresponding semiconductor absorption functions, it was found that the as-deposited nanocrystalline SnSe thin films are characterized by dipole-allowed indirect band-to-band electronic transition appearing 1.20 eV and a direct one (also allowed in dipole approximation) at 1.74 eV (Fig. 7).

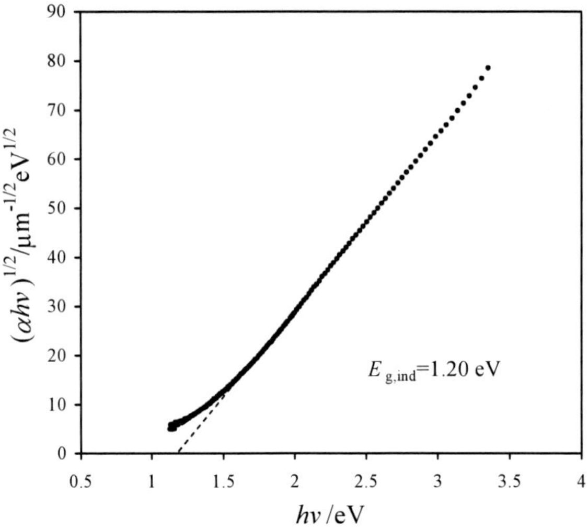

a)

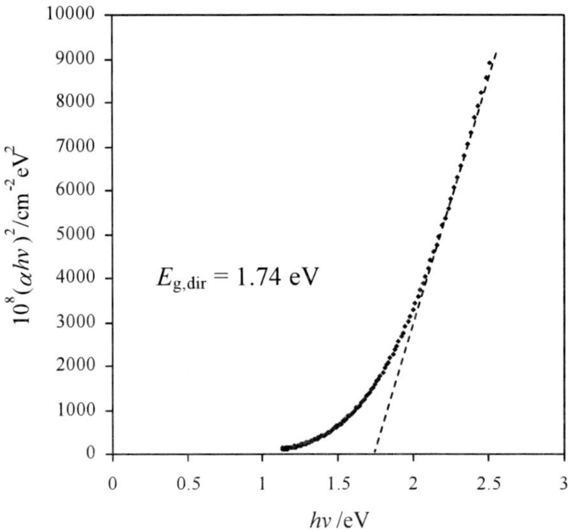

b)

Figure 7. The $(\alpha h\nu)^{1/2}$ and $(\alpha h\nu)^2$ *vs.* $h\nu$ dependencies of as-deposited SnSe thin film together with the extrapolations of linear parts to $\alpha h\nu = 0$.

Upon thermal treatment at 150 °C, the indirect transition exhibits a red shift to 1.10 eV, while the direct one shifts to 1.65 eV (Fig. 8). Both of the indirect band gap values are thus blue-shifted with respect to the bulk value of 0.90 eV [85,93].

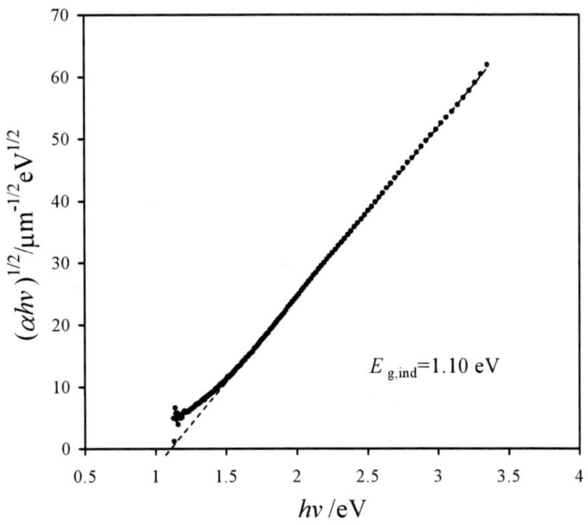

a)

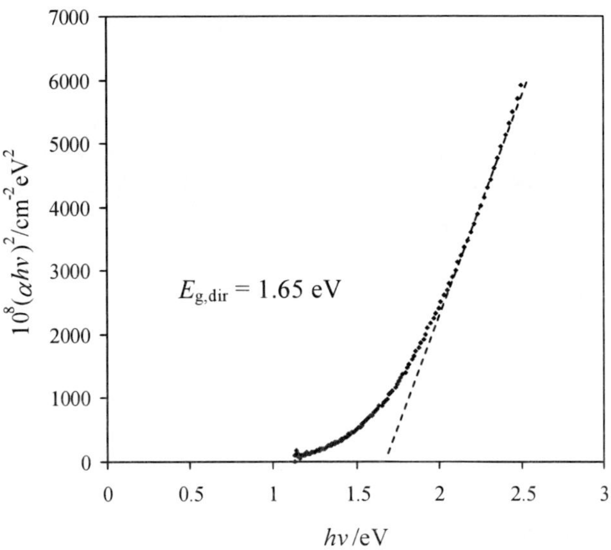

b)

Figure 8. The $(\alpha h\nu)^{1/2}$ and $(\alpha h\nu)^2$ *vs.* $h\nu$ dependencies of SnSe thin film annealed at at 150 °C together with the extrapolations of linear parts to $\alpha h\nu = 0$.

According to the literature data, the values for the indirect and direct band gap energies of a macrocrystalline SnSe specimen calculated by the tight-binding method [93] are 0.90 and 1.30 eV correspondingly (the ΔE value being 0.40 eV). In our case, both values are shifted to higher energies due to size quantization effects for both as-deposited and thermally treated films.

However, the energy differences between two minima of the conduction band, located at different points in the first Brillouin zone (*i.e.* the ΔE values ; $\Delta E = E_{g,dir} - E_{g,ind}$), are in very good agreement with the predicted ones (0.54 and 0.55 eV for as-deposited and annealed films correspondingly).

4.4. Optical Absorption of Binary and Ternary Metal Sulfide Nanocrystals Deposited as Thin Films

Analyses of the semiconductor absorption functions, constructed from the optical absorption spectra of nanocrystalline semiconducting metal sulfide films, led us to quantitative conclusions related to their optical properties, particularly the size quantization effects on the optical band gap energies and the possibilities for band gap engineering of these materials. It is important to note, in this context, that essentially all of the previously proposed literature methods for synthesis of bismuth(III) sulfide thin films have led to amorphous materials in thin film form which subsequently crystallized upon annealing [94-104]. Our chemical method thus allows following the band structure of ultrasmall nanocrystals of this material and its evolution upon thermal treatment [43]. In our previous papers, we also have reported systematic studies of the influence of ultrasonic irradiation on the optical properties of In_2S_3, $AgBiS_2$ and $CuInS_2$ nanocrystals deposited in thin film form [42,105,106]. Note that particularly $AgBiS_2$ is a material which has not been exploited much, besides its suitability for solar energy conversion applications. The most important results from our studies of optical properties of the title semiconducting compounds are summarized in Table 7.

Table 7. The types of electronic transitions and the corresponding band gap energies determined from the experimentally measured spectral dependencies of the absorption coefficient in the case of the synthesized nanocrystalline materials in thin film form

Material	post-deposition treatment	type of electronic transition	E_g / eV
ZnSe	as-deposited	direct, allowed	3.10
		direct, allowed	3.50
	annealed at 200 °C	direct, allowed	2.80
		direct, allowed	3.20
	annealed at 250 °C	direct, allowed	2.60
		direct, allowed	3.00
SnSe	as-deposited	indirect, allowed	1.20
		direct, allowed	1.74
	annealed at 150 °C	indirect, allowed	1.10
		direct, allowed	1.65
Bi_2S_3	as-deposited	direct, allowed	1.70
	annealed at 250 °C	direct, allowed	1.52
In_2S_3	as-deposited, chemical	direct, allowed	2.85
	as-deposited, sonochemical	direct, allowed	3.00
	annealed at 150 °C	direct, allowed	2.20
	annealed at 200 °C	direct, allowed	2.00
$AgBiS_2$	as-deposited, chemical	direct, allowed	1.10
	as-deposited, sonochemical	direct, allowed	1.40
	annealed at 200 °C	direct, allowed	1.00
$CuInS_2$	as-deposited, chemical	direct, allowed	1.82
	as-deposited, sonochemical	direct, allowed	1.87

4.5. Analysis of the Size Quantization Effects

The observed band gap energy blue shifts with respect to the bulk material values, as well as the red shifts of these quantities upon thermal annealing treatments of the films were rationalized and discussed in terms of various models for size quantization effects: the effective mass approximation model (Brus' model [107-110]), the hyperbolic band model [111] (for those systems for which it is applicable) and the Nosaka's model [112]. The most profound analysis was possible in the case of ZnSe, since most relevant data for such analyses are available for this material (such as described in details in our paper [40]).

Quantum confinement effects should be observable if the nanocrystal radius becomes less than (or at least comparable to) the Bohr excitonic radius r_B for the corresponding material, given by the following relation [107-110]:

$$r_B = \frac{4\pi\varepsilon_0\varepsilon_r\hbar^2}{e^2 m_0}\left(\frac{1}{m_e^*}+\frac{1}{m_h^*}\right) \tag{74}$$

where $\hbar = h/(2\pi) -$ h being the Planck's constant, m_0 is the electron mass, while m_e* and m_h* are the electron and hole relative effective masses, respectively (*i.e.* the effective masses expressed in units of electron mass m_0), e is the electron charge, ε_0 the permitivity of vacuum and ε_r is the relative dielectric constant of the semiconductor (usually taken as the bulk value $\varepsilon_{r,\infty}$ or the static value $\varepsilon_{r,s}$). Using the corresponding parameters for ZnSe, we have estimated a value of 3.2 nm for r_B in the case of this material, which is quite in line with our experimental data. The previously estimated literature values of about 9 nm [113] thus seem to be too high. First, the predicted band gap energy blue shifts according to Brus model were calculated by the expression:

$$\Delta E_g(R) = \frac{h^2}{8m_0R^2}\left(\frac{1}{m_e^*}+\frac{1}{m_h^*}\right) - \frac{1.8e^2}{4\pi\varepsilon_0\varepsilon_r R} - 0.248\frac{4\pi^2e^4m_0}{2(4\pi\varepsilon_0\varepsilon_r)^2h^2\left(\frac{1}{m_e^*}+\frac{1}{m_h^*}\right)} \tag{75}$$

for particular values of R corresponding to our experimental data, using the bulk material values for m_e*, m_h* and ε_r. The predicted blue shifts of E_g obtained in this way were too high in the whole range of studied R values. The kinetic quantum localization term (second term on the right-hand side of (75)) has a predominant contribution to overall $\Delta E_g(R)$, along with a negligible contribution of the spatial correlation term (third one in (75)). Such findings may be attributed to the relatively high ε_r value for bulk ZnSe of 8.1. A comparison of experimental data with the predictions of various theoretical models is given in concise form in Table 8. There are several reasons for the inadequacy of the Brus model to quantitatively predict the $\Delta E_g(R)$ values: the breakdown of the effective mass approximation upon crystal size decrease (*i.e.* the apparent non-parabolicity of energy bands), the reduced dielectric screening in small-sized semiconductor nanocrystals reflecting in a smaller value of ε_r (arising due to the inability of the lattice polarization to follow the more rapid electron and hole motions associated with a smaller crystal radius) and the existence of a finite barrier height at the nanocrystal boundary (allowing charge carrier leakage outside the quantum dot).

Application of the hyperbolic band model for analysis of our experimental data, based on the equation [111]:

$$E_g(R) = \sqrt{E_{g,\text{bulk}}^2 + \frac{2\hbar^2 E_g \pi^2}{m_0 R^2}\frac{1}{2}\left(\frac{1}{m_e^*} + \frac{1}{m_h^*}\right)} \tag{76}$$

did not lead to significantly improved agreement between theory and experiment, even upon inclusion of a term describing the Coulomb interaction between charge carriers in the manner of Brus model [107,108].

However, using modified Coulomb and spatial correlation terms in Brus eq. (75), with four times smaller value of ε_r than the bulk one, leads to an excellent agreement between the predictions of the Brus model and the experimental data. Finally, excellent agreement with the experimental data is obtained accounting for the finite potential barrier height at the nanocrystal/surrounding contact by Nosaka's approach. This was done using the equation [112]:

$$E_g(R) = E_{g,\text{bulk}} + E_{k,e}(R) + E_{k,h}(R) + E_{p,e-h}(R) \tag{77}$$

where $E_{k,e}(R)$ and $E_{k,h}(R)$ are the size-dependent kinetic energies of the electron and hole excited states respectively, while $E_{p,e-h}(R)$ is the size-dependent Coulomb interaction energy term. $E_{k,e}(R)$ and $E_{k,h}(R)$ may be calculated by the following formula derived by Nosaka [112]:

$$\frac{E_{k,e(h)}}{E_0} = a + \frac{b}{\left[\left(\frac{E_0 m_{e(h)}^*}{m_0}\right)^{1/2} R + c\right]^2} \tag{78}$$

The parameters a, b and c depend on the charge carriers' effective masses and may be taken from the plots giving the dependencies of these parameters on the ratio $m_{e(h)}^* / m_0$ given in Nosaka's paper [112]. E_0 is the potential energy (finite) depth (within the Brus model, $E_0 = \infty$). $E_{p,e-h}(R)$ in eq. (77) is in principle given by a relation of the form:

$$E_{p,e-h}(R) = -\frac{C_1 e^2}{4\pi\varepsilon_0\varepsilon_r R} \tag{79}$$

where C_1 is a parameter with a numerical value usually close to 1.8 (e.g. 1.786 in the papers by Kayanuma [114] and Schmidt and Weller [115]). As clearly seen from Table 8, two most probable reasons for the failure of the Brus model in the case of the presently studied ZnSe quantum dots in thin-film form are: (a) the change of relative dielectric constant value of a nanocrystal upon crystal size reduction and (b) the importance of electron leakage outside the nanoparticles. Which of these two effects is predominant remains still unclear, as both decreasing the value of ε_r in the Brus model and accounting for the finite depth of the potential within the semiconducting nanocrystal lead to essentially the same agreement with the present experimental data. However, the non-parabolicity of electron and hole bands seems to be of a very little importance in the case of the presently studied material.

Table 8. The experimentally determined band gap blue shifts ΔE_g (in eV, with respect to the bulk value) in the case of as-deposited and thermally annealed ZnSe quantum dot thin films together with the predictions of various models proposed in the literature; HB –Hyperbolic band model

ΔE_g	R = 1.4 nm	R = 1.84 nm	R = 2.0 nm
Experimental	0.52	0.32	0.02
Brus (ε_r = 8.1)	1.22	0.70	0.55
Brus (ε_r = 2.0)	0.52	0.16	0.06
HB	1.18	0.76	0.63
HB + Coulomb	0.95	0.59	0.47
Nosaka (E_0 = 1.5 eV)	0.39	0.25	0.21
Nosaka (E_0 = 3.0 eV)	0.52	0.33	0.27

5. The Transport Properties of Equilibrium Charge-Carriers in Ensembles of Closely Packed Semiconductor Nanocrystals in Thin Film Form

5.1. Electrical Resistivity of Nanocrystalline Semiconductor Thin Films and the Dominant Type of Charge Carriers

Electrical resistivity of the synthesized materials in thin film form was measured by the two-point probe, four-point probe (van der Pauw method) and also by the constant field method [74,78,116]. During these measurements, injection of charge carriers was allowed through suitable metal contacts. The non-rectifying character of the corresponding metal-semiconductor contacts was proved by measuring their current-voltage characteristics. In general, we have demonstrated that the electrical resistance of the studied thin film materials depends on a number of factors. Among these, the most significant are the experimental conditions during thin film deposition processes, the post-deposition treatment (e.g. annealing) and the average crystal size. In the case of all studied materials we have found that the average crystal size increase upon thermal annealing of the films is accompanied by decrease of thin films' resistivity. Since the charge-transport processes in ensembles of QD solids deposited in thin film form are barrier-dominated, we attributed this trend to both the average crystal size increase and decrease of the crystal boundary barrier height. The dependence of the conductivity of a thin film sample on various fundamental parameters which describe the surface band structure (Fig. 9) has the following form [116]:

$$\sigma = \frac{Ne^2\langle v \rangle}{4kT} L \exp\left(-\frac{e\phi}{k_B T}\right) \qquad (80)$$

where N is the number concentration of free charge carriers within the nanocrystals, $\langle v \rangle$ their average thermal velocity, L is the average distance between barriers, ϕ is the barrier potential, while e, k_B and T are the electron charge, Boltzmann constant and thermodynamic temperature, respectively. Employing eq. (80), we have calculated quantitatively the barrier potential changes upon annealing in the following manner, as discussed in details in our paper [41]. The ratio between film conductivities before and after annealing is, according to (80), given by:

$$\frac{\sigma_{as-dep.}}{\sigma_{annealed}} = \frac{L_{as-dep.}}{L_{annealed}} \exp\left[-\frac{(e\phi)_{as-dep.} - (e\phi)_{annealed}}{k_B T}\right] \qquad (81)$$

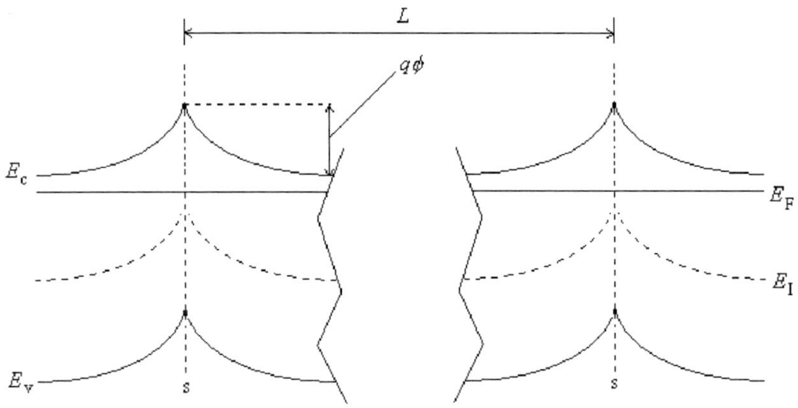

Figure 9. A schematic presentation of the existence of potential barriers at the intercrystalline (*i.e.* grain) boundaries due to band bending on the nanocrystal surfaces.

As discussed in our paper [41], the ratio $L_{as\text{-}dep.}/L_{annealed}$ is (to) a good approximation equal to the ratio of average crystal diameters before and after annealing (determined from the XRD studies). Therefore, the only remaining unknown quantity in (81) is $(e\phi)_{as-dep.} - (e\phi)_{annealed}$, *i.e.* the barrier height decrease. The calculated $(e\phi)_{as-dep.} - (e\phi)_{annealed}$ values for the materials studied by our group are given in Table 9.

Table 9. The decrease of intercrystalline barrier heights upon annealing in the case of the studied nanocrystalline materials in thin film form

sample	ZnSe	SnSe	Bi_2S_3	In_2S_3	$AgBiS_2$	$CuInS_2$
$\left(e\phi_S\right)_{as-dep.} - \left(e\phi_S\right)_{annealed}$ /eV	0.002	0.05	0.21	0.04	0.09	0.02

The predominant charge carriers' types in the case of the studied materials, determined by Seebeck effect studies (experimentally measuring *the sign* of the thermoelectric force of metal chalcogenide-metal thermocouple employing the hot point probe method), are summarized in Table 10.

Table 10. The dominant types of charge carriers in the case of the studied materials in thin film form

system	ZnSe	SnSe	Bi_2S_3	In_2S_3	$AgBiS_2$	$CuInS_2$
dominant type of charge carriers	e	h	e	e	e	h

5.2. Temperature Dependence of the Dark Electrical Resistivity of the Films ($R = f(T)$ function). Determination of Thermal Band Gap Energies and Impurity Levels Ionization Energies

Measurements of the temperature dependence of thin films' electrical resistivity were performed using the experimental setup presented in Fig. 10, which is discussed in details in our papers [27-43]. It consists of a special furnace basically designed for Franck-Hertz-type experiments (FH), a thermocouple with voltmeter (TC), variable transformer (Tr), a d.c. source and a standard resistor (R_s). To avoid chemical changes in the investigated films in the course of these measurements, they were carried out in inert (argon) atmosphere at P = 80 kPa. Slow heating of the furnace was achieved by the variable transformer (Tr). The film temperature was precisely measured directly on its surface using Fe-Constantane thermocouple, with a previously determined temperature dependence of thermoelectric force.

By mathematical analysis of the circuit containing the investigated thin film, it may be easily shown that:

$$R_f = \frac{V_b R_s}{V_s} - R_s \tag{82}$$

or, provided that $R_s \ll R_f$:

$$R_f = \frac{V_b R_s}{V_s} \tag{83}$$

During all measurements of temperature dependence of electrical resisitivity, the condition R_s<<R_f was provided by controlling the R_s value.

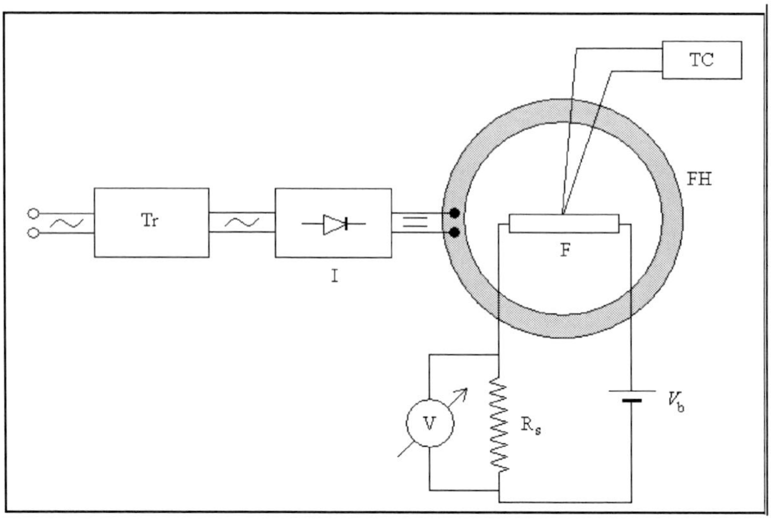

Figure 10. The experimental setup used for measurements of the temperature dependencies of thin films' electrical resistivity.

The experimentally obtained trends in the $R = f(T)$ dependencies were analyzed using the Arrhenius-type equation for the dependence of the conductivity on temperature, which holds in the most general case of a semiconductor containing n impurity levels present in the forbidden energy band [75]:

$$\sigma(T) = \sigma_0 \exp\left(-\frac{E_g}{2k_BT}\right) + \sum_{i=1}^{n} \sigma_{0,i} \exp\left(-\frac{\Delta E_i}{k_BT}\right) \tag{84}$$

In (84), E_g is the band gap energy of the studied semiconductor, ΔE_i is the ionization energy of the i-th impurity level, T is the thermodynamic temperature, while k_B is the Boltzmann constant. $\sigma_{0,i}$ -s are constants (preexponential factors) which correspond to the i-th-impurity level activation energies. The constant σ_0 , corresponding to the intrinsic temperature region, depends very weakly on T in comparison to the exponential multiplier $\exp[-E_g/(2k_BT)]$ and it may be regarded as a nearly constant term. Since each term on the right side of the equation (84) is expected to be "activated" within a different temperature interval, for each interval, thus, there will be a term in (84) which will predominate in the overall $\sigma(T)$ dependence. In the highest temperature interval, the $\sigma(T)$ will be dominated by the intrinsic conductivity mechanism (*i.e.* by band-to-band electronic transitions), provided that the band gap energy is not too high and the band gap energy was calculated by the following expression:

$$E_g = 2k_B \frac{\mathrm{d}\ln(R(T))}{\mathrm{d}(1/T)} \tag{85}$$

It may be relatively easily shown that regardless on the assumed explicit form of the function describing the temperature dependence of E_g, the measured thermal band gap energy is in fact an extrapolated value referring to 0 K. Having in mind that the band gap energy value is itself a temperature-dependent quantity [75] and assuming a linear $E_g = f(T)$ relationship (which holds only at temperatures significantly higher than the constant T_0 in a Varshni-type equation [117,118]):

$$E_g(T) = E_g(0) - \alpha T \tag{86}$$

it can be easily arrived to the following form of equation (84):

$$R(T) = R_0 \exp\left(-\frac{\alpha}{2k_B}\right)\exp\left(\frac{E_g(0)}{2k_B T}\right) \tag{87}$$

for the highest temperature region (corresponding to activation of intrinsic conductivity mechanism). Successive logarithming and differentiation of (87) leads to:

$$E_g(0) = 2k_B \frac{\mathrm{d}\ln(R(T))}{\mathrm{d}(1/T)} \tag{88}$$

The value obtained from the slope of $\ln(R(T))$ *vs.* T^1 dependence, thus, refers to 0 K within linear extrapolation approximation. We further demonstrate that even if one accounts a more realistic function describing temperature dependence of E_g, the obtained slope multiplied by $2k_B$ is essentially equal to $E_g(0)$. Assuming temperature dependence of E_g described by Varshni-type equation, *i.e.*:

$$E_g(T) = E_g(0) - \frac{\alpha T^2}{T + T_0} \tag{89}$$

and substituting in (84) one gets (in the case of predominance of intrinsic conductivity mechanism):

$$R(T) = R_0 \exp\left(-\frac{\alpha}{2k_B}\frac{T}{T+T_0}\right)\exp\left(\frac{E_g(0)}{2k_B T}\right) \tag{90}$$

Since the band gap energy is obtained from the $R(T)$ data in the higher temperature region (usually $T >> T_0$), the second multiplier on the right side in (90) is practically constant, *i.e.* equal to $\exp(-\alpha/(2k_B T_0))$, which leads to a conclusion that eq. (88) practically holds in this case as well. Another expression which has been used in the literature to describe the temperature dependence of band gap energy of semiconductors is a Bose-Einstein-type expression [119,120]. It has the form:

$$E_g(T) = E_g(0) - \frac{2a_B}{\exp\left(\frac{\Theta_B}{T}\right) - 1} \tag{91}$$

where the coefficient a_B is a measure of the strength of the electron-average phonon interaction, while Θ_B is the average temperature of phonons interacting with charge carriers in the semiconductor. By substitution of (91) in (84), the following equation arises:

$$R(T) = R_0 \exp\left(-\frac{a_B}{2k_B T} \frac{1}{\exp\left(\frac{\Theta_B}{T}\right) - 1} \right) \exp\left(\frac{E_g(0)}{2k_B T} \right) \tag{92}$$

In case when $T >> \Theta_B$, the exponent in the denominator of the second multiplier on the right side in eq. (92) may be expanded in McLaurin series, which leads to the following form of (92):

$$R(T) = R_0 \exp\left(-\frac{a_B}{2k_B T} \frac{1}{1 + \frac{\Theta_B}{T} + \frac{1}{2!}\left(\frac{\Theta_B}{T}\right)^2 + \cdots - 1} \right) \exp\left(\frac{E_g(0)}{2k_B T} \right) \tag{93}$$

or, neglecting the terms containing higher orders of (Θ_B/T):

$$R(T) = R_0 \exp\left(-\frac{a_B}{2k_B \Theta_B} \right) \exp\left(\frac{E_g(0)}{2k_B T} \right) \tag{94}$$

This result again leads to a conclusion that thermal band gap energy is essentially equal to $E_g(0)$.

In the temperature regions where extrinsic conductivity mechanism is predominant, the corresponding impurity level ionization energies were calculated by:

$$\Delta E_i = k_B \frac{\mathrm{d} \ln\left(R(T)_i\right)}{\mathrm{d}(1/T)} \tag{95}$$

A plot of ln(R) *vs.* $1/T$ constructed on the basis of experimentally measured temperature dependence of dark electrical resistance of SnSe quantum dot thin films in temperature interval from 25 °C to 310 °C is shown in Fig. 11. The thermal band gap energies and impurity levels ionization energies calculated in the previously described manner for all of the studied materials are summarized in Table 11.

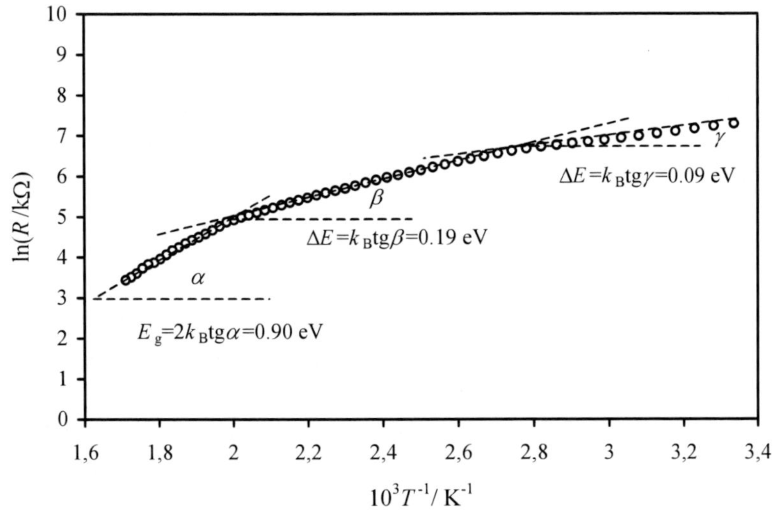

Figure 11. A plot of ln(R) *vs.* $1/T$ constructed on the basis of experimentally measured temperature dependence of dark electrical resistance of SnSe quantum dot thin films in temperature interval from 25 °C to 310 °C.

Table 11. The calculated thermal band gap energies and impurity levels ionization energies on the basis of experimentally measured $R = f(T)$ dependencies for the studied materials in thin film form

Material	E_g / eV	ΔE_1 / eV	ΔE_2 / eV
ZnSe	-	0.82	-
SnSe	0.90	0.19	0.09
Bi_2S_3	1.50	0.42	-
In_2S_3	2.22	0.82	-
$AgBiS_2$	1.05	0.45	-

The differences between the calculated thermal and optical band gap energies in the present cases may be rationalized either through the sign of the derivatives $\partial E_g / \partial T$ (since the optical values refer to room temperature, while the thermal ones are extrapolated values to 0 K) or on the basis of Franck-Condon principle reasoning. In fact, electronic transitions induced by interaction with external electromagnetic field occur rather fast (almost instantaneously), so that the new electronic state of the nanocrystal corresponds to higher-order vibrational state on a new (excited) potential energy hypersurface governing nuclear motions. Therefore, it is expected that the vertical electronic transition of this type would require higher energy than a transition that would occur (slowly) between ground vibrational states corresponding to the two electronic states. In the case of thermal excitation, it is the nuclear subsystem that firstly becomes excited and accumulation of energy allows subsequent occurrence of the electronic transition. It would thus be in general expected that optical band gap energies would be, due to this reason, higher than the thermal ones. However, in reality, one would expect that a complex interplay between these two physical reasons would determine the actual relation between the two values.

To check for the reproducibility of the $R = f(T)$ measurements in all of the studied cases, the studied samples were successively subjected to multiple heating-cooling cycles. Comparison of the data obtained from various cycles invariably showed that the measurements were highly reproducible, especially with respect to the calculated E_g and ΔE_i values.

6. Photophysical and Photoelectrical Properties of Low-Dimensional Semiconductor Systems

Two types of measurements were employed to characterize the photophysical properties of the synthesized materials which manifest photoconduction properties in our works: stationary ones and time-resolved studies. Studies of the stationary photoelectrical properties included measurements of the stationary photoconductivity of the samples and its spectral dependence, as well as extracting data on the corresponding band structures from these results. Time-resolved investigations included studies of photoresponse decay upon interaction of the sample with light impulses and modeling the kinetics of photoconductivity time-dependence.

6.1. Studies of the Stationary Photoconductivity Spectral Response

To measure the spectral dependence of stationary photoconductivity, we have constructed a set-up [27-43] containing a d.c. source serially connected to a standard resistor and the investigated photoconductive thin film sample (Fig. 12).

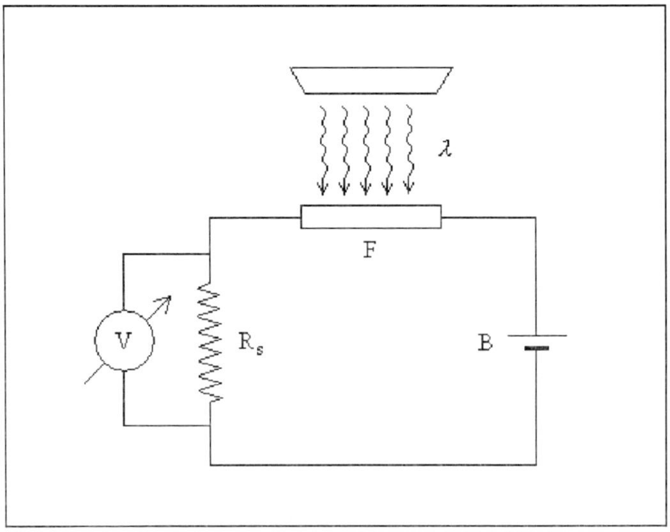

Figure 12. The experimental setup employed for measurements of the spectral dependence of stationary photoconductivity of the investigated samples.

As a source of illumination we have used a monochromator from spectrophotometer Beckman DU-2 within the wavelength range from 400 to 1200 nm. As can be shown by mathematical analysis of the employed electric circuit in these experimental measurements [74], the stationary photoconductivity ($\Delta\sigma_{st.}$) depends on dark electrical resistance of the investigated sample (r_0) and resistance of the standard resistor (R) according to the following equation at each sampled value of the incident light wavelength - λ:

$$\Delta\sigma_{st.}(\lambda) = \frac{v(\lambda)(R + r_o)^2}{r_o^2 VR - v r_o R(r_o + R)} \tag{96}$$

where V is the applied voltage and v is the voltage drop at the ends of the standard resistor. In the case of constant field method ($R << r_0$), equation (96) may be simply rearranged to the following form:

$$\Delta\sigma_{st.}(\lambda) = \frac{v(\lambda)}{RV} \tag{97}$$

Of all the studied nanocrystalline materials in thin film form, only those of Bi_2S_3, SnSe, $AgBiS_2$ and ZnSe manifested photoconductive properties. However, since the band gap value of ZnSe is quite high, as discussed before, the relevant part of its spectral photoresponse curve could not be measured by the available equipment in our laboratory. Therefore, only the results concerning Bi_2S_3, SnSe and $AgBiS_2$ will be discussed. The samples of Bi_2S_3 and $AgBiS_2$ manifest photoconductive properties only upon thermal annealing (*i.e.* the as-deposited films of these materials are non-photoconductive), while SnSe samples manifest photoconduction even before thermal treatment. Photoconductive properties of the last material are significantly degraded upon thermal treatment of the corresponding thin films. These findings were rationalized in terms of the relative importance of the grain boundary effects before and after annealing treatments. The details are given in our papers [39,41]. A typical stationary spectral photoresponse curve for as-deposited SnSe film is presented in Fig. 13. The general shape of the $\Delta\sigma_{st.} = f(h\nu)$ dependence implies the relevance of surface recombination phenomena, as shown quantitatively by Devore [74].

The quantity that was actually analyzed in our studies of photoconductive thin films of low-dimensional semiconductors is the stationary photoconductivity spectral response, usually defined as:

$$R(h\nu) = \frac{\Delta\sigma_{st.}(h\nu)}{\Delta\sigma_{st.,max.}} \tag{98}$$

We have first applied the simple Moss function [121,122] to model the photoconductivity response spectra of the studied films. This function is applicable for the ascending part of the experimental $R(hv)$ curve. Its analytical form is:

$$R(h\nu) = \frac{1}{1 + \exp[C(E_g - h\nu)]} \tag{99}$$

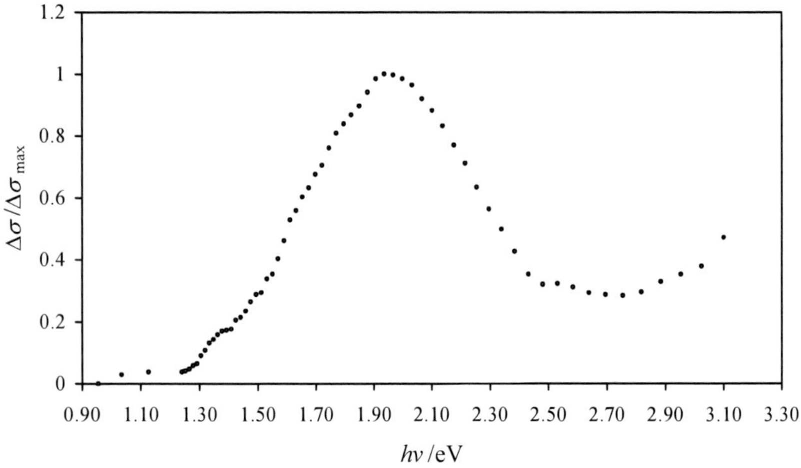

Figure 13. A typical stationary spectral photoresponse curve for as-deposited SnSe film

It may be easily seen that when R is equal to 0.5, then $h\nu = E_g$. This is the basis of the so-called Moss rule [121,122] for determination of band gap energies of photoconducting materials from their photoresponse spectral dependencies. The ascending parts of the $R(h\nu)$ curves in the case of all studied materials in our case could be excellently interpolated with functions of the form (99). An illustrative example is shown in Fig. 14 for Bi_2S_3. The resulting fitting parameters for this particular case are:

$$C = 10.917 \pm 0.317 \text{ eV}^{-1},\ E_g = 1.268 \pm 0.003 \text{ eV } (\chi^2 = 0.00052).$$

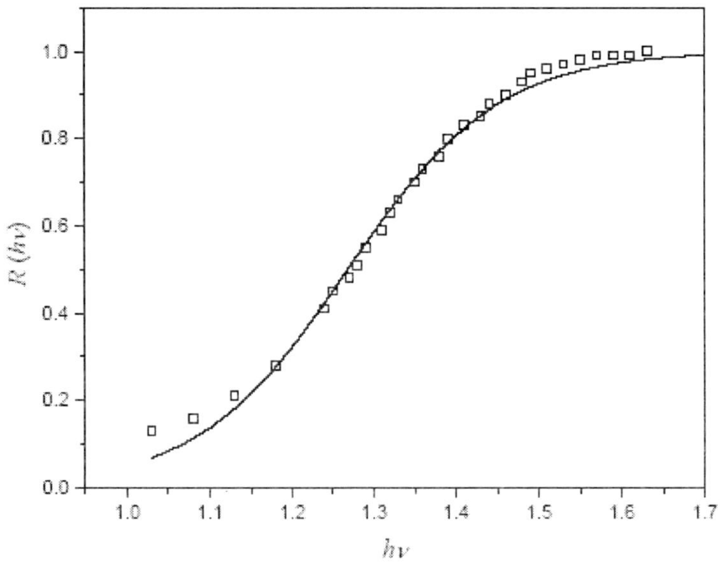

Figure 14. The ascending part of the $R(h\nu)$ curve for a photoconductive Bi_2S_3 thin film together with the interpolating Moss function of the form given by Eq. (99)

However, as the Moss function is an empirical one, we have proceeded with analysis of our experimental photoresponse spectra within the framework of some other models for this phenomenon. Basically all of the presently employed models that aim to quantitatively describe the photoconductivity response spectra are more or less based on solution of continuity equation [123-128]. In the case of polycrystalline thin films, in which the photoresponse is measured between two parallel planar contacts, the phenomenon of photoconductivity is due to enhancement of both free carrier concentration and their mobility [123-130]. While the first effect occurs due to absorption of light quanta, the last one is due to reduction of intercrystalline barrier height upon illumination. In the incident photon energy range $h\nu < E_g$ (i.e. $\alpha d \leq 1$), it may be shown [123,127]) that:

$$R(h\nu) \propto \left[1 - \exp(-\alpha(h\nu)d)\right] \tag{100}$$

Using the experimentally obtained $\alpha(h\nu)$ functional dependence from the optical spectroscopy data, we have obtained an excellent interpolation of the experimental photoconductivity spectra of the studied systems. Fig. 15 illustrates this for the case of a photoconductive Bi_2S_3 thin film.

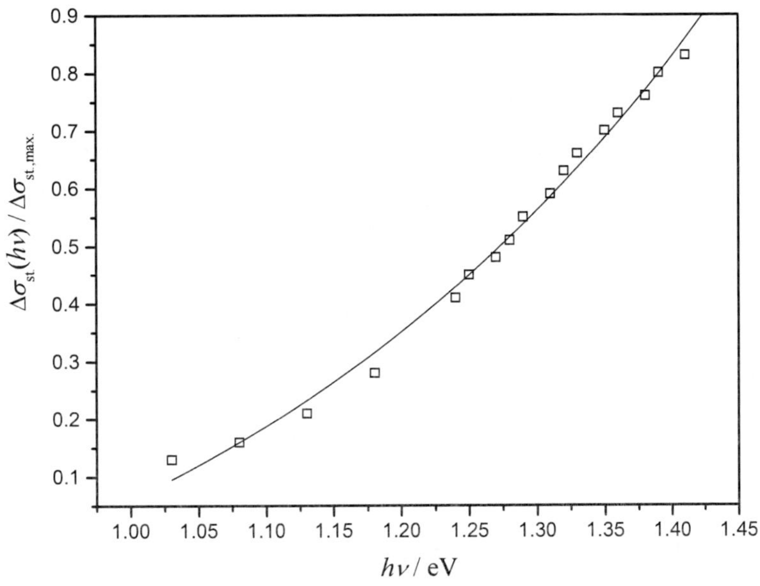

Figure 15. The experimental photoconductivity response spectrum of a photoconductive Bi_2S_3 thin film in the region $\alpha d \leq 1$, together with the interpolating function of the form $R(h\nu) \propto \left[1 - \exp(-\alpha(h\nu)\cdot d)\right]$, where $\alpha(h\nu)$ is experimentally obtained by interpolation of the electronic absorption spectrum with a function according to the modified Urbach rule.

Recently, Yano *et. al.* [124] have employed the following function to model the photoconductivity response in ZnS thin films:

$$R(h\nu) \propto [1-\exp(-\alpha(h\nu)d)] - \frac{\alpha(h\nu)L_d\{1-\exp[-d(\alpha(h\nu)+1/L_d)]\}}{(1+\tau_{sr}/\tau_{ph})[1+\alpha(h\nu)L_d]} \tag{101}$$

In the previous equation, L_d is the penetration depth of the incident light, while τ_{sr} and τ_{ph} denote the surface and bulk region relaxation times, correspondingly. Using the experimentally obtained data for $\alpha = f(h\nu)$ in the region below $E_{cr.}$ (according to the modified Urbach rule), we have obtained an excellent interpolation of the experimental data if the function (101) is used to model the photoconductivity response. This interpolation is presented on Fig. 16. The estimated ratio τ_{sr}/τ_{ph} from the least-squares fit to the experimental photoconductivity spectrum of Bi_2S_3 thin film is 1.87.

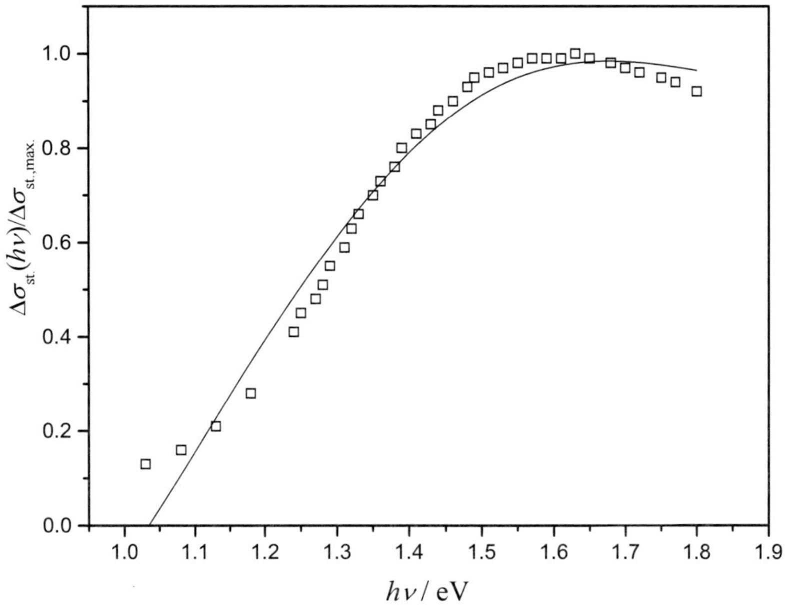

Figure 16. The experimental photoconductivity response spectrum of a photoconductive Bi_2S_3 thin film in the region below $E_{cr.}$, together with the interpolating function of the form (101), where $\alpha(h\nu)$ is experimentally obtained by interpolation of the electronic absorption spectrum with a function according to the modified Urbach rule.

Using the Urbach rule functional form of the dependence $\alpha = f(h\nu)$ is quite suitable in the present case, since this region includes the peak of the photoconductivity response function. Thus, if a suitable model function for interpolation of photoconductivity response spectrum is used, it is expected that it should reproduce its most prominent features (ascending trend and achievement of the maximal value of R, *i.e.* the peak) within the energetic interval in which the dependence of α on $h\nu$ is described by the modified Urbach rule.

Probably the most cited model aiming to provide a quantitative description of photoconductivity response functions is based on the Devore approach [126]. This model is based on solving the continuity equation for the photoresponse function and leads to the following functional form of the dependence of $\Delta\sigma$ on the incident photon energy:

$$\Delta\sigma(h\nu) \propto N = \frac{\tau I}{1-\alpha^2(h\nu)D\tau}\left\{\left[1-e^{-\alpha(h\nu)d}\right]-\frac{\left[\alpha S\tau\left(1+e^{-\alpha(h\nu)d}\right)+\alpha^2(h\nu)D\tau\left(1-e^{-\alpha(h\nu)d}\right)\right]}{1+S\left(\frac{\tau}{D}\right)^{1/2}\coth\left[\frac{l}{2(D\tau)^{1/2}}\right]}\right\} \tag{102}$$

In the last expression, N denotes the number of created carrier pairs upon interaction with the incident radiation, D is the carrier diffusion constant (governing the diffusion carrier current), S is the surface recombination velocity, τ the volume recombination lifetime, while all other terms have the usual meanings. It is convenient to introduce the following dimensionless parameters in (102):

$$\lambda = d/(D\tau)^{1/2} \tag{103}$$

$$\zeta = \alpha\,(D\tau)^{1/2} \tag{104}$$

$$\xi = S\,(\tau/D)^{1/2} \tag{105}$$

$$Z = \alpha d \tag{106}$$

In case when surface recombination is large compared to the volume recombination, the function (102) reduces to a simpler form, *i.e.*:

$$R(Z) = \frac{1-e^{-Z}}{\xi\coth\left(\frac{\lambda}{2}\right)}\left\{1+\frac{\xi\lambda\left[\lambda\coth\left(\frac{\lambda}{2}\right)-Z\coth\left(\frac{Z}{2}\right)\right]}{\lambda^2-Z^2}\right\} \tag{107}$$

On the other hand, in cases when surface recombination is negligible compared to the volume recombination ($S << D/\tau$, i.e. $\xi << 1$):

$$R(Z) \propto \left(1-e^{-Z}\right) \tag{108}$$

Also, in case of small sample thickness:

$$R(Z) \propto \frac{\lambda}{2\xi}\left(1-e^{-Z}\right) \tag{109}$$

In other words, in the last two considered cases, the Devore function basically gets the form (100), which we have already mentioned before. However, although the general trends of the photoconductivity response spectrum are basically reproduced by functions (102) and (107) arising from the Devore model, we could not get a satisfactory good interpolation of our experimental data in the whole energy (or αd) region. Similar results were also obtained by Kebbab and co-workers in a recent study devoted to Bi_2S_3 thin films obtained by spray pyrolysis technique [123].

In this context, we note that besides with the functions of the previously mentioned forms, we have also tried to interpolate the experimental $R(h\nu)$ dependencies using some other functions which were expected to resemble the form of our experimental photoconductivity spectrum on a purely intuitive basis. Thus, the photoconductivity response spectrum can be almost excellently interpolated by a Fano-type function ([70]) and also by the log-normal distribution function. However, up to now these interpolation functions have not been given a physical significance for this particular purpose (interpolation of the photoconductivity spectra). Therefore we will not devote more attention to them in the present review. Despite that, we consider that the knowledge that the mentioned functions actually closely resemble the form of the experimentally measured photoconductivity response spectra may be rather valuable for a more profound understanding of this phenomenon at fundamental level. This knowledge may also be helpful in deriving new theoretical models for description of photoconductivity phenomena, especially employing statistical methods.

The stationary photoconductivity spectral dependencies were employed in our works as an additional probe to the band structure of the synthesized materials in thin film form. To determine the band gap energy of the photoconductor, as well as the type of electronic transition being responsible for (*i.e.* contributing to) the internal photoelectric effect, we have employed the approach outlined in details in our papers [27-43], based on the proportionality of α on $\Delta\sigma_{st.}$ in the limit of low absorption of the films [123].

In fact, the photoconductivity, *i.e.* the $\Delta\sigma$ value is proportional to the number of photons absorbed by the investigated thin film. The last value is, on the other hand, proportional to the decrease of the intensity of incident light upon absorption by the film of thickness d:

$$\int_0^d |\mathrm{d}I|\,\mathrm{d}x \tag{110}$$

However, it is known that:

$$|\mathrm{d}I| = \alpha\, I\,\mathrm{d}x \tag{111}$$

or:

$$|\mathrm{d}I| = \alpha\, I_0 \exp(-\alpha\, x)\mathrm{d}x \tag{112}$$

Therefore, we may write:

$$\Delta\sigma \propto \int_0^d \alpha\, I_0 \exp(-\alpha\, x)\mathrm{d}x = \alpha\, I_0 \int_0^d \exp(-\alpha\, x)\mathrm{d}x \tag{113}$$

The last integral is easily solvable:

$$\int_0^d \exp(-\alpha x)dx = \frac{1}{\alpha}(1 - \exp(\alpha d)) \quad (114)$$

The term in parentheses in the last expression reduces to αd in the limit of low values of the argument in the exponential function (the limit of low absorption *i.e.* $\alpha d << 1$):

$$(1 - \exp(\alpha d)) \approx \alpha d \quad (115)$$

In the case of low absorption limit the following relation holds:

$$\Delta\sigma \propto \alpha d \quad (116)$$

This proportionality was employed to calculate the band gap energy of photoconductive semiconductor materials employing the Fermi's golden rule for band-to-band transitions, within the framework of parabolic approximation for the dispersion relation. In the expressions, arising from the Fermi's golden rule, the function $\alpha(h\nu)$ is simply replaced by $\Delta\sigma_{st.}(h\nu)$ or $\Delta\sigma_{st.}(h\nu)/\Delta\sigma_{st.,max.}$. For example, when the dependencies of $((\Delta\sigma_{st.}/\Delta\sigma_{st.,max.})h\nu)^n$ on $h\nu$ for various values of the exponent n in the case of SnSe films were constructed, clearly manifested linear trends were observed only for $n = 2$ and 1/2. This finding suggests that both indirect (the case of $n = 1/2$) and direct ($n = 2$) band-to-band electronic transitions contribute to the internal photoelectrical effect in these films. The determined values for $E_{g,dir.}$ and $E_{g,ind.}$ according to the described procedures were 1.54 and 1.23 eV correspondingly (Fig. 17). These values are both slightly lower than those determined from optical absorption spectroscopy data outlined before in this review (1.74 and 1.20 eV correspondingly). However, they are still larger than thermal band gap energy value of this material which was also determined in our studies [105].

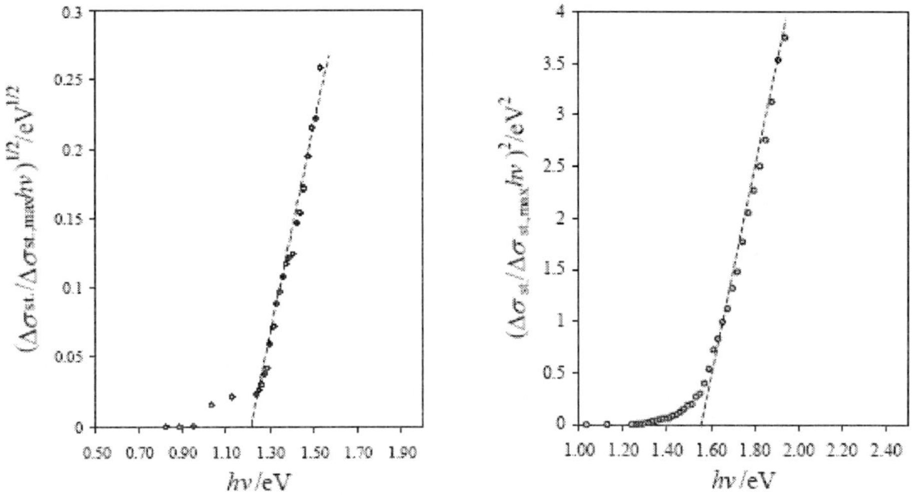

Figure 17. The $((\Delta\sigma_{st.}/\Delta\sigma_{st.,max.})h\nu)^{1/2}$ and $((\Delta\sigma_{st.}/\Delta\sigma_{st.,max.})h\nu)^2$ *vs.* $h\nu$ dependencies for as-deposited SnSe thin film together with the extrapolations of linear parts to $(\Delta\sigma_{st.}/\Delta\sigma_{st.,max.})h\nu = 0$.

This may be mainly attributed to size quantization effects, as the band gap energies determined from photoconductivity response measurements refer to as-deposited samples, while the thermal band gap refers to annealed sample. The band gap energies of Bi_2S_3 and $AgBiS_2$ samples determined by the outlined approach are 1.30 and 1.18 eV (the transitions contributing to the photoelectric effect being of a direct dipole-allowed type), in rather good agreement with the values derived from optical spectroscopy data.

6.2. Study of the Relaxation Dynamics Phenomena of Non-Equilibrium Charge-Carriers in Nanocrystalline Metal Chalcogenide Thin Films

The experimental set-up, which was employed to study the kinetics of the photoconductivity decay process, is schematically depicted in Fig. 18 and is essentially based on the constant-field circuit [27-43,74]. Photoexcitation of the investigated material in thin film form was achieved by illumination with a flash lamp, while the time dependence of the voltage drop at the ends of the standard resistor was followed by transferring this signal to an oscilloscope with previously calibrated time and voltage axes. From the digitalized $\Delta v = f(t)$ function, the non-equilibrium conductivity $\Delta\sigma(t)$ of investigated thin film at each sampled value of t was calculated by the constant field method.

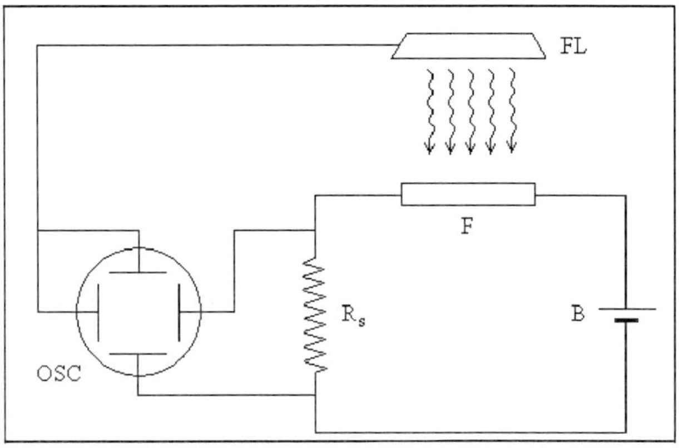

Figure 18. The experimental set-up employed for studies of photoconductivity relaxation dynamics.

A typical oscillogram for as-deposited SnSe thin film is presented in Fig. 19. The experimentally obtained dependencies $\Delta\sigma / \Delta\sigma_{max} = f(t)$ were further analyzed by interpolation with functions governing the time dependence of photocarriers' concentration in the case of linear and quadratic recombination process regimes.

If one considers photoexcitation with Heaviside[1] - type rectangular light impulses, the overall dynamics of photoexcited non-equilibrium charge-carriers is governed by the following differential equation [74]:

[1] The Heaviside approximation for the light impulse is quite valid in cases when time constant of the lamp signal decay is significantly smaller than the time constant of the non-equilibrium charge carrier decay.

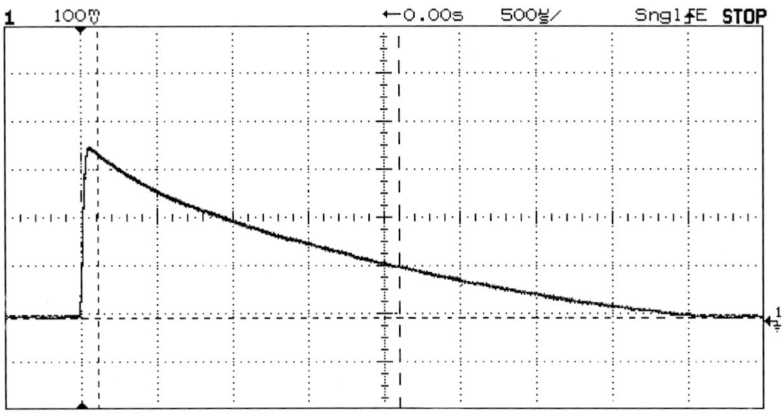

Figure 19. A typical oscillogram for an as-deposited nanocrystalline SnSe thin film.

$$\frac{\mathrm{d}(\Delta N)}{\mathrm{d}t} = v_{\text{gen.}} - v_{\text{rec.}} \tag{117}$$

In Eq. (117), ΔN denotes the excess number concentration of charge carriers with respect to the equilibrium value (*i.e.* the concentration of photo-generated charge carriers), $v_{\text{gen.}}$ denotes the rate of charge-carriers generation, while $v_{\text{rec.}}$ is the rate of their recombination. In the case of linear recombination mechanism, only one type of charge-carriers is relevant for the overall recombination kinetics and the rate of recombination is proportional to their actual concentration, *i.e.*:

$$v_{\text{rec.}} = \frac{\Delta N}{\tau} \tag{118}$$

where τ denotes the average lifetime of photogenerated charge carriers. On the other hand, the rate of generation is given by:

$$v_{\text{gen.}} = \beta \alpha I \tag{119}$$

In the last equation, I is the intensity of the incident light, α is the absorption coefficient of the investigated material, while β is the quantum yield. The differential equation (117) thus takes the form:

$$\frac{\mathrm{d}(\Delta N)}{\mathrm{d}t} = \beta \alpha I - \frac{\Delta N}{\tau} \tag{120}$$

Under initial condition $\Delta n(t = 0) = 0$ it may be shown that the solution of the ordinary linear differential equation (120) has the form:

$$\Delta N(t) = \tau \beta \alpha I(1 - e^{-t/\tau}) \tag{121}$$

Denoting the constant before the parentheses in (121) as $\Delta N_{\text{st.}}$ (the stationary value of the photogenerated charge carriers concentration), the dependence of ΔN on t in the case when both excitation and recombination processes are present takes the form:

$$\Delta N(t) = \Delta N_{\text{st.}} (1 - e^{-t/\tau}) \tag{122}$$

When the light impulse is switched off than $v_{\text{gen.}} = 0$. In such case, eq. (120) gets the form:

$$\frac{\mathrm{d}(\Delta N)}{\mathrm{d}t} = -\frac{\Delta N}{\tau} \tag{123}$$

The solution of (123) is:

$$\Delta N(t) = \Delta N_{\text{st.}}\, e^{-t/\tau} \tag{124}$$

Since $\Delta\sigma$ is proportional to ΔN, for the first-order relaxation kinetics the $\Delta\sigma$ (t) dependence is given by:

$$\Delta\sigma(t) = \Delta\sigma_{\text{st.}}\, e^{-t/\tau} \tag{125}$$

If the relaxation process is kinetically of a second order, then:

$$v_{\text{rec.}} = \gamma\, \Delta N_{\text{e}}\, \Delta N_{\text{h}} \tag{126}$$

where ΔN_{e} and ΔN_{h} denote the number concentrations of photogenerated electrons and holes respectively, while γ is the recombination coefficient. When $\Delta N_{\text{e}} = \Delta N_{\text{h}} = \Delta N$:

$$v_{\text{rec.}} = \gamma(\Delta N)^2 \tag{127}$$

In such case, eq. (117) takes the following form:

$$\frac{\mathrm{d}(\Delta N)}{\mathrm{d}t} = \beta\alpha I - \gamma(\Delta N)^2 \tag{128}$$

Under initial condition $\Delta n(t = 0) = \Delta n_0$ ($= \tau\beta\alpha I$) the solution of (128) has the form:

$$\Delta N(t) = \sqrt{\frac{\alpha\beta I}{\gamma}}\,\mathrm{tgh}(t\sqrt{\alpha\beta\gamma I}) \tag{129}$$

When the light impulse is switched off, the time dependence of ΔN is determined by:

$$\Delta N(t) = \sqrt{\frac{\alpha \beta I}{\gamma}} \frac{1}{t\sqrt{\alpha \beta \gamma I} + 1} \tag{130}$$

Similar expressions govern the rise or decay of $\Delta\sigma$, *i.e.* the function $\Delta\sigma(t)$.

Thus, in the case of linear recombination regime ($v_{rec.} \propto \Delta N$), $\ln(\Delta\sigma)$ should depend linearly on t, while in the case of quadratic regime ($v_{rec.} \propto (\Delta N)^2$), it is $1/\Delta\sigma$ that should depend linearly on t. Upon analyzing the $\ln(\Delta\sigma) = f(t)$ and $1/\Delta\sigma = f(t)$, dependencies, linear trend in a wide t-interval was observed only in the first case for all of the studied photoconducting films (a typical $\ln(\Delta\sigma)$ *vs.* t plot is presented in Fig. 20). This indicates that the non-equilibrium charge carrier recombination occurs according to linear kinetics (linear relaxation mechanism), *i.e.* following the exponential decay law: $\Delta\sigma = \Delta\sigma_0 \exp(-t/\tau)$, where τ is the average lifetime of the photocarriers (*i.e.* relaxation time). This parameter was calculated on the basis of the experimental data from the slope of $\ln(\Delta\sigma)$ as a function of t ($d\ln(\Delta\sigma)/dt = -1/\tau$). In the case of SnSe thin films, $\tau = 1.78$ ms, while for Bi_2S_3 and $AgBiS_2$ this parameter is 1.58 ms and 1.67 ms respectively. Such relatively high values for the relaxation time suggest that the studied thin film materials have a potential application in solar cell engineering.

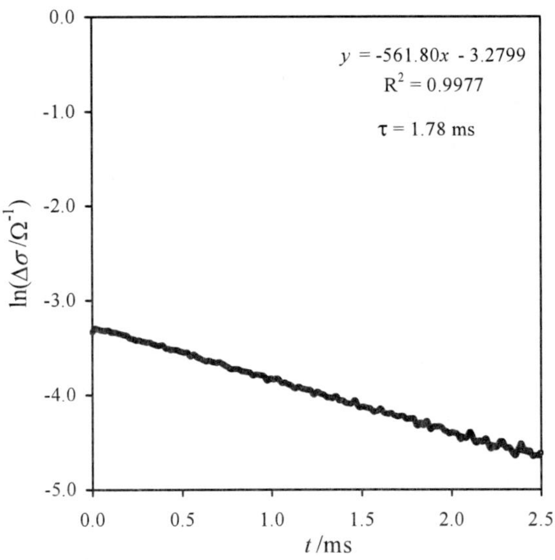

Figure 20. A plot of $\ln(\Delta\sigma)$ *vs.* t for an as-deposited nanocrystalline SnSe thin film as calculated from oscilloscopic data

As the recombination process is kinetically of a linear (first-order) type, charge carrier relaxation processes most probably occur via local trapping centers, describable within the Schockley-Read-Hall model [74,76,77]. Due to the nanocrystalline character of the investigated films, also the possibility for recombination via surface states as alternative channels cannot be in principle excluded. Photoconductivity of residual type is not manifested in the studied cases. Within the framework of the model of Schockley-Read-Hall, for relatively deep recombination centers located in the forbidden band (*i.e.* the band gap) in the case of a semiconductor with holes as predominant charge carriers, the measured relaxation time corresponds approximately to the average lifetime of minority charge carriers [76,77] - electrons. Opposite conclusion may be derived in the case when the electrons are predominant type of charge carriers.

6.3. Lux-Ampere Characteristics of the Photoconductive Thin Films

In a most general case of photoconductor, on the basis of equations governing the dynamics of non-equilibrium charge carriers, it may be shown that photoconductivity ($\Delta\sigma$) is a power function of the intensity (I) or the flux of the incident light (Φ) [74,131-133]: $\Delta\sigma \sim I^k$, i.e. $\Delta\sigma \sim \Phi^k$. The value of the exponent k should be strictly equal to 1 or 1/2, in cases of linear and quadratic relaxation mechanisms correspondingly. It has been, however, shown that in the case of amorphous materials, k often lies in the interval between 0.5 and 1. This finding is often referred to as a sublinear behavior [131-133]. The proposed models for such lux-ampere regime are based on the concept of distribution of impurity levels in the forbidden band which is often assumed to be of exponential type. As the properties of semiconducting materials synthesized in quantum dot form often fall somewhere between those of a macrocrystalline sample and an amorphous sample, it was of certain interest to study the lux-ampere characteristics of the synthesized photoconductive thin films in our works.

For that purpose, we carried out measurements of the non-equilibrium conductivity as a function of the intensity of incident light in the case of all studied samples manifesting photoconduction properties. A source of non-monochromatic and unpolarized light was used for these measurements. Constant field circuit-based methodology was implemented for measurements of the non-equilibrium conductivity, while the incident light intensity (i.e. its flux and consequently the sample illumination) was modulated by changing the distance between light source and the investigated thin film. Sample illumination was measured by a MASTECH MS 6610 luxmeter.

On the basis of least-squares linear interpolation of the experimental data for the dependence of $\ln(\Delta\sigma)$ on $\ln\Phi$ the values of the exponent k were calculated for all of the studied cases. The results are summarized in Table 12. As can be seen the experimental value of k is rather close to 1 in all of the studied systems. These findings are in accordance with those from the photoconductivity relaxation measurements, indicating a linear recombination regime of the non-equilibrium charge-carriers. It thus seems that a single-trap level governs the dynamics of the photoconductivity decay (by a simple Schockley-Reed-Hall mechanism), rather than a quasi-continual distribution of trap levels within the forbidden energy band of the nanocrystals.

Table 12. The experimentally determined values of *k* in the expression $\Delta\sigma \sim \Phi^k$ from the measured lux-ampere characteristics of the studied photoconductive materials in thin film form

system	*k*
SnSe	0.980
Bi_2S_3	0.993
$AgBiS_2$	0.985

REFERENCES

[1] D. Yoffe, *Adv. Phys.* 50 (2001) 1.

[2] D. Yoffe, *Adv. Phys.* 51 (2002) 799.

[3] Burda, X. Chen, R. Narayanan, M. A. El-Sayed, *Chem. Rev.* 105 (2005) 1025.

[4] T. Trinade, P. O'Brien, N. L. Pickett, *Chem. Mater.* 13 (2001) 3843.

[5] R. F. Khairrutdinov, *Russ. Chem. Rev.* 67 (1998) 109.

[6] O'Regan, M. Grätzel, *Nature* 353 (1991) 737.

[7] V. L. Colvin, M. C. Schlamp, A. P. Alivisatos, *Nature* 370 (1994) 354.

[8] Hagfeldt, N. Vlachopoulos, M. Gräitzel, *J. Electrochem. Soc.* 141 (1994) L82.

[9] T. S. Kang, , D. Kim, K. J. Kim, *J. Electrochem. Soc.* 145 (1998) 1982.

[10] Y. Lin, A. Böker, J. He, K. Sill, H. Xiang, C. Abetz, X. Li, J. Wang, T. Ernrick, S. Long, Q. Wang, A. Balazs, T. P. Russell, *Nature* 434 (2005) 55.

[11] Z. L. Wang, J. S. Yin, *Mater. Sci. Eng.* A 286 (2000) 39.

[12] W. U. Dittmer, F. C. Simmel, *Appl. Phys. Lett.* 85 (2004) 633.

[13] G. U. Kulkarni, P. J. Thomas, C. N. R. Rao, *Pure Appl. Chem.* 74 (2002) 1581.

[14] F. Landes, S. Link, M. B. Mohamed, B. Nikoobakht, M. A. El-Sayed, *Pure Appl. Chem.* 74 (2002) 1675.

[15] Y. A. Vlasov, N. Yao, D. J. Norris, *Adv. Mater.* 11 (1999) 165.

[16] Rabani, D. R. Reichman, P. L. Geissler, L. E. Brus, *Nature* 426 (2003) 271.

[17] F. Klingshirn, *Semiconductor Optics*, Springer, Berlin, 1997.

[18] Nesheva, in *Handbook of Surfaces and Interfaces of Materials* (H. S. Nalwa, Ed.), Academic Press, New York, 2001.

[19] K. S. Suslick, G. J. Price, *Annu. Rev. Mater. Sci.* 29 (1999) 295.

[20] T. G. Leighton, *The Acoustic Bubble*, Academic Press, London, 1994.

[21] Y. T. Didenko, K. S. Suslick, *Nature* 418 (2002) 394.

[22] W. B. McNamara, Y. T. Didenko, K. S. Suslick, *Nature* 401 (1999) 772.

[23] Dantsin, K. S. Suslick, *J. Am. Chem. Soc.* 122 (2000) 5214.

[24] W. B. McNamara, Y. T. Didenko, K. S. Suslick, J. *Phys. Chem.* B 107 (2003) 7303.

[25] M. Ohring, *The Materials Science of Thin Films,* Academic Press, New York, 1992.

[26] X. Gao, Y. Cui, R. M. Levenson, L. W. K. Chung, S. Nie, *Nature Biotechnology* 22 (2004) 969.

[27] B. Pejova, M. Najdoski, I. Grozdanov, S. K. Dey, *J. Mater. Chem.* 9 (1999) 2889.

[28] B. Pejova, M. Najdoski, I. Grozdanov, S. K. Dey, *Mater. Lett.* 45 (2000) 2694.
[29] B. Pejova, T. Kocareva, M. Najdoski, I. Grozdanov, *Appl. Surf. Sci.* 165 (2000) 271.
[30] B. Pejova, M. Najdoski, I. Grozdanov, A. Isahi, *J. Mater. Sci.: Materials in Electronics* 11 (2000) 405.
[31] B. Pejova, I. Grozdanov, *Appl. Surface Sci.* 177 (2001) 152.
[32] B. Pejova, A. Isahi, M. Najdoski, I. Grozdanov, *Mater. Res. Bull.* 36 (2001) 161.
[33] B. Pejova, I. Grozdanov, *J. Solid State Chem.* 158 (2001) 49.
[34] B. Pejova, I. Grozdanov, *Thin Solid Films* 408 (2002) 6.
[35] B. Pejova, A. Tanuševski, I. Grozdanov, *J. Solid State Chem.* 172 (2003) 381.
[36] B. Pejova, A. Tanuševski, I. Grozdanov, *J. Solid State Chem.* 174 (2003) 276.
[37] B. Pejova, I. Grozdanov, *Mater. Lett.* 58 (2004) 666.
[38] B. Pejova, I. Grozdanov, A. Tanuševski, *Mater. Chem. Phys.* 83 (2004) 245.
[39] B. Pejova, A. Tanuševski, I. Grozdanov, *J. Solid State Chem.* 177 (2004) 4785.
[40] B. Pejova, I. Grozdanov, *Mater. Chem. Phys.* 90 (2005) 35.
[41] B. Pejova, A. Tanuševski, I. Grozdanov, *J.Solid State Chem.* 178 (2006) 1786.
[42] B. Pejova, I. Grozdanov, Czech. *J. Phys.* 56 (2006) 75.
[43] B. Pejova, I. Grozdanov, *Mater. Chem. Phys.*, 2006, in press.
[44] Ph. Carlach, *Online Nanotechnologies Journal* 2 (2001) 1.
[45] C. Zettlemoyer, *Nucleation*, Marcel Dekker, New York, 1969.
[46] Y. Mastai, G. Hodes, *J. Phys. Chem.* B 101 (1997) 2685.
[47] C. A. Estrada, P. K. Nair, M. T. S. Nair, R. A. Zingaro, E. A. Meyers, *J. Electrochem. Soc.* 141 (1994) 802.
[48] D. Lokhande, P. S. Patil, A. Ennaoui, H. Tributsch, *Appl. Surf. Sci.* 123/124 (1998) 294.
[49] C. D. Lokhande, P. S. Patil, H. Tributsch, A. Ennaoui, *Solar Energy Mater. Solar Cells* 55 (1998) 379.
[50] *Handbook of Chemistry and Physics*, CRC Press, 64th edition, 1983-1984.
[51] J. M. Doña, J. Herreo, *J. Electrochem. Soc.* 142 (1995) 764.
[52] N. N. Greenwood, A. Earnshaw, *Chemistry of the Elements*, Second Edition, Buterworth and Heinemann, London, 1997.
[53] P. Pramanik, S. Bhattacharya, *J. Mater. Sci. Lett.* 7 (1988)1305.
[54] P. Atkins, T. Overton, J. Rourke, M. Weller, F. Armstrong, *Inorganic Chemistry*, Fourth edition, Oxford University Press, Oxford, 2006.
[55] K. S. Suslick, Sonochemistry in: *Kirk-Othmer Encyclopedia of Chemical Technology*, Fourth edition, Vol. 26, John WileyandSons,Inc.: New York, 1998.
[56] K. S. Suslick, Sonoluminiscence and Sonochemistry in: *Encyclopedia of Physical Science and Technology*, Third edition, Academic Press, Inc.: San Diego, 2001.
[57] K. S. Suslick, Y. Didenko, M. M. Fang, T. Hyeon, K. J. Kolbeck, W. B. Mcnamara III, M. M. Mdleleni, M. Wong, Phil. Trans. R. *Soc. Lond.* A 357 (1999) 335.
[58] T. J. Mason, J. P. Lorimer, *Sonochemistry: Theory, Applications and Uses of Ultrasound in Chemistry,* Chichester, UK, 1988.
[59] K. S. Suslick, M. Fang, T. Hyeon, *J. Am. Chem. Soc.* 118 (1996) 11960.
[60] T. Prozorov, R. Prozorov, K. S. Suslick, *J. Am. Chem. Soc.* 126 (2004) 13890.
[61] R. West, *Basic Solid State Chemistry*, Second Edition, John WileyandSons, LTD, New York, 2000.
[62] M. Weller, *Inorganic Materials Chemistry*, Oxford Science Publications, Oxford, UK, 1994.

[63] P. Coppens, A. M. Glazer, J. P. Glusker, A. Hordvik, K. Kuchitsu, J. H. Robertson, V. I. Simonov, *International Union of Crystallography Texts on Crystallography*, Oxford University Press, Oxford, 1995.
[64] STATISTICA, 6.0, 1984-2001 Statsoft, Inc.
[65] J. F. Nye, *Physical Properties of Crystals*: Their Representation by Tensors and Matrices, Clarendon Press, Oxford, 1986.
[66] J. A. Venables, *Introduction to Surface and Thin Film Processes*, Cambridge University Press, Cambridge, 2000.
[67] R. Stokes, in (H. S. Peiser *et al.*, editors) *X-Ray Diffraction by Polycrystalline Materials,* Chapman and Hall, London, 1960.
[68] W. D. Callister, *Materials Science and Engineering – an Introduction*, John Wiley and Sons, New York, 1997.
[69] S. Velumani, Sa. K. Narayandass, D. Mangalaraj, *Semicond. Sci. Technol.* 13 (1998) 1016.
[70] Kuzmany, *Solid-State Spectroscopy: An Introduction*, Springer-Verlag, Berlin Heidelberg, 1998.
[71] P. Y. Yu, M. Cardona, *Fundamentals of Semiconductors*, Second, Edition, Springer-Verlag, Berlin, 1999.
[72] Urbach, *Phys. Rev.* 92 (1953) 1324.
[73] W. Martienssen, *J. Phys. Chem. Solids* 2 (1957) 257.
[74] S. M. Rifkin, *Fotoelektrichesskiye yavleniya v poluprovodnikhah,* GI Fiz-mat. literaturi, Moskva, 1963.
[75] N. W. Ashcroft, N. D. Mermin, *Solid State Physics*, Holt, Rinehart and Winston, New York, 1976.
[76] S. M. Sze, *Semiconductor Devices, Physics and Technology*, Wiley, New York, 1985.
[77] R. Dalven, *Introduction to Applied Solid State Physics*, Plenium Press, New York, 1990.
[78] K. Seeger, *Semiconductor Physics*, Springer-Verlag, New York, 1997.
[79] M. P. Marder, *Condensed MatterPhysics*, John Wiley and Sons, New York, 2000.
[80] J. S. Blakemore, *Solid State Physics*, W. B. Sauders Company, London, 1974.
[81] R. J. Elliott, *Phys. Rev.* 108 (1957) 1384.
[82] S. Adachi, *Phys. Rev.* B 12 (1991) 9569.
[83] N. Chestnoy, R. Hull, L. E. Brus, *J. Chem. Phys.* 85 (1986) 2237.
[84] Baldereschi, N.G. Lipari, *Phys. Rev.* B 3 (1971) 439.
[85] B. Novoselova (editor), *Physical and chemical properties of semiconductors*, Handbook (Moscow, 1978).
[86] Y. Wang, N. Herron, *J. Phys. Chem.* 95 (1991) 525.
[87] D. I. Kim, M. A. Islam, L. Avilla, I. P. Herman, *J. Phys. Chem.* B 107 (2003) 6318.
[88] M. V. Artemyev, A. I. Bibik, L. I. Gurinovich, S. V. Gaponenko, H. Jaschinski, U. Woggon, *Phys. Stat. Solidi* B 224 (2001) 393.
[89] O. I. Micic, S. P. Ahrenkiel, A. J. Nozik, *Appl. Phys. Lett.* 78 (2001) 4022.
[90] M. V. Artemyev, U. Woggon, H. Jaschinski, L. I. Gurinovich, S. V. Gaponenko, *J. Phys. Chem.* B 104 (2000) 11617.
[91] S. Kim, L. Avila, L. E. Brus, I. P. Herman, *Appl. Phys. Lett.* 76 (2000) 3715.
[92] M. V. Artemyev, A. I. Bibik, L. I. Gurinovich, S. V. Gaponenko, U. Woggon, *Phys. Rev.* B 60 (1999) 1504.

[93] Lefebvre, M. A. Szymanski, J. Olivier-Fourcade, J. C. Jumas, Phys. Rev. B 58 (1998) 1896.
[94] V.V. Killerdar, C. D. Lokhande, C. H. Bhosale, *Thin Solid Films* 289 (1996) 14.
[95] N. Benramdane, M. Latreche, H. Tabet, M. Boukhalfa, Z. Kebbab, A. Bouzidi, *Mater. Sci. Eng.* B64 (1999) 84.
[96] Julien, N. Benramdane, J. P. Guesdon, *Semicond. Sci. Technol.* 5 (1990) 905.
[97] N. S. Yesugade, C. D. Lokhande, C. H. Bhosale, *Thin Solid Films* 263 (1995) 145.
[98] D. Lokhande, C. H. Bhosale, *Bull. Electrochem.* 6 (1990) 622.
[99] P. Pramanik, R. N. Bhattacharya, *J. Electrochem. Soc.* 129 (1982) 332.
[100] R. S. Mane, B. R. Sankapol, C. D. Lokhande, *Mater. Chem. Phys.* 60 (1999) 196.
[101] R. S. Mane, B. R. Sankapol, C. D. Lokhande, *Mater. Res. Bull.* 35 (2000) 587.
[102] D. Desai, C. D. Lokhande, *Mater. Chem. Phys.* 41 (1995) 98.
[103] M. E. Rincon, J. Campos, R. Suarez, *J. Phys. Chem. Solids* 60 (1999) 385.
[104] Grozdanov, M. Ristov, Gj. Sinadinovski, M. Mitreski, *Chemtronics* 5 (1991) 71.
[105] B. Pejova, I. Grozdanov, submitted.
[106] B. Pejova, I. Grozdanov, submitted.
[107] E. Brus, *J. Chem. Phys.* 80 (1984) 4403.
[108] E. Brus, *J. Chem. Phys.* 90 (1986) 2555.
[109] M.L. Steigerwald, L.E. Brus, *Annu. Rev. Mater. Sci.* 19 (1989) 471.
[110] M.L. Steigerwald, L.E. Brus, *Acc. Chem. Res.* 23 (1990) 183.
[111] Y. Wang, A. Suna, W. Mahler, R. Kasowski, *J. Chem. Phys.* 87 (1987) 7315.
[112] Y. Nosaka, *J. Phys. Chem.* 95 (1991) 5054.
[113] Kumbhojkar, S. Mahamuni, V. Leppert, S. H. Risbud, *Nanostruct. Mater.* 10 (1998) 117.
[114] Y. Kayanuma, *Phys. Rev.* B 38 (1988) 9797.
[115] H.M. Schmidt, H. Weller, *Chem. Phys. Lett.* 129 (1986) 615.
[116] Green, R. E. Miles, *J. Phys. D: Appl. Phys.* 6 (1973) L45.
[117] Y. P. Varshni, *Physica* 34 (1967) 149.
[118] M. Fernández, P. Prete, N. Lovergine, A. M. Mancini, R. Cingolani, L. Vasanelli, M. R. Perrone, *Phys. Rev.* B 55 (1997) 7660.
[119] Lautenschlager, P. B. Allen, M Cardona, *Phys. Rev.* B 33 (1986) 5501.
[120] Lautenschlager, M. Garriga, M Cardona, *Phys. Rev.* B 35 (1987) 9174.
[121] T. S. Moss, *Photoconductivity in the Elements*, Butterworths, London, 1952.
[122] M. Elkorashy, *J. Phys.: Condens. Matter 2* (1990) 6195.
[123] Z. Kebbab, N. Benramdane, M. Medles, A. Bouzidi, H. Tabet-Derraz, *Solar Energy Mater. and Solar Cells* 71 (2002) 449.
[124] S. Yano, R. Schroeder, B. Ullrich, H. Sakai, *Thin Solid Films* 423 (2003) 273-276.
[125] C. Bouchenaki, B. Ullrich, P. Zielinger, H. N. Cong, P. Chartier, *J. Opt. Soc. Am.* B 8 (1991) 691.
[126] H. B. Devore, *Phys. Rev.* 102 (1956) 86.
[127] E. Rakhshani, *J. Phys.:Condens. Matter* 12 (2000) 4391.
[128] M. Balkanski, *Handbook of Semiconductors*, Vol. 2, Moss Series Editor, 1980.
[129] H. Bube, *Photoelectric Properties of Semiconductors*, Cambridge University Press, Cambridge, 1992.
[130] J. W. Orton, B. J. Goldsmith, J. A. Chapman, M. J. Powell, *J. Appl. Phys.* 53 (1982) 1602.

[131] F. Qasrawi, N. M. Gasanly, *Phys. Stat. Solidi* (A) 194 (2002) 81.
[132] P. Padiyan, S. J. Ethilton, R. Murugesan, *Phys. Stat. Solidi* (A) 185 (2001) 231.
[133] M. Hammam, G. J. Adrianssens, W. Grevendonk, *J. Phys.* C: *Solid State Phys.* 18 (1985) 2151.

In: Progress in Solid State Chemistry Research
Editor: Ronald W. Buckley, pp. 117-164
ISBN 1-60021-313-8

Chapter 3

PRECURSORS ROUTES FOR THE PREPARATION OF Nb-BASED MULTIMETALLIC OXIDES

Daisy A. Bayot and Michel M. Devillers*
Université catholique de Louvain, Unité de Chimie des Matériaux Inorganiques et Organiques, Place Louis Pasteur 1, B-1348 Louvain-la-Neuve, Belgium

ABSTRACT

In the past decades, Nb-containing multimetallic oxide materials have generated considerable interest because of their attractive physical properties such as catalysis, ferroelectricity, ion conductivity and luminescence. Most of the time, the preparation method is the classical ceramic route; because of the serious limitations of this high-temperature route, alternatives based on precursors have also been considered. The development of such processes in the case of niobium-containing materials constitutes a great challenge because of the cruel lack of appropriate sources of niobium(V). Many efforts have been made recently on the syntheses of novel precursors that display versatile properties.

The purpose of this review is to cover the open literature concerning precursors routes for the preparation of Nb-containing multimetallic oxides. It provides a classification of the different kinds of such methods developed up to now and offers a broad view of the possible ways to form niobium-based oxide materials.

CHEMICALS ABBREVIATIONS

acac	acetylacetonate
bis-dmap	bis(dimethylamino)propanolate
dmae	dimethylaminoethoxide
dpm	dipivaloylmethanate
EG	ethyleneglycol

* Corresponding author.

EGMEE	2-ethoxyethanol(ethylene glycol monoethyl ether)
gu^+	guanidinium
H_3BHMP	bis(hydroxymethyl)propionic acid
H_4cit	citric acid
H_5DTPA	diethylenetriaminepentaacetic acid
H_4EDTA	ethylenediaminetetraacetic acid
H_3NTA	nitriloacetic acid
H_2ox	oxalic acid
H_3sal	salicylic acid
H_4tart	tartaric acid
H_3THME	tris(hydroxymethyl)ethane
H_6TTHA	triethylenetetraaminehexaacetic acid
HMT	hexamethylenetetramine
pdb	1-phenyl-1,3-butanedione
PVA	polyvinylalcohol
TEA	triethanolamine
thd	2,2,6,6-tetramethylheptane-3,5-dionate

1. Introduction

In recent years, multimetallic Nb(V)-containing oxides have generated considerable interest in many fields, because of their attractive physical properties. These oxides are widely studied as ferroelectric and/or piezoelectric materials like the so-called PMN ($Pb(Mg_{1/3}Nb_{2/3})O_3$) [1-5], PZN ($Pb(Zn_{1/3}Nb_{2/3})O_3$) [6], SBN ($Sr_xBa_{1-x}Nb_2O_6$) [7-9] and LN ($LiNbO_3$) [10-12] oxides, and also phases from the Bi-Nb-O system [13-17]. Nb-based multimetallic oxide materials are also described as conductors (i) of oxygen ions (fluorite-like structured phases, Y_3NbO_7 [18] or Bi_3NbO_7 [14]), (ii) of lithium ions (oxides from the Li-La-Nb-O system [19]), or (iii) of electrons (VNb_9O_{25} [20]). They are also reported as mixed electronic-ionic conductors like the Sr_2FeNbO_6 phase [21]. Nb-containing luminescent or phosphorescent oxides have also been discovered like the yttrium-based oxides, $YNbO_4$ or Bi-doped $YNbO_4$ [22-24]. Another important feature of Nb-containing mixed oxides is their promising catalytic behaviour in several highly challenging industrial processes, like alkane oxidation or ammoxidation. Pure niobium pentoxide itself is a very selective catalyst for propane oxidative dehydrogenation (ODH) even though its catalytic activity is only moderate [25]. In recent years, it has been shown that associating niobium with elements such as vanadium, molybdenum or antimony could considerably improve the catalytic activity while maintaining a high selectivity in such reactions. Bulk or supported oxides from multicomponent systems like the Nb-V-Mo [26-28], Nb-V-Mo-Sb [29,30] and Nb(-V)-Mo-Te [31-36] systems are mainly described in such a context and these catalysts are generally multiphasic. Moreover, some Nb-containing mixed oxides are also reported to be efficient catalysts in the photodecomposition of water into molecular hydrogen and oxygen. Nb and Bi-based phases, Bi_2MNbO_7, with M corresponding to iron, yttrium, a rare earth or a group 13 element, have been described recently as photocatalysts [37-39]. Some tantalum-containing solid solutions, $(Ta_xNb_{1-x})_2O_5$ and $Ta_xNb_{1-x}BiO_4$, have also shown an important catalytic activity in that process [40-43].

Most of the time, the preparation method of Nb-based multimetallic oxides reported in the literature is the conventional ceramic route, which involves reacting an appropriate mixture of oxides or carbonates at elevated temperatures. Because of the serious limitations of this high-temperature route in terms of phase purity and specific surface area essentially, alternatives based on precursors, often called soft chemical routes, have been developed for many years. The advantages of these precursors methods in general are numerous [44-46] : (i) because of the frequent solubilization step required by these processes, their use provides homogeneous materials due to the mixing at a molecular level of the metals involved; (ii) they also allow forming crystalline oxides under conditions significantly milder than those employed in conventional solid state syntheses, (iii) they sometimes enable trapping of metastable phases and finally, (iv) they generally result in materials with particular textural properties and relatively high specific surface areas, which are critical parameters in several domains like catalysis.

As far as Nb-containing multimetallic oxides are concerned, progresses in the soft chemistry direction have been hindered because of the cruel lack of appropriate sources of niobium, whether commercial or not. The development of such routes thus constitutes a great challenge in that context and many efforts have been made recently on the syntheses of novel Nb precursors that display versatile properties (solubility, stability, ...) and can be engaged in precursors methods. Moreover, there is a great diversity among the methods developed in the literature and the authors must generally adapt well-known routes to answer the strict requirements of niobium (solvent, atmosphere,...).

This paper will review the precursors routes for the preparation of multimetallic Nb(V)-containing oxides, whether they are mono- or multiphasic, and in a bulk or supported form. It presents a general overview of the different methods used as well as their numerous variants, with a particular attention to the precursors themselves and to the formulation of the aimed materials. The preparation methods engaging niobium under its oxide form, like the ceramic, hydrothermal or combustion routes, will not be described here. The present review will not cover the literature on Nb-doped oxides, nor the field of Nb-containing silicates and zeolites, because the preparation methods of these mesoporous materials are generally very different from those of stoichiometric oxides containing niobium [47-49]. The first section is dedicated to a brief description of the most currently used commercial niobium(V) sources, their properties, advantages and drawbacks.

2. Commercial Sources of Niobium(V)

Table 1 lists the niobium(V) sources commercially available and most currently engaged as precursors. Three main situations are encountered : (i) These precursors are used as such, (ii) they can be modified in situ without clear identification of the species involved or (iii) they are used as reactants for the synthesis of derived precursors well characterized at the molecular level and often pre-isolated. Most of these commercial compounds display unfortunately several drawbacks that restrict their applicability.

Table 1. Most currently used commercially available niobium(V) sources

Common name	formulation	aspect
Niobium pentoxide	Nb_2O_5	white crystalline powder
Niobic acid or niobium oxide hydrate	$Nb_2O_5.nH_2O$	white amorphous powder
Niobium pentachloride	$NbCl_5$	yellow crystalline powder
Niobium ethoxide	$Nb(OC_2H_5)_5$ or $Nb(OEt)_5$	liquid
Niobium iso-propoxide	$Nb(OCH(CH_3)_2)_5$ or $Nb(O^iPr)_5$	liquid
Niobium ter-butoxide	$Nb(OC(CH_3)_3)_5$ or $Nb(O^tBu)_5$	liquid
Niobium normal-butoxide	$Nb(O(CH_2)_3CH_3)_5$ or $Nb(O^nBu)_5$	liquid
Niobium ammonium oxalate salt	$NH_4H_2[NbO(C_2O_4)_3].3H_2O$	white crystalline powder

Niobium pentoxide, Nb_2O_5, is a very insoluble and very unreactive compound. It is most of the time used in the high-temperature ceramic method, in the mechanochemical activation and sometimes in hydrothermal syntheses. Its hydrated version, commonly called niobic acid and noted $Nb_2O_5.nH_2O$, is poorly defined but more reactive than Nb_2O_5. It dissolves by reaction in water with diverse additives like chelating ligands or hydrogen peroxide. Niobic acid constitutes a more convenient Nb(V) starting material than Nb_2O_5 for synthesis involving aqueous solutions. Nevertheless, it is quite unstable with time and looses reactivity. It is generally freshly precipitated prior to use, from hydrolysis in basic aqueous medium of niobium pentachloride, niobium oxalate or Nb_2O_5 dissolved in HF.

Niobium pentachloride and some pentalkoxides are the most used sources of niobium. Both kinds of compounds are moisture sensitive and must be stored and handle under inert atmosphere. There are soluble in most organic solvent, provided they are first dried to prevent hydrolysis. The major drawbacks that can be pointed out is the expensiveness of alkoxides and, when using $NbCl_5$, the possible chloride contamination of the final material and liberation of toxic HCl vapours during chemical processing or upon calcination.

The niobium oxalate ammonium salt, also called "niobium oxalate soluble complex", is generally noted $NH_4H_2[NbO(C_2O_4)_3].3H_2O$ even if other stoichiometries have also been cited in the literature. This compound is yet poorly defined and characterized. It is soluble in water but the resulting solution is not stable with time towards hydrolysis and addition of oxalic acid allows its stabilization.

3. Precursors Methods for the Preparation of Nb-Containing Multimetallic Oxides

3.1. Sol-Gel Route

The sol-gel process is commonly used for the preparation of multimetallic oxides, in a bulk or supported form. It is also often considered for the formation of thin oxide films on diverse substrates, mainly using the dip- or spin-coating methods. In its usual version (hydrolytic route), the sol-gel method involves hydrolysis and polycondensation of ionic precursors, allowed by at least one molecular alkoxide in the medium. This process leads to the formation of an oxide network directly from the solution. A non-hydrolytic variation also exists and should be more properly named non-hydrolytic condensation, because it does not involve hydroxylation.

Niobates of alkali metals were widely prepared by the sol-gel method. Potassium niobate, $KNbO_3$ was obtained in a bulk and pure form from the controlled hydrolysis of metal ethoxides or propoxides [50]. The single alkoxides were reacted to form in situ the "mixed metal" alkoxides, i.e. $K[Nb(OEt)_6]$ and $K[Nb(O^iPr)_6]$, evidenced by ^1H, ^{93}Nb-NMR and mass spectrometry. The hydrolysis rate, depending on the amount of added water, is a critical parameter in this method to obtain the single $KNbO_3$ phase. The sol-gel method has been mainly used up to now to synthesize that kind of niobates as thin films, sometimes with particular properties. Highly oriented $K(Ta,Nb)O_3$ (Ta:Nb = 65:35) (KTN) thin films of perovskite structure were prepared successfully by dip-coating on Pt(1000)/MgO(100) substrates from the controlled hydrolysis of a solution of metal alkoxides [51]. The coating solution, containing $Nb(OEt)_5$, $Ta(OEt)_5$ and KOEt in ethanol, was analyzed by ^1H, ^{13}C and ^{93}Nb-NMR. The KTN precursor consists in a homogeneous mixture of $K[M(OEt)_6]$ (M = Ta or Nb) units interacting at the molecular level, as in the structure proposed in Figure 1 [51]. The perovskite KTN films, crystallized on the substrates at 700°C, were found to orient preferentially through calcination in water vapour/oxygen gas mixture. Homogeneous $LiNbO_3$ thin films were also prepared on Si(100) and sapphire(0001) substrates, at temperatures as low as 250°C, by dip-coating using a partially hydrolyzed solution of a double alkoxide similar to that used for the KTN oxide [52]. The concentration of the solution influences both thickness and quality of the films. Films crystallized on sapphire(0001) showed the preferred orientation along the *c* axis, while a preferred orientation could not be observed on Si(100) substrates. The same phase was also obtained in a bulk form above 400°C, by sol-gel from the controlled hydrolysis of an ethanolic solution containing LiOEt and $Nb(OEt)_5$ [53]. The appearence of the evolved products during heat-treatment of the gel, mainly water and ethanol, was followed by mass spectrometry and this allows selecting the most appropriate thermal conditions to obtain the oxide (above 400°C).

The sol-gel method is also often used to prepare niobates containing alkaline-earth metals. $Pb(Mg_{1/3}Nb_{2/3})O_3$ oxides (PMN) can be obtained from alkoxide-type reactants, which allow the formation of the desired pure perovskite phase at temperature ranging from 800 to 900°C [5,54]. The Mg and Nb precursors are most of the time the double metal ethoxide, $Mg[Nb(OEt)_6]_2$, formed in situ by progressive dissolution of the magnesium ethoxide into the niobium ethoxide. The lead precursor is obtained either (i) by reacting the organometallic compound, $Pb[N(SiMe_3)_2]$ with *t*-butanol to give $Pb(O^tBu)_2$ [5] or (ii) by reacting lead acetate with sodium isopropoxide [5] or 2-methoxyethanol [54]. The organic mixture of the lead precursor and the double alkoxide is then hydrolyzed to produce a gel which is dried and then calcined under appropriate conditions. The preparation of the $0.4Pb(Mg_{1/3}Nb_{2/3})O_3$-$0.3Pb(Mg_{1/2}W_{1/2})O_3$-$0.3PbTiO_3$ ceramic, also noted 0.4PMN-0.3PMW-0.3PT, was investigated by the sol-gel method [55]. Figure 2 illustrates the experimental procedure followed by the authors. The starting materials were lead acetate trihydrate, titanium isopropoxide, niobium ethoxide, magnesium ethoxide and tungsten ethoxide. About 89% of the desired perovskite phase is obtained at 1050°C. A slight contamination of the pyrochlore phase is always detected at that temperature.

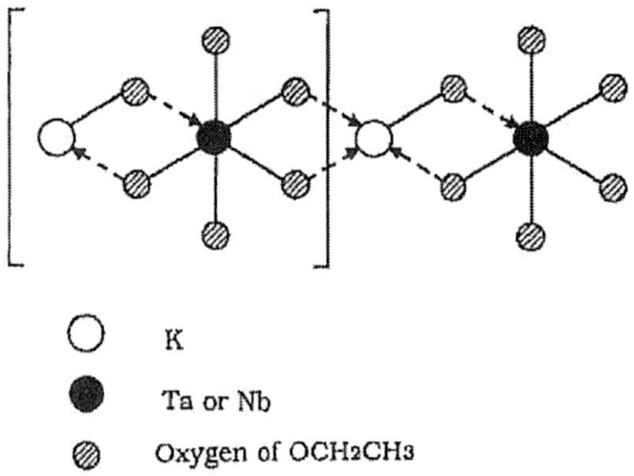

Figure 1. Structure of the KTM precursor consisting of $K[M(OEt)_6]$ (M = Ta or Nb) units interacting at the molecular level. Reproduced from Ref. [51] with permission from copyright owners.

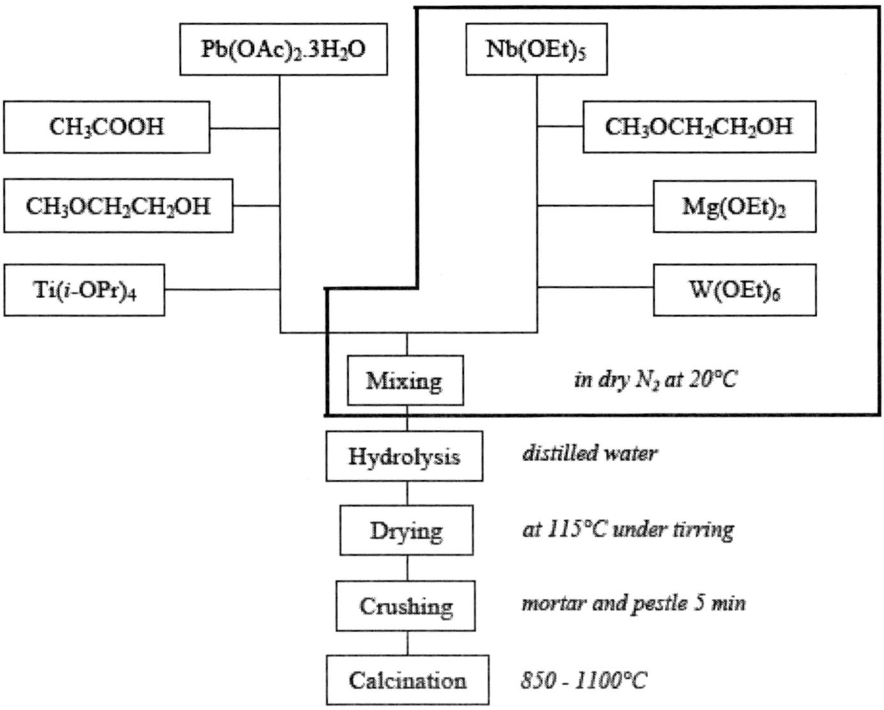

Figure 2. Flow chart showing the preparation procedure of $0.4Pb(Mg_{1/3}Nb_{2/3})O_3$-$0.3Pb(Mg_{1/2}W_{1/2})O_3$-$0.3PbTiO_3$ ceramics [55].

Thin films containing strontium and barium or bismuth were obtained through metal alkoxides. Highly oriented $Sr_{0.5}Ba_{0.5}Nb_2O_6$ (SBN50) films were prepared from a homogeneous precursor solution prepared via the controlled reaction of niobium ethoxide and metallic strontium and barium in absolute ethanol [8]. The SBN precursor solution was then stabilized by addition of 2-ethoxyethanol(ethylene glycol monoethyl ether) (EGMEE). The effect of addition of this ligand was studied by ^1H and ^{13}C-NMR, which revealed the coordination of EGMEE to metals via its 2-ethoxyethanol group. Based on spectroscopic data, the SBN50 precursor consists of $M[Nb(OR)_6]$ units (M = Sr or Ba, R = CH_2CH_3 (Et) or $CH_2CH_2OCH_2CH_3$). SBN50 thin films on MgO(100) and Pt(100)/MgO(100) were obtained by a repeated dip-coating-calcination process, and crystallized to a mixture of orthorhombic and tetragonal phase at 700°C and then transformed completely to the tetragonal oxide at 1000°C [8]. Thin films of the $SrBi_2Nb_2O_9$ (SBN) oxide can be prepared from a precursor occurring from the reaction of an in situ formed double Sr-Bi (2:2) methoxyethoxide and of niobium ethoxide in methoxyethanol, followed by partial hydrolysis [56]. By a deep interpretation of the results of ^1H, ^{13}C , ^{93}Nb-NMR and FTIR, the molecular structure of this precursor was found to be very similar to the sublattice of the layered-structured perovskite crystal of $SrBi_2Nb_2O_9$. A comparison of the structure of the SBN oxide with the one proposed for the SBN precursor is allowed in Figure 3. As a result of this structural similarity, the crystallization of the SBN thin films, prepared by spin-coating on Si(100) and Pt-passivated silicon, occurs at the relatively low temperature of 550°C [56]. An identical study, leading to the same conclusions, was carried out on $Sr_2Bi_2Nb_3O_{12.5}$ (SBN223) oxides [57].

The $Pb(Fe_{1/2}Nb_{1/2})O_3$ oxide (PFN) was prepared by sol-gel starting from niobium ethoxide, $Nb(OEt)_5$, iron(III) acetylacetonate, $Fe(acac)_3$, and lead acetate, $Pb(OAc)_2.3H_2O$ [58]. An acetic solution of $Fe(acac)_3$ was added to an ethanolic solution of $Nb(OEt)_5$. The mixture was then refluxed and according to the authors, some chemical reactions must occur during this step. They would correspond to the nucleophilic substitution of alkoxy or acac groups by acetate ligands. The lead acetate solution was then added. A red gel is formed rapidly without addition of water because it is formed in situ from esterification reactions between alcohol and acetic acid, as well as from the crystallization water of the hydrated lead acetate. Analysis of the gel was carried out by TGA and DTA measurements and the crystallization of the PFN phase was followed by XRD at different temperatures. A two-phase ceramic, containing the pyrochlore and the perovskite phases, is obtained at 650°C. The desired perovskite oxide is obtained in a pure form at 700°C, a temperature much lower than in the conventional ceramic method.

Group five-containing oxides, from the binary Nb-V or ternary Nb-V-Ta systems, were prepared by sol-gel techniques. In 1990, Yamaguchi et al carried out a systematic investigation of the niobium-vanadium oxide system in the entire Nb/V composition range [59]. Amorphous precursors were obtained by the simultaneous hydrolysis with ammonia of an isopropanolic solution of niobium isopropoxide and vanadyl ethoxide. The powders, with compositions ranging from pure Nb_2O_5 to pure V_2O_5, were analyzed thoroughly by differential thermal analysis and calcined at different temperatures. Four defined phases crystallized in the Nb-V system : $23Nb_2O_5.2V_2O_5$ (1250°C), $9Nb_2O_5.V_2O_5$ or Nb_9VO_{25} (1050°C), $9Nb_2O_5.2V_2O_5$ or $Nb_{18}V_4O_{25}$ (525-630°C), and the $NbVO_5$ (420-470°C) oxides [59]. Amarilla et al synthesized the bulk MVO_5 phases (M = Nb or Ta) and $Ta_xNb_{1-x}VO_5$ solid solutions ($0 < x < 1$), from the niobium and/or tantalum pentachorides and from vanadyl *t*-butoxide, $VO(O^tBu)$, in isopropanol [60-62]. They have first used a standard sol-gel method to obtain $Ta_xNb_{1-x}VO_5$ oxides. The precursor solution was transformed by aging into a gel, which was dried and the resulting powder, washed until elimination of chlorides, was finally calcined in air at 600°C [60]. The authors then focused on the MVO_5 phases and $Ta_xNb_{1-x}VO_5$ solid solutions to establish the influence of the different experimental parameters on this standard sol-gel method [61,62]. They also investigated the thermal behaviour of the dried gels by thermal analyses. Amorphous Nb-V mixed oxides (Nb:V = 6:1) were also prepared starting from $Nb(OEt)_5$ and $VO(O^iPr)_3$ in an ethanol/HCl medium [63]. The H_2O/V ratio was varied and the clear transparent gels obtained under hydrolytic conditions were dried to provide amorphous powders which display relatively high specific surface areas reaching 160 $m^2.g^{-1}$ [63].

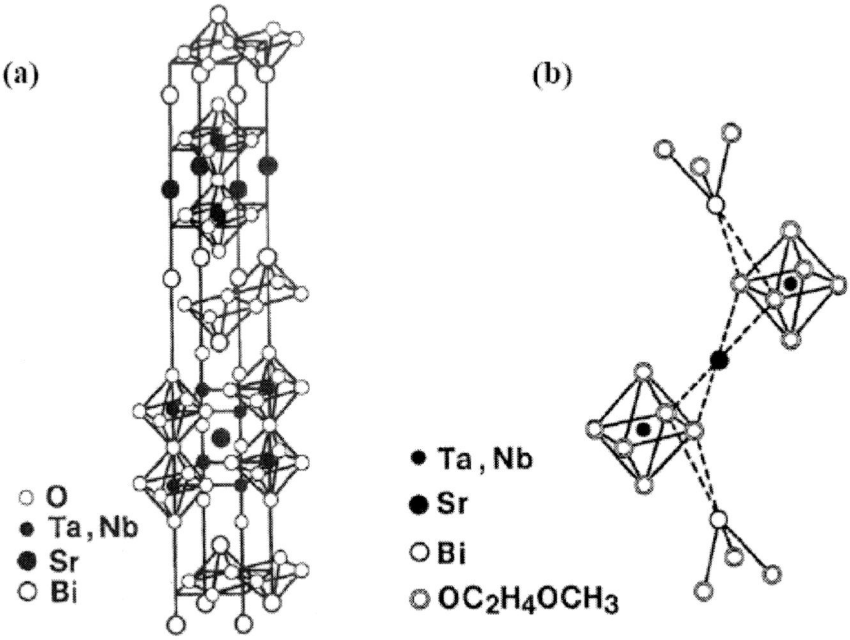

Figure 3. Comparison of (a) the crystal structure of the layer-structured perovskite $SrBi_2Nb_2O_9$ (SBN) or $SrBi_2Ta_2O_9$ (SBT) oxide and (b) the proposed molecular structure of SBN or SBT precursors in solution. Reproduced from Ref. [56] with permission from copyright owners.

Thermal analyses of these powders were carried out. Their further calcination up to 550°C strongly decreases their surface areas due to crystallization. The H_2O/V ratio was shown to influence the catalytic features of the product obtained after calcination. A comparison was made between samples prepared by sol-gel and those having the same composition but prepared by an impregnation of Nb_2O_5 with an aqueous metavanadate solution [63]. The niobium-rich Nb_9VO_{25} bulk phase was obtained from the sintering at 680°C of the precursor resulting from the controlled hydrolysis of $Nb(OEt)_5$ and $VO(OEt)_3$ [20]. This temperature is much lower than that reported by Yamaguchi et al for the formation of this phase [59]. Another variant of the sol-gel method was also used: a Nb_2O_5/V_2O_5 mixture was transformed into the corresponding alkoxides with the use of absolute ethanol and benzene as solvent. The solution was then hydrolyzed by adding water [20]. Very recently, Nb-V mixed oxides were prepared via the hydrolytic sol-gel procedure. The effects of various Nb precursors ($Nb(OEt)_5$ or $NbCl_5$), acid promoters (HCl, HNO_3) and/or complexing agents (citric or oxalic acid) and the influence of the Nb/V ratio (1:1, 4.5:1 or 9:1) on the sol-gel process were investigated [64]. The vanadium precursor engaged is vanadyl isopropoxide, $VO(O^iPr)_3$. Table 2 summarizes the Nb/V molar ratios, the nature of the added promoter and the results of analyses carried out on the final oxides.

Table 2. XRD results of Nb-V samples prepared by modified hydrolytic sol-gel methods [64]

Nb/V molar ratio	Nb precursor	additive	Crystalline phases
1:1	$Nb(OEt)_5$	/	$NbVO_5$; $Nb_{18}V_4O_{55}$; V_2O_5
1:1	$Nb(OEt)_5$	HCl	$NbVO_5$; $Nb_{18}V_4O_{55}$; V_2O_5
1:1	$Nb(OEt)_5$	HNO_3	$NbVO_5$; $Nb_{18}V_4O_{55}$
1:1	$Nb(OEt)_5$	citric acid	$NbVO_5$
1:1	$Nb(OEt)_5$	oxalic acid	$NbVO_5$; $Nb_{18}V_4O_{55}$
1:1	$NbCl_5$	/	$NbVO_5$; $Nb_{18}V_4O_{55}$; V_2O_5
4.5:1	$Nb(OEt)_5$	/	Nb_2O_5
9:1	$NbCl_5$	/	Nb_2O_5

The addition of citric acid as complexing agent was shown to stabilize the Nb-V precursor solution after addition of water, favouring the direct formation of gel. As a result, a homogeneous 1:1 Nb-V system was obtained by this method. In all other cases investigated with Nb/V = 1:1, precipitation of a solid occurred, leading after calcination to heterogeneous multiphasic systems. In the case of the Nb-richer systems (Nb/V = 4.5:1 or 9:1), solutions are stable without additive, and turned rapidly to gels. Nevertheless, there is no evidence of V-based crystalline phases and only crystalline Nb_2O_5 is detected by XRD. Raman spectroscopy evidenced the presence of $Nb_{18}V_4O_{25}$ in the 4.5:1 Nb-V sample [64]. Pure and silica-supported Nb-V and Nb-V-M (M = Sb or Mo) oxides were also prepared by hydrolytic sol-gel routes [65,66]. This process involves addition of a mixture of water and isopropanol to mixtures of $Nb(OEt)_5$, $VO(O^iPr)_3$, $Sb(O^iPr)_3$ and $Mo(O^iPr)_5$. In the case of the preparation of supported samples, $Si(OEt)_4$ is added to this mixture [66]. The metal precursors were mixed in order to obtain the following atomic ratios : 1:1 for the binary Nb-V system; 1:1:1, 1:2:1 and 1:1:5 for the ternary Nb-V-Mo, Nb-V-Sb and Nb-V-Si systems, respectively. A non-hydrolytic route was conducted to prepared pure and silica-dispersed Nb-V oxides by adding dropwise $VO(O^iPr)_3$ to $NbCl_5$, solid or suspended in hexane (V:Nb = 1:1, 5:3). $SiCl_4$ and/or $Si(OEt)_4$ is/are further added when supported samples are desired (V:Nb:Si = 1:1:5, 1:1:20) [65,66]. All formulations, obtained by hydrolytic and non-hydrolytic processes, were thermally treated at 550°C. The supported samples are all practically amorphous and characterized by quite high surface areas. The crystalline $NbVO_5$ phase is identified in the 1:1 Nb-V sample while mixture of mixed oxide phases are detected in the 1:1:1 Nb-V-Mo and 1:2:1 Nb-V-Sb samples. The experimental parameters characterizing the preparation procedure appear to be really critical to determine the physical properties of the final oxides (surface, morphology,...) but also their catalytic properties, which were evaluated in the propane ODH reaction [65,66].

3.2. Methods Based on in Situ Modified Precursors

The preparation modes of multimetallic oxides based on metal precursors modified in situ have been widely used up to now and are still under constant consideration. The precursors engaged in such routes are mostly obtained in solution from commercial metal sources, in the presence of additive(s), which are mostly chelating agents like oxalic, citric, tartaric, EDTA acids and also hydrogen peroxide. Sometimes, gelifying (citric acid) or polymerizing (ethyleneglycol) additives are also engaged, like in the citrate and Pechini routes, respectively. This section is divided according to the nature of the additives engaged in the preparation method to modify in situ the metal sources.

3.2.1. The Citrate Route and Variations

The citrate method constitutes a polymeric approach developed thirty years ago, in which different metals are trapped in a gelified citric acid network. In this process, citric acid acts as a complexing agent but the in situ formed metal citrato complexes have never been characterized by a given stoichiometry. The metal precursors engaged are most of the time nitrates and the solvent is water. Some modifications in the usual approach are sometimes necessary to answer the requirement of particular elements (solubility, precursors available, gel stabilization by additives, …).

Ultrafine powder of single-phased PMN was prepared by this method [5]. The starting reactants are lead and magnesium nitrates, dissolved together in nitric acid, and niobium pentachloride, dissolved in an aqueous 30wt% hydrogen peroxide solution. Citric acid is added to the nitrates solution and the chloride solution is added to the resulting mixture at pH 1.5. The water is slowly evaporated from the resulting clear Pb-Mg-Nb solution by heating at 80°C. A gel product with a high viscosity is obtained. It is calcined at 500°C and directly heated up to 900°C during a short period of time (<1h). This treatment is called "the thermal shock method" and is quite effective to exclude undesired pyrochlore phase and decreasing PbO vaporization.

Very recently, for synthesizing the $BaNb_2O_6$ phase, $BaCl_2.6H_2O$ and $NbCl_5$ were engaged as metal sources in a citrate method [67]. Barium chloride was dissolved in water while niobium chloride was dissolved in a minimum amount of HF while heating in a hot water bath for 20h in order to obtain soluble niobium fluoride species, $NbOF_5^{2-}$ and NbF_7^{2-}. An excess amount of HCl is added to the mixture of both solutions in order to dissolve barium fluoride formed in situ. Citric acid is finally added and a yellowish gel is obtained after water evaporation. It is subsequently decomposed at various temperatures ranging from 200 to 800°C. The citrate precursor crystallized in orthorhombic $BaNb_2O_6$ at 600°C. For comparison, the ceramic route requires a more elevated temperature (1300°C) and repeated grinding and calcining steps to obtain the same oxide. Zhao et al also reported an adapted citrate method to prepare $Sr_xBa_{1-x}Nb_2O_6$ (SBN, x = 0.4, 0.5 and 0.6) oxides, which are solid solutions between $BaNb_2O_6$ and $SrNb_2O_6$ [9,68]. The starting materials engaged are $Ba(NO_3)_2$, $Sr(NO_3)_2$ and $Nb_2O_5.nH_2O$. The flow chart followed to prepared these oxides is detailed in Figure 4. For the preparation of stock-solutions of Ba and Sr, the corresponding nitrates are separately dissolved in water in the presence of ethylenediaminetetraacetic acid (EDTA) as complexing agent. The stock solution of Nb is obtained from freshly precipitated niobic acid dissolved in an aqueous solution of citric acid at 60°C. Stoichiometric amounts of the Ba-EDTA, Sr-EDTA and Nb-citrate solutions are mixed, followed by the addition of citric acid. The gelatinous precursor, obtained by drying the solution at 80°C, is then calcined between 400 and 800°C. The crystallization of $Sr_xBa_{1-x}Nb_2O_6$ begins at 500°C but the single phase is completely crystallized at 800°C. Thermal analyses were carried out on the gel before calcination to understand its thermal behaviour and the influence of pH and citric acid:Nb molar ratio on the formation of homogeneous Sr-Ba-Nb precursors gels has also been studied [9]. Strontium bismuth niobate, $SrBi_2Nb_2O_9$ was prepared using an aqueous acetate-citrate gel precursor [69,70].

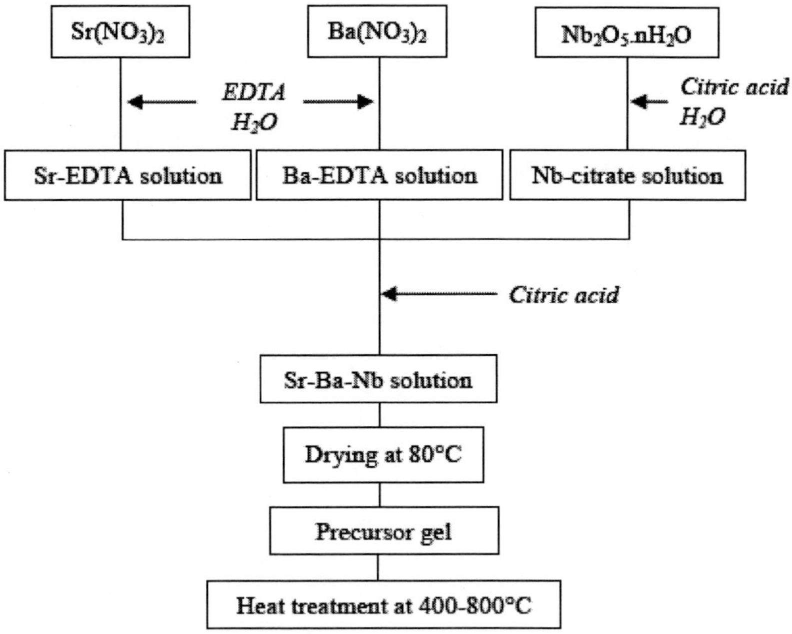

Figure 4. Flow chart showing the preparation procedure of $Sr_xBa_{1-x}Nb_2O_6$ oxides (x = 0.4, 0.5 and 0.6) by the citrate route [9].

This method starts from solutions of bismuth acetate in acetic acid, strontium acetate in water, and from a niobium-peroxo-citrato precursor, which is prepared by the dissolution of niobium ammonium oxalate in an aqueous solution of citric acid and H_2O_2. In the latter case, the oxalate ions, not very suitable for gel formation, are decomposed by excess hydrogen peroxide. The preparation and thorough characterization of the Nb-peroxo-citrato solution were reported by Narendar et al and will be described in section 3.2.4 [71]. The three metallic solutions are mixed in the appropriate amount with regard to the stoichiometry of the desired oxide and citric acid is added with a citric acid:metal molar ratio equal to 3:1. The pH is then adjusted to 4.5 with ammonia [69,70] or ethylenediamine [69] and the solvent is evaporated at 100°C leading to the formation of a homogeneous transparent gel. This gel was thoroughly studied by complementary thermoanalytical techniques (TGA-MS, TGA-FTIR) to evidence its decomposition pathway, and by SEM or TEM to evaluate its homogeneity. The crystallization of the precursor was followed in situ by high-temperature-XRD analyses : the fluorite phase starts to form at 425°C and it transforms into the perovskite phase above 625°C. Pure perovskite SBN oxide is detected at 675°C [69,70]. This temperature is lower than that required by the ceramic process (above 1000°C) which causes significant Bi losses, leading to non-stoichiometric compositions. The $SrBi_2Nb_2O_9$ phase can also be prepared via a slightly different citrate method, using aqueous solutions including Sr-EDTA, Bi-EDTA and Nb-citrate complexes [72]. The solutions are prepared according the same procedures used by Zhao et al for the preparation of Sr-Ba-Nb oxides and described above [9,68], from $Sr(NO_3)_2$, $Bi(NO_3)_3$ and niobic acid, freshly prepared from hydrolysis of $NH_4H_2[NbO(C_2O_4)_3].3H_2O$ with NH_3. Stoichiometric amounts of the Bi-EDTA, Sr-EDTA and Nb-citrate solutions are mixed, followed by the addition of citric acid. Drying the solution at 80°C gives rise to a gel which is first pyrolyzed at 300°C and then heated between 500 and 800°C. The fluorite phase forms in the 500-600°C range and appears to act as an intermediate phase prior to the formation of the perovskite one, which crystallizes at 650°C. This temperature is higher than that reported by Kato et al for the same phase prepared by sol-gel (550°C) [56]. However, the authors emphasized the important practical and environmentally beneficial aspects of their aqueous way, with regard to the alkoxide sol-gel route [72]. A β-$BiNbO_4$ oxide powder with nanometer sized particles (~65 nm) was also obtained from a mixture of Bi-citrate and Nb-citrate solutions, following a classical procedure [17]. The crystalline phase is obtained at the low temperature of 600°C, in comparison with that required in the case of the ceramic route (>1020°C) [73].

3.2.2. The Pechini Method

The Pechini route, also called polymerizable complex method, is a variant of the citrate process. It consists in using ethylene glycol (EG) together with citric acid, which leads to polyesterification reactions and the formation of a polymeric resin.

Thin films of $LiNbO_3$ were prepared by the Pechini method, and deposited by dip or spin-coating on several substrates [74-78]. The coating solution is prepared by the mixture of lithium citrate and niobium citrate solutions. The citrate complex solutions were obtained (i) from the separate dissolution in aqueous citric acid of niobic acid and Li_2CO_3 [74-76] or (ii) from the simultaneous dissolution of Nb_2O_5 and Li_2CO_3 in EG and citric acid at 90°C (molar ratio metal:citric acid:EG = 1:4:16) [77,78]. The subsequent addition of EG and the heating of the mixture leads to a polymeric precursor solution, the viscosity of which is then adjusted by adding a controlled amount of water. Films were deposited by spin or dip coating on sapphire [74,76,78], or silicon [75,77] substrates and, after each deposition, heat-treated first at 300°C and then at temperatures in the range 500-600°C. The $LiNbO_3$ phase was obtained as multi-layered films (i) on sapphire with preferred orientation [74,76,78], and (ii) on silicon with no preferred orientation and with the presence of small amounts of the Li-deficient phase, $LiNb_3O_8$ [75,77]. This oxide is originated from an interface reaction between silicon oxidized surface and the lithium niobate according to the following reaction:

$$3\ LiNbO_3 + SiO_2 \rightarrow LiNb_3O_8 + Li_2SiO_3$$

The effect on crystallinity, morphology and optical properties of Mg doped-$LiNbO_3$ thin films prepared by the Pechini method was also investigated [78]. Ultrafine $NaNbO_3$ powder was synthesized from aqueous solutions of niobium ammonium oxalate, $NH_4H_2[NbO(C_2O_4)_3].3H_2O$ and sodium nitrate, $NaNO_3$ in the presence of citric acid [79]. The citrate complex solutions are mixed considering the 1:1 Na:Nb molar ratio and EG is added. The polymerization is carried out in two stages, at 120°C for water elimination, and at 250°C for the formation of a solid resin with high porosity. This resin is ball-milled and calcined at different temperatures. The authors investigated the relative crystallinity of the $NaNbO_3$ phase with the calcination temperature. As illustrated in Figure 5, the crystallinity of the material, after calcination at 500°C for 3h, is 90% and, after calcination at 700°C, reaches 100%.

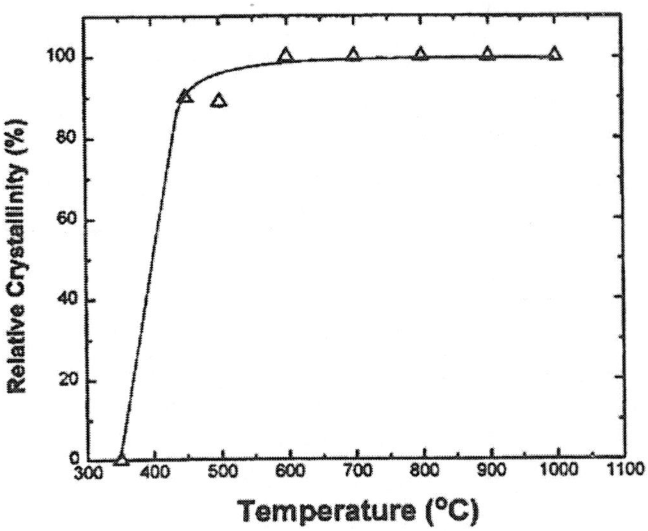

Figure 5. Relative crystallinity of the $NaNbO_3$ powder prepared by the Pechini method as a function of calcination temperature. Reproduced from Ref. [79] with permission from copyright owners.

These results suggest the nucleation of $NaNbO_3$ to occur between 400 and 500°C, temperatures below those required for the solid state reaction between Nb_2O_5 and Na_2CO_3 (900-1200°C). Thin films of $KNbO_3$ were obtained by Simoes et al via a similar procedure as they used for $LiNbO_3$ [77,78], from Nb_2O_5 and replacing the lithium carbonate by potassium hydrogenophtalate [80]. Films deposited by spin coating on MgO(100) substrates and annealing at 600°C revealed the presence of the pure phase. Powders of $KNbO_3$ and $KNbO_3$-$BaTiO_3$ solid solution (KN-BT) were prepared by the Pechini route, from $NbCl_5$ in methanol, K_2CO_3, $Ti(O^iPr)_4$ and $Ba(NO_2)_2$ [81]. The detailed preparation scheme is shown in Figure 6. The resin precursor obtained at the end of the procedure is calcined at temperatures between 460 and 900°C. Its thermal behaviour was followed by TG-DTA beforehand. The crystallization of KN proceeds via the $K_4Nb_6O_{17}$ intermediate phase, which appears at 460°C. The perovskite KN is obtained after heating at ~ 600°C and the KN-BT solid solution is formed at temperatures below 900°C. The layered $K_4Nb_6O_{17}$ phase was prepared as thin films from a niobium-chelated precursor, which is obtained from the dissolution at 80°C of freshly precipitated niobic acid into an EG/citric acid solution. Anhydrous potassium acetate is then added [82]. Before spin coating on a quartz glass plate substrate, the resulting sol is aged for 24h in a sealed glass. Calcination of the films at different temperatures was carried out. After heat treatment at 600°C, thin films of nanostructured $K_4Nb_6O_{17}$ with a preferred orientation are obtained.

Precursor solutions for $Pb(Mg_{1/3}Nb_{2/3})O_3$ oxides (PMN) were obtained by the Pechini method [83]. In the case of niobium, the authors considered two procedures to obtain the starting solution : from the dissolution of niobic acid into an EG/citric acid solution or from the mixture of niobium ethoxide into the same solution. In the case of lead and magnesium, they used lead acetate and magnesium carbonate as such or solubilized in a solution of EG and citric acid. The influence of the concentration of organic materials (EG-citric acid) on the phase formation was studied. The mixed Pb-Mg-Nb solutions were deposited by dip coating onto a Si(100) substrate. The films were annealed between 600 and 900°C for 1h. The perovskite phase formed at 600°C when the precursor solution was prepared with a lower amount of organic materials and starting from niobic acid. By increasing the temperature to 800 or 900°C, only the formation of the pyrochlore phase was observed. With the solution prepared from niobium ethoxide, the presence of the pyrochlore oxide was detected independently of the calcination temperature. The same authors also prepared PMN oxides, growth on $BaTiO_3$ seed particles, from a Nb-EG-citric acid solution, lead acetate and magnesium carbonate [84]. Evidence of hetero-epitaxial growth of perovskite PMN on $BaTiO_3$ was observed after calcination at temperatures as low as 450°C. Pyrochlore-free PMN could also be obtained by a combination of the Pechini method and a partial oxalate route [85]. The method starts from a high-reactive $MgNb_2O_6$ precursor, synthesized from a niobium-citric acid aqueous solution in the presence of $MgCO_3.Mg(OH)_2.5H_2O$ and EG. The appropriate heat treatment of the resulting resin at 800°C gives rise to the $MgNb_2O_6$ oxide. This material is dispersed in an oxalic acid solution in the presence of $Pb(NO_3)_2$ at pH 11. The precipitated lead oxalate on the surface of the oxide particles are filtered off and washed to eliminate nitrate ions. Calcination is carried out between 600 and 900°C. At 700°C, the single perovskite phase is detected without the use of lead excess or a closed system, as required by conventional solid state methods. By increasing the temperature up to 800°C, the formation of the unwanted pyrochlore oxide is observed, as a result of the lead loss by volatilization.

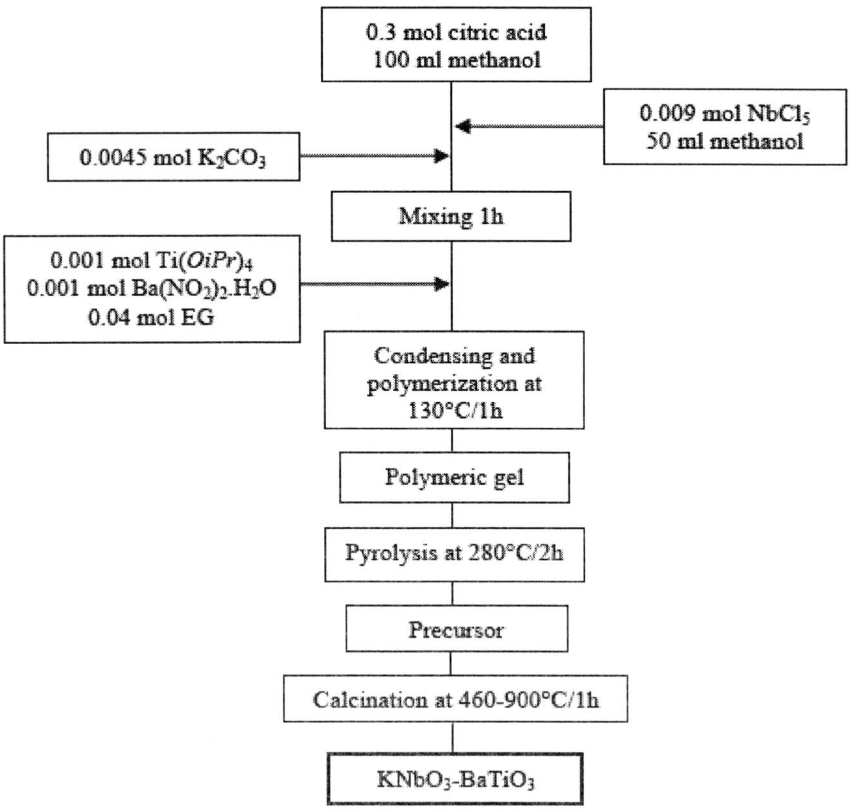

Figure 6. Flow diagram for the preparation of the $KNbO_3$-$BaTiO_3$ solid solution by the Pechini method [81].

Recently, Aguilar et al reported the synthesis of $SrBi_2Nb_2O_9$ (SBN) powders by a modified Pechini route [86]. A solution of Bi-EDTA, similar to that described above [9], is prepared from Bi_2O_3. A Sr-Nb viscous suspension is obtained from the reflux of $SrCO_3$ and Nb_2O_5 in an aqueous citric acid and EG medium. The Bi solution and Sr-Nb suspension are mixed and heated at 100°C to promote polyesterification. The resulting gel is then pyrolyzed at 300°C and finally, calcined at temperatures ranging from 600 to 900°C. A comparative study between the Pechini method and the ceramic route showed that pure SBN oxide is obtained at 820°C and 1000°C by these two methods, respectively. A very similar process, but without the use of EDTA as additive, was also implemented to prepare the same phase, as powders or thin films [87]. $SrCO_3$ and a solution of Bi_2O_3 in HNO_3 are added to a Nb-citrate solution, prepared from the dissolution of niobic acid in citric acid. EG is then added dropwise and the resulting Sr-Bi-Nb resin is used to prepared thin films on Si(100) substrates by dip coating or powders, by heat treatment between 300 and 800°C. The pure perovskite SBN phase is obtained via this method, in a bulk form or as crack-free thin films, at ~700°C. Single-phased bismuth titanate niobate nanoparticles of $Bi_7Ti_4NbO_{21}$ stoichiometry were obtained by a Pechini-type reaction method [88]. The required amount of bismuth nitrate and niobium pentachloride are separately dissolved into HNO_3 and HCl, respectively, and then mixed together to obtain a Bi-Nb solution, which is added to a solution of $Ti(OBu)_4$ and citric acid in EG. Esterification occurs with heating at 130-180°C for several hours. The resin so-obtained is pyrolized at 280°C, pre-calcined at 350°C and then calcined between 400 and 800°C. It completely converts into homogeneous and finely divided $Bi_7Ti_4NbO_{21}$ powders at 500°C, a temperature much lower than that required by the ceramic method (1100°C).

The Y_3NbO_7 phase was prepared at reduced temperatures by a modified Pechini process, from $NbCl_5$ and $Y_2(CO_3)_3$ dissolved in methanol, in the presence of citric acid and EG [18,89]. The resulting clear solution, stable with respect to hydrolysis, is heated at 130°C to remove methanol and subsequently promote esterification reactions, leading to a resin which chars at 300°C. The powder precursor obtained is heat-treated between 500 and 900°C. Its thermal decomposition and crystallization were investigated by means of thermal analyses (TG-DTA). The sample heated above 500°C exhibits a pure cubic Y_3NbO_7 phase with a fluorite structure. This Pechini method allows obtaining this oxide at reduced temperatures, as opposed to the conventional solid-state reaction methods requiring temperatures higher than 1400°C for a nearly single-phase material. This success in lowering the crystallization temperature down to 500°C may indicate, according to the authors, an improved level of mixing of cations in the powder precursor.

3.2.3. Carboxylate-Based Methods

The carboxylate-based methods are close to the citrate or Pechini method because of the presence of chelating ligands in the medium, which stabilize cations in solution. Nevertheless, they differ from these polymeric approaches by the absence of polymerization process and thus gel formation.

Ceramics of $NaNbO_3$ and sodium-rich $Na_{1-x}Li_xNbO_3$ solid solution (x = 0.12) were prepared by an oxalate route [90,91]. Sodium nitrate is reacted with oxalic acid in aqueous medium adjusted to pH 3 with ammonia. $NH_4H_2[NbO(C_2O_4)_3].3H_2O$ is then added to this solution and the resulting mixture is refluxed at 80°C for 3h. The solvent is completely evaporated and the hygroscopic mixed precursor obtained is calcined between 250 and 900°C for 5h. The diffraction peaks of the calcined powder reveal the single niobate phase to crystallize near 350°C but the crystallinity is higher at 450 or even 600°C. Lithium or sodium niobate powders can be prepared by a method based on water-soluble niobium malato complexes, according to the flow chart illustrated in Figure 7 [10,92].

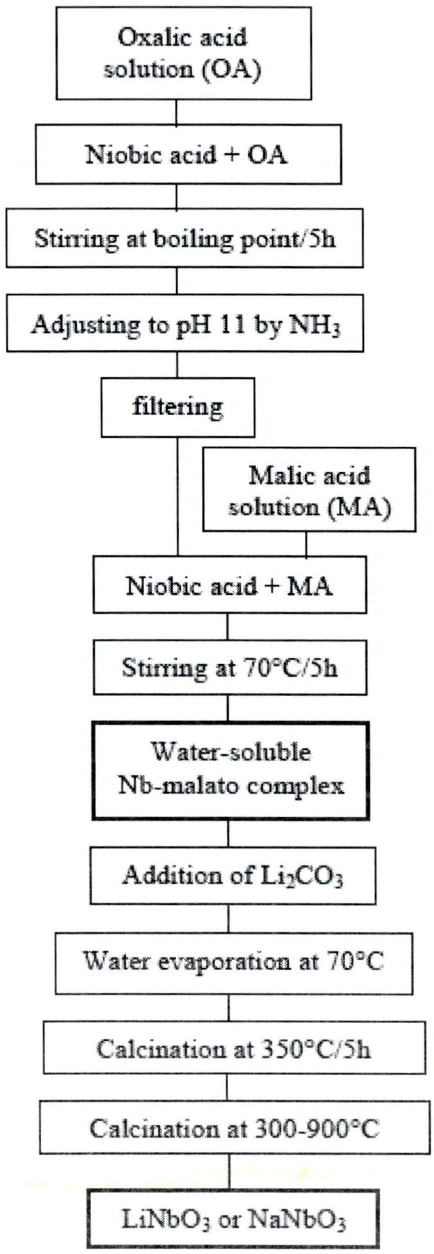

Figure 7. Flow chart for the synthesis of $LiNbO_3$ or $NaNbO_3$ powders based on a water-soluble niobium malato complex solution [10,92].

A freshly precipitated niobic acid, prepared from hydrolysis of a niobium-oxalate solution, is dissolved in a malic acid solution at 70°C. Addition of Li_2CO_3 or Na_2CO_3 to the niobium malato solution following by water evaporation leads to the formation of a solid precursor which is then calcined to give the final oxide. A small amount of the Nb-malato complex was isolated and characterized by IR spectroscopy and TG-DTA. Its molecular composition should correspond to a dinuclear $H[Nb_2O(H_3C_4O_5)_3]$ complex. Crystalline sodium and lithium niobates were identified at 450 and 500°C, respectively, although the total elimination of the inorganic carbon required higher temperatures (500 and 650°C, respectively).

Lead magnesium niobates (PMN) can be obtained from the freeze-drying of tartrate stabilized solutions of lead acetate, magnesium acetate and niobium-ammonium-oxalate in acetic acid medium [93]. The PMN acetate-oxalate solid precursor has been characterized by thermal analyses and both the mechanism of formation and the properties of oxide products were investigated and elucidated. Thermodiffractometry measurements reveal the pyrochlore phase to crystallize near 450°C. From 700 to 900°C, the perovskite phase dominates and above 1000°C pyrochlore reappears as a result of PbO volatilization. A similar preparation route was used by Narendar et al to prepare the same materials, engaging a Nb-peroxo-citrato solution, lead acetate and magnesium acetate, in the presence of EDTA as additional chelating agents [94]. This route allows the perovskite formation in a temperature window between 700 and 750°C, while pyrochlore formation is favoured below 650°C and PbO volatilization occurs above 800°C. Das et al used, for the synthesis of PMN powders, a method starting from niobium-tartrate, lead-triethanolamine (TEA), and magnesium-polyvinylacohol (PVA) complex solutions [95,96]. The procedure used to prepare the Nb-tartrate complex solution is detailed in Figure 8. After complete water evaporation of a stoichiometric mixture of the Nb, Pb and Mg solutions, the resulting solid precursor, first analyzed by TG-DTA, provides the perovskite PMN in a single-step calcination in a nearly pure form at 850°C for 2h [96], and as pyrochlore-free material at 1000°C for 1h [95]. Some $(1\text{-}x)Pb(Mg_{1/3}Nb_{2/3})O_3\text{-}xPb(Fe_{1/2}Nb_{1/2})O_3$ ($x = 0.1$, 0.2 and 1) ceramics were also prepared by the same method, engaging iron in stoichiometric amounts as aqueous $Fe(NO_3)_3$ [96]. These PMN-PFN oxides crystallize in the pure perovskite phase at 800°C. Powders prepared via this tartrate-modified precursor method have shown better dielectric properties than those obtained by the standard solid-state method. Moreover, the authors emphasized the advantages of using the tartrate complex prepared from niobic acid as the niobium source, i.e. avoiding the major problems of moisture sensitivity and expensiveness of pentachloride and alkoxides. The same authors also used a Nb-citrate solution, obtained from the dissolution of niobic acid in aqueous citric acid with catalytic amounts of H_2O_2, together with Pb-TEA and Mg-PVA or Fe-PVA complex solutions, to prepare finely divided ferroelectric niobate powders as PMN, PFN or $PbTiO_3$-stabilized PMN materials [2]. The addition of H_2O_2 in the case of the niobium solution facilitates the breaking of polymer chain of the hydrated niobium oxide and promotes the dissolution of niobic acid in carboxylic acid by peroxo complexes formation. Table 3 gives the percentage of perovskite phase detected on the XRD patterns of the calcined solid precursor for each system investigated. The same tartrate precursor, as used for the preparation of PMN materials by Das et al and described above, was used to prepare strontium barium niobate (SBN), $Sr_xBa_{1-x}Nb_2O_6$ ($x = 0.4$, 0.5 and 0.6) [3].

Table 3. Percentage of perovskite phase detected by XRD with calcination temperature of the precursor obtained using a citrate complex as niobium source [2].

sample	750°C/2h	850°C/2h	950°C/2h	1050°C/2h
PFN	99%	-	-	-
PMN	-	59%	76%	88%
PMN-PT (60/40)	-	-	53%	100%

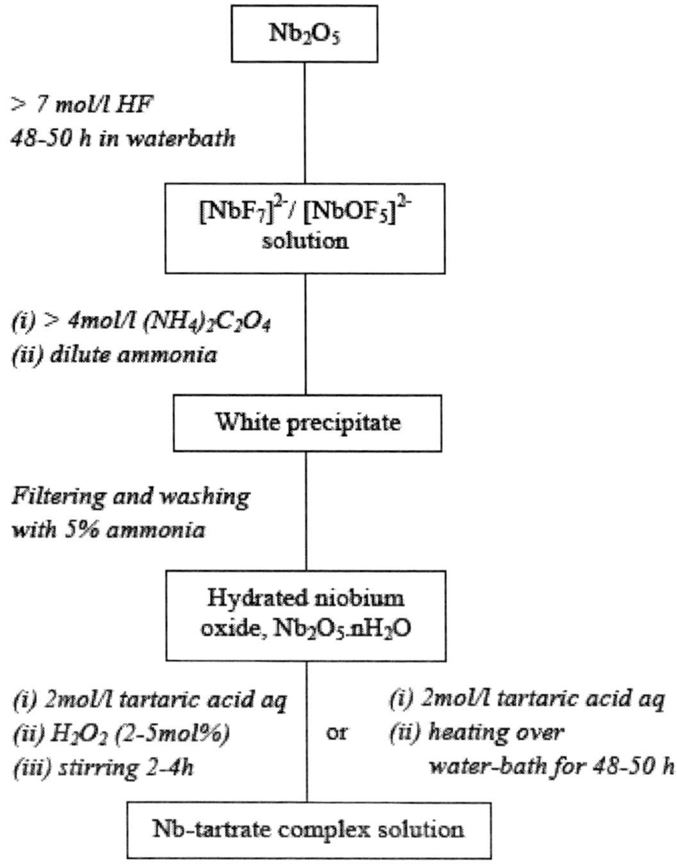

Figure 8. Schematic procedure for the preparation of a stable niobium tartrate complex aqueous solution [95].

The method consists in a single step process starting from a mixture of solutions containing niobium-tartrate, EDTA complexes of strontium and barium, and triethanolamine in excess as chelating additive. These aqueous solutions are prepared separately from niobic acid, $Sr(NO_3)_2$ and $Ba(NO_3)_2$, respectively. The complete dehydration of the mixed precursor solution at 200°C yields a black organic-based solid which is then calcined to directly produce the desired nanocrystalline single phase of SBN powders at 725-750°C for 2h, without passing through any intermediate phase. It was mentioned that the heat generated, due to the combustion of the carbonaceous residues obtained from the decomposition from organic materials (EDTA, TEA), reduces the external temperatures required for the formation of the SBN phase. The same material with x = 0.5 was also prepared by a combustion synthesis starting from $Sr(NO_3)_2$, $Ba(NO_3)_2$, and a Nb-oxalate solution prepared from the dissolution of freshly precipitated niobic acid in aqueous citric acid with catalytic amounts of H_2O_2 [7]. The metal solutions are placed in a silica crucible together with fuel (urea, HMT or glycine) and NH_4NO_3 as oxidizer, and inserted for 5 minutes in a pre-heated furnace (700-800°C) for spontaneous ignition. The pure tetragonal SBN tungsten bronze is obtained at 700°C when urea or HMT are used as fuel or at 800°C when engaging glycine. The dielectric properties of the oxides were shown to vary according to the nature of the fuel.

Langbein et al reported the preparation of copper niobates by thermal decomposition of freeze-dried metal oxalate solutions, prepared by dissolving freshly precipitated niobic acid and commercial copper oxalate in aqueous ammonium oxalate solution [97]. The thermal decomposition of the solid precursor obtained after a freeze-drying step was analyzed by TG-DTA, coupled with a mass spectrometer. In the case of the precursor with a Cu:Nb ratio of 1:2, the pure monoclinic form of $CuNb_2O_6$ crystallizes upon thermal treatment at 600°C, while the orthorhombic phase is formed above 900°C. In the case of a Cu:Nb ratio of 3:2, monoclinic $CuNb_2O_6$ appears in addition to CuO. The solid state reaction between both oxides begins between 700 and 800°C and leads to $Cu_3Nb_2O_8$. The reaction is complete after heating at 900°C for 8h. Other Cu:Nb compositions, 0:10, 1:9, 2:8, 5:5, 8:2 and 10:0, were investigated.

A lot of catalytically relevant materials can be prepared by carboxylate-based methods, mainly engaging oxalate modified precursors. The oxide systems containing niobium and vanadium are the most investigated in that context. The preparation of Nb-V supported catalysts were reported by Resini et al, and prepared by impregnation of a silica support [98]. The impregnation solution consists in a mixture of ammonium metavanadate and niobium ethoxide in aqueous oxalic acid. Vanadium-doped Nb bulk oxides (0.25-10 mol% V), investigated in the alkane selective oxidation context, were prepared by the so-called "melt" method [99,100]. A solid mixture of NH_4VO_3, niobic acid and oxalic acid in large amount is heated overnight at 120°C in air. The oxalic acid dissolves in its own crystallization water and then reacts with niobium and vanadium to form oxalate complexes. The excess oxalic acid evaporates slowly and the oxalate complexes decompose partially, leading to a fine powder which is finally calcined at 600-630°C for 5h. In order to investigate the effect of various parameters on the properties of the final materials prepared by this method, the authors synthesized Nb-V catalysts by varying the evaporation temperature, the amount of oxalic acid and the rate of flowing air. The influence of the calcination temperature on the performances of the catalysts was also investigated [99]. Niobium pentoxide promoted with 5 mol% of P, V, Cr, Ge or Mo were also prepared by Smits et al by evaporation of water from an oxalic acid solution of niobium and the second element [100]. The resulting solids were filtered, washed, dried at 80°C and finally calcined at 630°C, which is the optimal temperature for unpromoted Nb_2O_5. The precursors engaged were $(NH_4)_2HPO_4$, NH_4VO_3, $Cr(NO_3)_3.9H_2O$, GeO_2 or $(NH_4)_6Mo_7O_{24}.4H_2O$. In these catalysts, the *TT*-Nb_2O_5 oxide (hexagonal structure) was found to be the major phase in all cases. For the Nb-Ge system, the pure dopant oxide was detected. Nb-Mo bulk oxide catalysts could be obtained by a simply way : the evaporation of an oxalic acid solution of the oxalate salts of metals, followed by a calcination of the solid at the appropriate temperature [99]. This method was also applied to Nb-V and Nb-Cr oxides. Titania-supported Nb-V or Nb-V-Mo oxide catalysts were obtained by an impregnation method of the support with an oxalic acid solution containing the desired metal salts: NH_4VO_3, $(NH_4)_6Mo_7O_{24}.4H_2O$ and $NH_4H_2[NbO(C_2O_4)_3].3H_2O$, with the appropriate metal ratio [101,102]. Burch et al prepared bulk or supported Nb-Mo, Nb-V and Nb-V-Mo oxide materials from a solution containing the metal precursors, ammonium heptamolybdate, ammonium metavanadate and niobium oxalate. The source of niobium used is obtained from the dissolution of niobic acid, freshly precipitated from $NbCl_5$, in aqueous oxalic acid. The solid mixed precursor is obtained either by a freeze-drying process [27] or by slow evaporation of water at 80°C [26,27]. Samples supported on γ-alumina were obtained by wet impregnation of the precursors solution, followed by the same sequence of drying [26]. In order to obtain the pure $Nb_2Mo_3O_{14}$ phase (wrongly formulated as $Nb_2Mo_3O_{11}$, see ref. [103]), the thermal treatment of the solid precursor was carried out above 800°C in an evacuated sealed quartz tube, to prevent loss of molybdenum by volatilization. The $NbVO_4$ phase was produced from the calcination of the precursor at 200°C, followed by a thermal treatment under hydrogen up to 900°C. Multiphasic Nb-V-Mo materials were obtained at 400°C and they showed the presence of crystalline MoO_3, $Nb_2Mo_3O_{14}$ and $Mo_6V_9O_{40}$ oxides. Other phases were detected by energy dispersive spectroscopy, but were not observed by XRD and are therefore at least to a large part amorphous or present in low amounts. An identical preparation method as that used by Burch et al was considered by Desponds et al to synthesize bulk Nb-Mo, Nb-V and Nb-V-Mo oxide catalysts [104]. More complicated oxide systems, involving four different metal atoms, were also investigated and often prepared from oxalate-based metal sources [29,30,32,33,105]. Mo-V-Nb-Te oxides with an orthorhombic structure can be prepared by mixing two solutions in desired amounts, one containing NH_4VO_3, $(NH_4)_6Mo_7O_{24}.4H_2O$ and H_6TeO_6, and the other containing niobic acid dissolved in oxalic acid [32,33,105]. The water is then evaporated at 150°C and the resulting solid is calcined in two steps in order to obtain the desired phase : (i) a calcination under air at 275°C for 2h and then, (ii) a calcination for another 2h in a quartz reactor in flowing argon at either 550°C or 600°C. An almost pure orthorhombic $Mo_{7.5}V_{1.5}NbTeO_{29}$ phase was obtained via this preparation route [32,33]. Mo-V-Sb-Nb oxide catalysts, investigated in the context of selective oxidation of propane to acrylic acid, were also prepared by an oxalate-based method by varying the metal composition [29,30]. These materials were prepared from the mixture in appropriate amounts of stock solutions of (i) ammonium heptamolybdate in water, (ii) ammonium metavanadate and antimony trioxide dissolved by reflux in water and (iii) niobic acid dissolved at 80°C in an oxalic acid aqueous solution. The drying of the mixed solution at 120°C followed by the calcination of the resulting slurry under nitrogen at 200°C and then at 600°C, provide the oxide material. A mixture of many phases, involving MoO_3, $SbVO_4$, $Mo_6V_9O_{40}$ or $Nb_2Mo_3O_{14}$, was observed by X-ray diffraction or micro-Raman spectroscopy for each composition considered.

Several papers reported the use of a Nb-oxalate solution, mainly prepared from the dissolution of niobic acid in excess aqueous oxalic acid, as precursor of niobium in coprecipitation methods [3,99,100,106]. These preparation routes will be described in section 3.5.

3.2.4. Peroxo-Based Methods

The niobium precursor solution engaged in peroxo-based methods consists either in a "peroxo niobic acid sol" or in soluble peroxo species of niobium, complexed by ancillary chelating ligands as citric acid or EDTA. Some peroxo complexes of niobium have been isolated and characterized in the solid-state prior to their use as precursors. The literature concerning them will be exposed in the following section dedicated to molecular precursors.

Modified methods engaging a Nb-peroxo-citrato complex solution in the presence of citric acid as gelifying additive [69,70], or in the presence of chelating EDTA [94], were already described above. This peroxo-citrato solution is obtained from the decomposition of niobium oxalate, $NH_4H_2[NbO(C_2O_4)_3].3H_2O$ by acidified H_2O_2 at 65°C in the presence of citric acid, followed by increasing pH to 7.5 by adding ammonia. This stable precursor was first described by Narendar et al and thoroughly characterized in aqueous solution but also as a precipitated form, and engaged as metal source for the formation of Nb_2O_5 [71]. On the basis of IR and Raman spectroscopies, a hypothetical structure was proposed (Figure 9): the aqueous complex would correspond to a dimeric species with each niobium atom bearing an oxo group and a peroxo ligand. Two citrato ligands are also present and both are tridentate and bridging. UV spectrophotometry confirms the Nb:peroxo mole ratio of 1:1 in the complex in solution. The degradation behaviour of the precipitated precursor was also investigated by TG-DTA measurements.

Lithium, sodium or potassium niobate oxides were often obtained via peroxo-based solution routes. Kallinikov et al used a "peroxo niobic sol", formed by the dissolution of freshly precipitated niobic acid in the presence of hydrogen peroxide and a stoichiometric amount of lithium or sodium hydroxide at room temperature [11]. The resulting clear solution was boiled in the range 80-100°C until precipitation of the intermediate hydrated compound, $LiNbO_4.nH_2O$ or $NaNbO_4.nH_2O$. Calcining this precipitate at 800°C, which corresponds to the temperature optimized by DTA, leads to the single-phase $LiNbO_3$ or $NaNbO_3$. The same procedure was also used to prepare $Li_xNa_{1-x}NbO_3$ solid solutions and the $LiTaO_3$ tantalate. Moreover, a similar peroxo niobic acid sol was reported for the low-temperature preparation of niobium oxide nanoparticles [107]. $LiNbO_3$ and $NaNbO_3$ powders can also be prepared from an aqueous tetraperoxoniobate species, $[Nb(O_2)_4]^{3-}$ formed in situ by the reaction of niobic acid in excess H_2O_2 in ammonia medium, according to the following equation [108]:

$$Nb_2O_5 + 8\ H_2O_2 + 6\ NH_3 \rightarrow 2\ (NH_4)_3[Nb(O_2)_4] + 5\ H_2O$$

To the clear solution, Na_2CO_3 or $LiOH.H_2O$ is added. Heating the mixture at 70-80°C leads to the apparition of a white precipitate, which is filtered, washed with water and dried in vacuum. The powder obtained is treated at different temperatures for 2h. The crystallization initiation of the $Li(Na)NbO_3$ phase occurs at 300°C and the pure crystalline oxide is obtained at 400°C. The synthesis of $LiNbO_3$ at this very low temperature is significant because of the severe loss of lithium by volatilization during calcination steps above 600°C. Moreover, the obtained powders show small particles size (from 25 to 30 nm) and relatively high specific surface areas, ranging from 18 to 25 m^2/g. The tetraperoxo niobate species $(NH_4)_3[M(O_2)_4]$, isolated in the solid form, was used to prepare high purity and fluorine-free Nb_2O_5 [109,110]. Cheng et al reported the preparation of lithium or potassium niobate from H_2O_2-modified alkoxides solutions [111,112]. In this route, schematized in Figure 10, lithium or potassium ethoxide and niobium ethoxide in equivalent molar ratio are dissolved in ethanol/hydrogen peroxide medium. After refluxing the resulting colloidal mixture for several hours at 100°C and decomposing excess H_2O_2 by platinum foil at the same temperature, a clear stable peroxo solution is obtained. It is either used as such to coat films or dried into powder at 120°C in air. Spin-coated films and powders are heat-treated from 350 to 800°C with a step heating process. Pure $KNbO_3$ crystallizes at 650°C while $LiNbO_3$ at a lower temperature of 400°C, identical to that reported by Dey et al with a similar method [108].

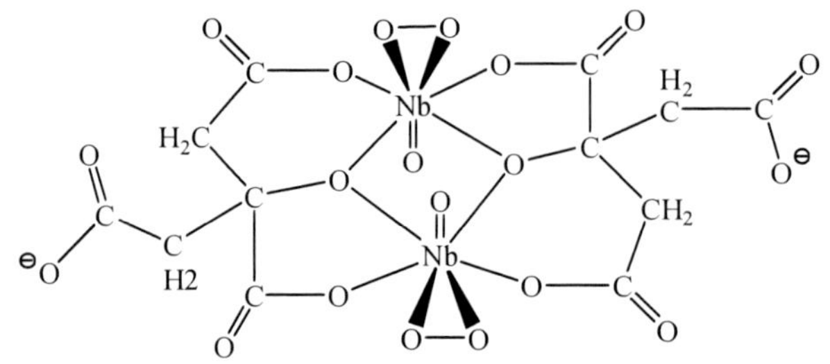

Figure 9. Proposed molecular structure for the bridged peroxo-citrato niobium complex in aqueous solution [71].

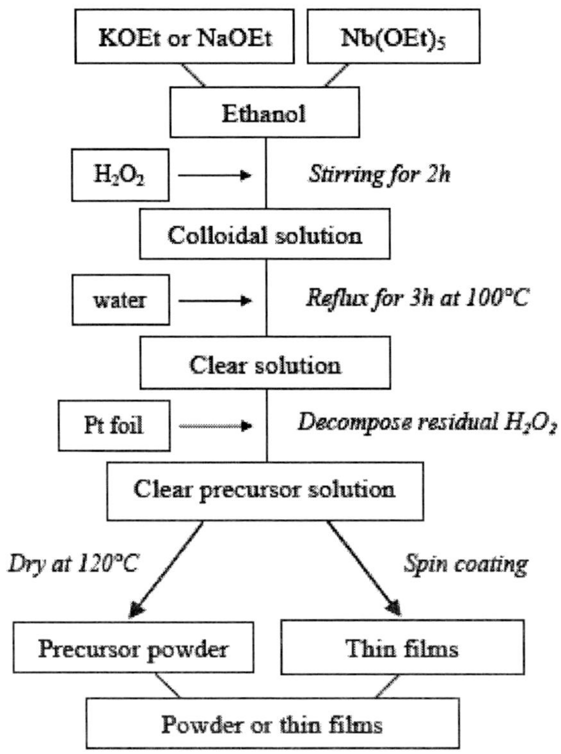

Figure 10. Flow chart of the preparation of $KNbO_3$ or $NaNbO_3$ thin films and powders by a peroxo-based method [111,112].

PMN samples were synthesized by a spray-drying process of a solution of metal precursors [1]. Magnesium and lead nitrate were engaged in such a route together with an acid solution of ammonium-peroxo-niobium complex, prepared from niobium-ammonium oxalate dissolved in an aqueous nitric acid/hydrogen peroxide solution. The Pb-Mg-Nb solution is spray-dried and the resulting powder was calcined in the 350-900°C temperature range for 1h. The XRD patterns evidenced the formation of the pyrochlore phase at 450°C, and the perovskite ones together with pyrochlore and unreacted PbO in the 700-800°C region. The PbO and the pyrochlore phases are almost completely transformed into the perovskite oxide at 800°C, while above 800°C, PbO partially volatilizes and the lead-deficient pyrochlore is formed again. This peroxo-based method thus requires temperatures lower to form perovskite-type PMN, in comparison with those required by the classical multi-step sintering "columbite" ceramic method. Moreover, this route is further improved by the spray-drying process providing particular morphology, homogeneous microstructure, and increasing final density of the precursor powder. Other peroxo-based methods modified by adding chelating agents were reported for the preparation of PMN oxides. Uniform submicrometer particles of lead niobate can be obtained (i) by adding slowly a lead nitrate solution to a solution of niobium peroxo complex chelated with EDTA, providing a precipitate by ligand replacing reactions [113]. The niobium solution is prepared by dissolving $NbCl_5$ in H_2O_2/HCl medium, followed by the addition of an EDTA solution and ammonia up to a final pH of 3. These particles can also be prepared (ii) by introducing a calcium nitrate solution in the mixture of nitriloacetic acid(NTA)-chelated Pb and NTA-chelated Nb complexes solutions, which causes the precipitation of lead niobium-peroxo-hydroxide at pH 9. The so-obtained precursor powders (Pb:Nb = 0.75) were converted to the $Pb_3Nb_4O_{13}$ oxides by heat treatment at ~ 500°C. Figure 11 displays the TG and DTA of the Pb-Nb precursor powder obtained from method (i). The differential analysis shows three exothermic peaks, the highest accounting for oxide crystallization, according to the authors. Mechanisms of the replacement reactions were suggested, based on the chemical, thermal and XRD analyses.

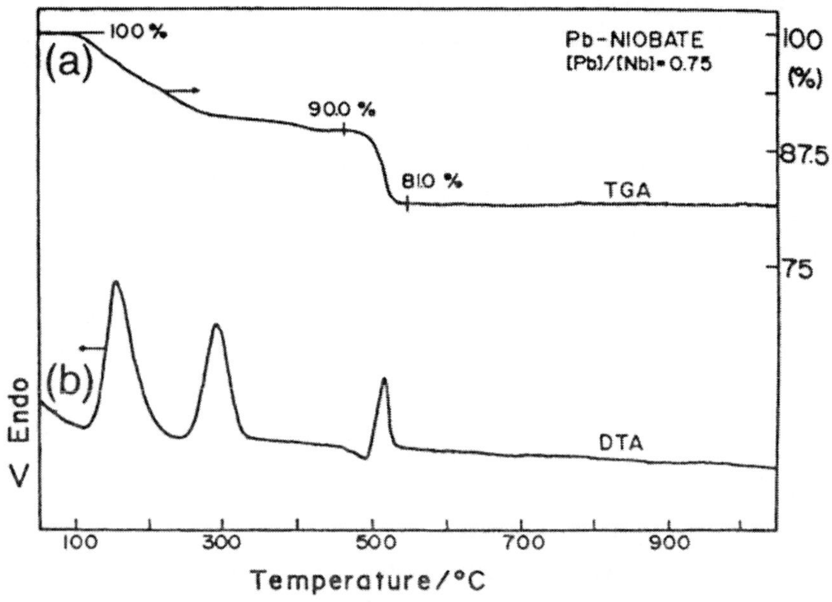

Figure 11. (a) TGA and (b) DTA of the Pb-Nb precursor powder (Pb:Nb = 0.75) obtained, via a peroxo-based method, by adding lead nitrate to a solution of niobium peroxo complex chelated with EDTA. Reproduced from Ref. [113] with permission from copyright owners.

3.3. Methods Based on Pre-Isolated Molecular Precursors

Preparation methods of multimetallic oxides based on molecular precursors start from metal compounds which are first isolated and characterized by a well-defined stoichiometry. These molecular precursors correspond to coordination complexes in which the metal is surrounded by organic ligands, mainly chelating molecules. The *multiple molecular precursors route* consists to start from a mixture of homometallic complexes of each metal to engage in the material, while the *single molecular precursor method* uses a heterometallic single-source complex, containing all the metals (in general, two) to be incorporated in the oxide. A much greater control of the stoichiometry of the final oxide can be achieved via this second way.

Three approaches of such methods exist, according to the treatment imposed to the molecular precursors. The *thermolytic approach* consists in the thermal decomposition of the solid precursor at the appropriate temperature to provide the final crystalline oxide. The second method consists in a *hydrolytic route*, in which the ligands in the complexes, most of the time alkoxide groups, are hydrolized in basic aqueous medium. The resulting hydrolysis product can be either heated further to crystallize oxides or just dried to obtain an amorphous fine powder. The third approach, mainly used when preparing thin films, corresponds to the *metal-organic chemical vapor deposition* (MOCVD). It consists in the vaporization of the precursor in a closed system, which is then deposited on a substrate. The oxide is obtained either directly by heating the substrate, or by a further calcination step. Precursors engaged in this MOCVD route must be volatile at a moderate temperature to avoid their decomposition and are almost exclusively alkoxide-type complexes.

Metal alkoxides or oxo-alkoxides are versatile molecular precursors of oxides. Their attractive features include a high purity, a large solubility in a wide variety of solvents but mainly, their ability to form heterometallic compounds that can be engaged as single-source in oxides syntheses. Many efforts are still made to design these kinds of molecules in order to reach novel materials for the technological demands. New compounds are often elaborated by varying the set of ligands that are able to provide suitable properties such as solubility, decomposition pathways, control of stoichiometry,... Alkoxides are often volatile and are therefore adequate in the MOCVD process but because of this particular property, they are rarely engaged in a thermolytic approach. The alternative based on hydrolysis has been developed to bypass volatilization problems.

3.3.1. Thermolytic Approach

PMN oxides were prepared by the multiple molecular precursors method from modified niobium alkoxides [114]. A more sterically demanding analogue of $Nb(OEt)_5$ was isolated as $[Nb(\mu\text{-}OEt)(ONp)_4]_2$ (Np = $CH_2C(CH_3)_3$) by alcohol exchange. Moreover, multidentate ligands such as the bis(hydroxymethyl)propionic acid (H_3BHMP) and tris(hydroxymethyl) ethane (H_3THME) were reacted with $Nb(OEt)_5$ to reduce the number of terminal ligands, yielding $[(\mu\text{-}BHMP)Nb_2(\mu\text{-}O)(OEt)_5]_2$ and $[(\mu\text{-}THME)Nb(OEt)_2]_2$, respectively. The characterization, structural aspects and solution properties of the three modified niobium alkoxides reported are described in detail. These complexes were used separately as niobium precursor together with lead neo-pentoxide and $Mg(n\text{-}Bu)_2$ in hot toluene. The mixture solution turned rapidly to a dark brown colour and the precipitated Pb(0) is filtered. Thin films of $Pb(Mg_{1/3}Nb_{2/3})O_3$ were obtained by spin-coating of these precursors solution (Pb:Mg:Nb = 1.1:0.33:0.66) onto Pt-coated SiO_2/Si substrates following by firing at 700°C. The surface uniformity of the resultant films followed the general trend of $Nb(OEt)_5$ < BHMP complex < THME complex, which is consistent with the decreasing number of readily hydrolyzable ligands. The same process was also used for the preparation of Nb-doped $Pb(Zr_{1-x}Ti_x)O_3$ (PZT) oxides and the authors came to similar conclusions according to the preference of sterically hindered precursors to existing $Nb(OEt)_5$ for materials application.

The so-called single molecular precursor method was used by Nakano et al in the context of preparation of Ba-Nb ceramics [115]. A bimetallic alkoxide compound of formula $Ba_5Nb_4(OEt)_{30}.xC_6H_6$ was obtained in ethanol from $Nb(OEt)_5$ and $Ba(OEt)_2$. This Ba_5-Nb_4 alkoxide precursor was heated at temperatures up to 900°C. At temperatures above 700°C, the trigonal $Ba_5Nb_4O_{15}$ phase appears. Its amount increases with increasing temperatures but small quantities of the monoclinic $BaNb_2O_6$ are always present. After heat treatment at 500°C, particles with size of 2-3 nm are formed while at 900°C, sizes of 70-150 nm are detected.

The heterometallic single-source method was also envisaged for the preparation of mixed oxides of rhenium, niobium and tantalum [116]. The $Nb_2(OMe)_2(ReO_4)_2$ and $M_4O_2(OMe)_{14}(ReO_4)_2$ complexes (M = Nb or Ta) were obtained via interaction between $Nb_2(OMe)_{10}$ or $Ta_2(OMe)_{10}$ with Re_2O_7 in toluene. Their thermal behaviour was investigated and the decomposition of these compounds in air at 900°C gave rhenium-niobium or rhenium-tantalum mixed oxide phase, supposedly based on the monoclinic Nb_2O_5 and orthorhombic Ta_2O_5 phases, respectively.

Thurston et al reported two heterometallic complexes of niobium and bismuth, involving salicylate ligands together with alkoxo groups [117,118]. The reaction under argon between triphenylbismuth, salicylic acid (H_3sal) and niobium ethoxide in toluene provides $Bi_2Nb_2(\mu$-O)(sal)$_4$(Hsal)$_4$(OEt)$_2$ and Bi_2Nb_2(sal)$_4$(Hsal)$_4$(OEt)$_4$. The presence of isopropanol in the reaction medium gives rise to the $BiNb_4$(sal)(Hsal)$_4$(O^{i}Pr)$_4$ and $BiNb_4(\mu$-O)$_4$(sal) (Hsal)$_3$(O^{i}Pr)$_4$ complexes. Similar heterometallic complexes with tantalum and bismuth were also reported as precursors of $BiTaO_4$. The thermal decomposition of $Bi_2Nb_2(\mu$-O)(sal)$_4$(Hsal)$_4$(OEt)$_2$ was explored by TG-DTA and powder X-ray diffraction. The heterometallic precursor is converted to the triclinic form of $BiNbO_4$ at 650°C as the major crystalline oxide phase. Heating at higher temperatures (850°C) favours conversion of the material to the orthorhombic form as well as disproportionation into $Bi_5Nb_3O_{15}$ and Nb_2O_5. Bimetallic oxides are also achievable through hydrolysis of the precursor in the presence of Et_4NOH in refluxing water.

Very recently, several water-soluble peroxo complexes of niobium, and also tantalum, were prepared, characterized and engaged in molecular precursors methods [119-122]. The use of binuclear compounds containing niobium and tantalum was reported in the context of preparation of bulk Nb-Ta mixed oxides [121,122]. Heterobimetallic peroxo complexes with tartaric acid, H_4tart and high denticity polyaminocarboxylic acids, H_5DTPA and H_6TTHA were prepared from an equimolar aqueous mixture of the tetraperoxo complexes, $(CN_3H_6)_3[Nb(O_2)_4]$ and $(CN_3H_6)_3[Nb(O_2)_4]$ ($CN_3H_6^+$ = guanidinium cation, gu$^+$), in the presence of the ligand. Compounds of stoichiometries $(CN_3H_6)_5[NbTa(O_2)_4$(tart)(Htart)]$.4H_2O$, $(CN_3H_6)_3[NbTa(O_2)_4(DTPAO_3)].5/2H_2O$ and $(CN_3H_6)_3[NbTa(O_2)_4$(HTTHAO$_4$)]$.2H_2O$ ($CN_3H_6^+$ = guanidinium cation, gu$^+$) were obtained and characterized by classical techniques like NMR, mass spectrometry, IR spectroscopy, TGA and sometimes monocrystal X-ray diffraction. The in situ oxidation of the nitrogen atoms of the polyaminocarboxylate ligands into N-oxide groups was evidenced in the complexes [123]. The thermal treatment of these Nb-Ta precursors in their solid state, between 700 and 800°C, provided in one step the solid solution, $NbTaO_5$. This study also investigated the thermal decomposition of the corresponding homobimetallic complexes into Nb_2O_5 or Ta_2O_5 and an equimolar mixture of these precursors were also engaged in the same method. BET and SEM measurements revealed these Nb-Ta mixed oxides to possess relatively high specific surface areas and to display a mesoporous disordered character. A comparison between the single and multiple molecular precursors methods is given. Nb-Ta, Nb-Ta-V and Nb-Ta-Bi mixed oxides were prepared by the same authors via a multiple molecular precursors route based on water-soluble peroxo-EDTA complexes [120]. Solid solutions of general formula $(Ta_xNb_{1-x})_2O_5$, $Ta_xNb_{1-x}VO_5$ and $BiTa_xNb_{1-x}O_4$ ($0 < x < 1$) were obtained from a homogeneous aqueous solution of the pre-isolated metal complexes, $(NH_4)_3[V(O)_2(EDTA)].H_2O$, (gu)[Bi(EDTA)($H_2O$)] $(gu)_3[Nb(O_2)_2(EDTAO_2)].2H_2O$ and $(gu)_3[Ta(O_2)_2(EDTAO_2)].2H_2O$. The organization of the chelating ligands around the metal atoms in these precursors is shown in Figure 12. The aqueous mixed solution of these precursors (Nb+Ta or Nb+Ta+M) was freeze-dried to give an amorphous powder which was heated first at 300°C to remove most of the organic materials and then at temperatures between 700 and 800 °C for the Nb-Ta system, or at 600°C for the Nb-Ta-M (M = V, Bi) systems. The formation of a solid solution over the entire range of composition ($0 < x < 1$) was evidenced, for the three systems studied, by Raman spectroscopy but also by DTA/DSC analyses and powder X-ray diffraction. The incorporation of small amounts of tantalum (1-3 mol%) in the Nb-V-O and Nb-Bi-O systems was shown to stabilize phases difficult to obtain in a pure form, like the $NbVO_5$ and the triclinic $BiNbO_4$ phases. The solid solutions were obtained via this method under quite mild thermal conditions and with particular morphology and textural properties.

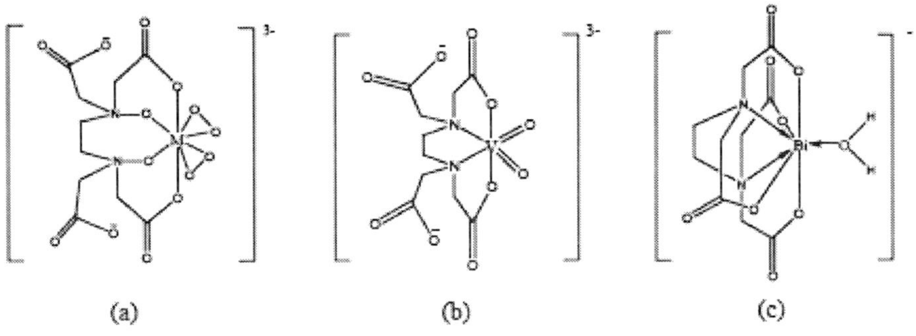

Figure 12. Chemical structures of the various anionic EDTA complexes engaged in the preparation of the Nb-Ta, Nb-Ta-V or Nb-Ta-Bi oxides [120]. (a) $[M(O_2)_2(EDTAO_2)]^{3-}$ (M = Nb or Ta), (b) $[V(O)_2(EDTA)]^{3-}$ and [Bi(EDTA)(H_2O)].

Catalytically relevant bulk and silica-supported oxides containing niobium and molybdenum were prepared by the thermolytic approach of the multiple molecular precursors method [103,119]. Within that context, water-soluble peroxo complexes of niobium with carboxylato ligands like oxalic, tartaric and citric acid were isolated and characterized from a spectroscopic and structural point of view [119]. The complexes, of general formula

$C^I_3[Nb(O_2)_x(H_yL)].nH_2O$ (C^I = NH_4^+ or $CN_3H_6^+$, L = oxalate, tartrate or citrate), were combined in aqueous solution with analogous molybdenum(VI) peroxo-carboxylato precursors, $(NH_4)_2[MoO(O_2)_2(H_yL)].nH_2O$ (L = oxalate, tartrate, citrate). In this preparation mode, the silica support was impregnated by the resulting Nb-Mo solution and the heat treatment of the dried precursor, carried out at 800°C for 6h in a closed quartz tube to prevent loss of molybdenum by volatilization, allowed the formation of the supported $Nb_2Mo_3O_{14}$ in a pure form. Unfortunately, the authors deplored the very low specific surface areas of the supported samples as a result of the severe experimental conditions required for the formation of the desired ternary phase : while the initial specific surface of the silica support engaged was 291 m^2/g, the prepared catalysts display surfaces lower than 5 m^2/g. Nevertheless, the authors emphasized the high solubility of such peroxo-carboxylato complexes and the high stability of their aqueous solutions, focusing therefore on their usefulness as metal precursors in the preparation of oxide materials. This method was later optimized by the same authors for the formation of the pure $Nb_2Mo_3O_{14}$ bulk or supported on silica [103]. Oxo-oxalato complexes, $(NH_4)_3[NbO(ox)_3].H_2O$ and $(NH_4)_2[MoO_3(ox)].H_2O$ were dissolved in water according to the appropriate Nb/Mo proportion of 2/3. In the case of supported samples, silica was added with different metal (Nb+Mo)/Si loadings (5, 10 and 20 mol%). The solvent was removed by a freeze-drying process and the mixed solid precursor obtained was first analyzed by TG and calcined at different temperatures in flowing air. The thermogram of the mixed Nb-Mo oxo-oxalato precursor obtained is illustrated in Figure 13. After the dehydration step around 100°C, the precursor decomposes into oxide until a final temperature of 600°C. When the temperature reaches ca 750°C, a progressive weight loss is observed and continues to 850°C.

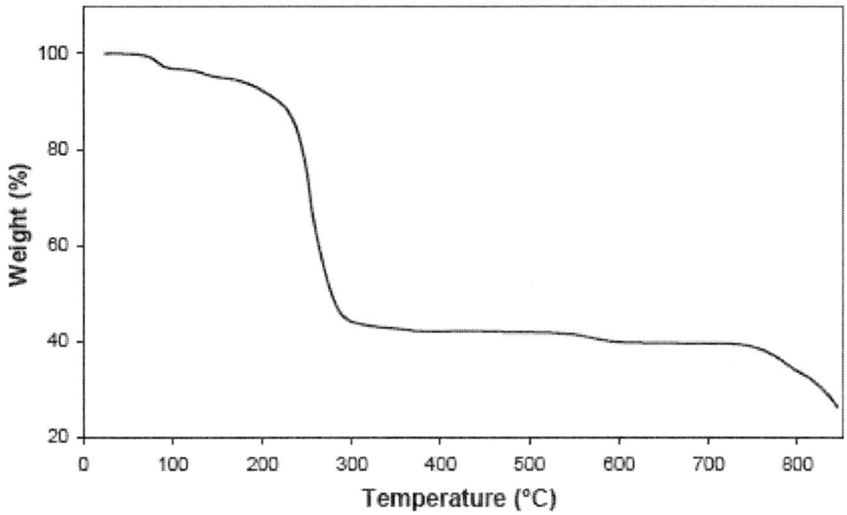

Figure 13. Thermogram under air of the Nb-Mo mixed oxo-oxalato precursor used in the preparation of the bulk $Nb_2Mo_3O_{14}$ oxide. Reproduced from Ref. [103] with permission from copyright owners.

This phenomenon is due to the volatilization of molybdenum oxide in this temperature range. Consequently, the calcination of the mixed precursors was carried out at temperatures between 600°C and 750°C. The $Nb_2Mo_3O_{14}$ phase was shown to be generated in a pure form at 700°C for 2h, i.e. 100°C below the lowest temperature reported so far for the formation of the phase by the ceramic method. A characterization of the bulk phase obtained via this conventional ceramic route in a closed system was also carried out and comparison was made with the precursors route. Moreover, the supported $Nb_2Mo_3O_{14}$ samples displayed appreciable specific surface areas of 60-70 $m^2.g^{-1}$, much larger than those reached in previous attempts under vacuum in sealed vials. The SEM and EDX analyses revealed a high dispersion of the desired phase on the silica support.

3.3.2. Hydrolytic Approach

Homogeneous and stoichiometric $LiNbO_3$ films were synthesized by the single precursor method from the double alkoxide, $LiNb(OEt)_6$ [124]. Controlled partial hydrolysis of this bimetallic compound, prepared under refluxing conditions in ethanol of LiOEt and $Nb(OEt)_5$, gave a clear solution, which was concentrated before dip-coating. The thickness of the films obtained by one-dip run was related to the concentration of the dip solution. Films crystallized at temperatures as low as 250°C and showed preferred orientation at 400°C on sapphire substrates.

The group of L. Hubert-Pfalzgraf is working since a long time in the design of homo- and heterometallic molecular alkoxide-type precursors to Nb-containing high-tech oxides [125-131]. This group has mainly used the hydrolytic approach of the molecular precursors route and has reported results in the context of Nb-containing multimetallic oxides with lead, magnesium or cadmium. Heterometallic aggregates involving alkoxide and acetate groups, $MNb_2(OAc)_2(O^iPr)_{10}$ (M = Mg, Cd or Pb) were described as precursors of the corresponding MNb_2O_6 oxides [126-128]. The complexes were obtained from the reaction under argon between $M(OAc)_2$ and $Nb(O^iPr)_5$ in hexane, according to the following reaction :

$$M(OAc)_2 + 2\ Nb(O^iPr)_5 \rightarrow MNb_2(OAc)_2(O^iPr)_{10}$$

The hydrolysis products of these different heterometallic sources were analyzed by TG-DTA and heated between 400 and 800°C to provide oxides. A comparison was made between this alkoxide-acetate route and the more classical one using homoleptic alkoxide. As an example, Figure 14 illustrates the results of thermal analyses and XRD measurements carried out on the Pb-Nb sample. The TG-DTA curve of the hydrolysis product of $PbNb_2(OAc)_2(O^iPr)_{10}$ reveals a nearly continuous mass loss between 80 and 420°C, which indicates the presence of residual carboxylate ligands towards hydrolysis. The DTA curve shows an exotherm around 600°C suggesting the crystallization of oxide. This is confirmed by XRD which displays pure crystalline $PbNb_2O_6$ phase. Traces of $PbNb_4O_{11}$ are observed at 800°C as a result of PbO volatilization. By contrast, the $[PbNb_2O(O^iPr)_{10}]_m$ alkoxide provided after thermal treatment at 600°C no $PbNb_2O_6$ but crystalline $Pb_3Nb_4O_{13}$ together with $PbNb_4O_{11}$ at 750°C. These observations suggest, according to the authors, that the acetate groups assist the building up of ordered, crystalline material in the course of the hydrolysis process, thus favouring the formation of single-phase materials at a quite low temperature. The same authors obtained the pure $BaNb_2O_6$ phase from the calcination at 700°C of the hydrolysis product of the heterometallic complex, $BaNb_2(O^iPr)_{12}(iPrOH)_2$ [125,130]. This species was prepared under argon by a reaction between BaI_2 and $KNb(O^iPr)_6$ in refluxing toluene and crystallization in the presence of isopropanol. The isolated mixed alkoxide was characterized thoroughly by usual techniques as elemental analysis, IR, NMR and crystal X-ray diffraction.

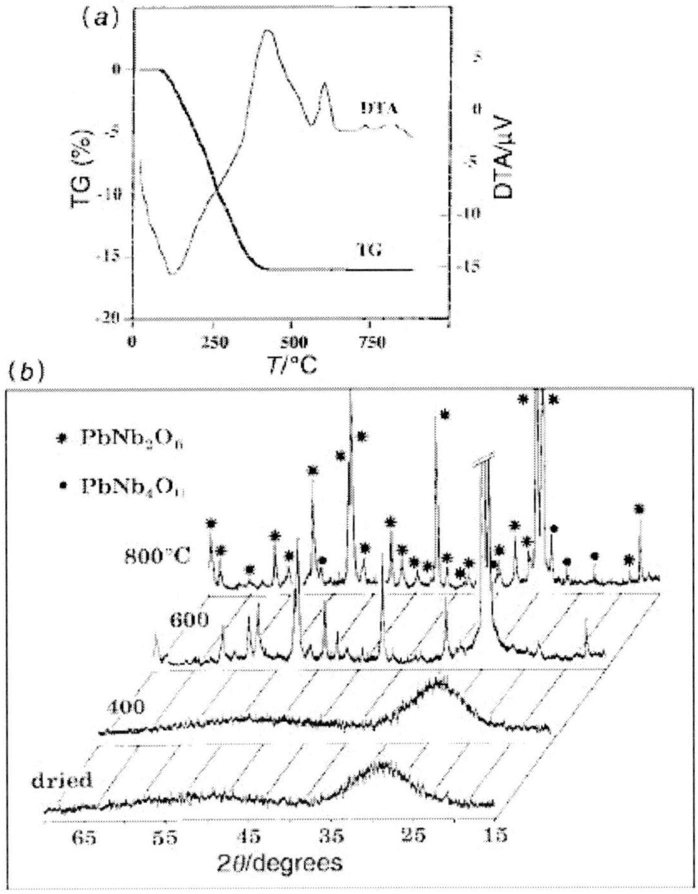

Figure 14. (a) Thermal analyses and (b) XRD patterns for various calcination temperatures of the powder resulting from the hydrolysis of $PbNb_2(OAc)_2(O^iPr)_{10}$. Reproduced from Ref. [126] with permission from copyright owners.

A mixed alkoxide was synthesized from a mixture of $Pb(OAc)_2$, $Nb(OEt)_5$ and Mg metal, and engaged in the preparation of PMN oxides [132]. The chemical structure of the alkoxide was described on the basis of the change in solubility in organic solvents, and IR and Raman spectra. The alkoxide was assumed to be a polymer produced by ester-exchange reactions of $Pb(OAc)_2$ and $Nb(OEt)_5$, in which Mg atoms are coordinated to acetate groups. The powder precursor was obtained by hydrolysis of the mixed alkoxide solution with aqueous ammonia in the presence of acetone. The crystallization of the hydrolysis product of this precursor was examined. It transforms into the perovskite phase at the very low temperature of 300°C with a 70% yield, but a pyrochlore oxide coexists. The authors attributed this observation to the chemical structure of the obtained precursor powder.

Hydrolytic decomposition of the $Bi_2Nb_2(\mu\text{-}O)(sal)_4(Hsal)_4(OEt)_2$ complex in the presence of $(Et_4N)OH$ in refluxing water was shown to provide amorphous nanometer scale bimetallic oxide particles, while its thermal decomposition, which is described above, produces porous and crystalline $BiNbO_4$ oxide [117]. Because the nanoparticles synthesized by this hydrolytic approach show a significant tendency to aggregate once dried and become difficult to redisperse, the authors suggest that this method might be an accessible way to stable thin films of ferroelectric materials.

3.3.3. Chemical Vapor Deposition Approach

The metal-organic chemical vapor deposition (MOCVD) technique, mainly used for the preparation of layered thin films, requires the availability of precursors with appropriate physical properties (volatility, stability in solution for liquid injection applications,...) and decomposition characteristics (large temperature window between evaporation and degradation, clean decomposition without residual impurities). In order to exploit the full potential of this method, it is sometimes necessary to tailor the precursors properties to optimize process parameters such as evaporation temperature, deposition temperature, thin films purity and homogeneity. The multiple molecular precursors route has been used in this context but the engaged precursors are often incompatible which each other, having different physical properties and decomposition behaviour which can lead to poor film uniformity. For this reason, the MOCVD process has mainly been applied with single-source precursors containing the elements required for the oxide film. Moreover, precursors engaged in this route are almost exclusively alkoxide-type complexes. Some examples of the use of chloride compounds like the volatile $NbCl_5$ were reported, essentially for the preparation of Nb_2O_5 thin films [133,134].

In the classical MOCVD approach, the precursors are stored in stainless steel container (bubbler) held at moderate temperatures (50-150°C) and transported into the gas phase using a carrier gas delivered through a dip-pipe. Although this approach has been successfully used for the growth of a number of oxides, the majority of the engaged precursors has insufficient thermal stability to withstand heating for long periods and decompose in the bubbler or in the inlet pipe-work leading to poor homogeneity of the oxide layer. However, the problems of premature thermal decomposition associated with low volatility precursors can be largely avoided by the use of liquid injection or spray MOCVD. In this technique, a solution containing the precursors dissolved in an inert solvent, usually an ether (e.g. tetrahydrofuran) or a hydrocarbon (e.g. heptane or nonane) is delivered at precisely controlled rate and quantity into a heated evaporator and then transported by a carrier gas into the reactor zone. In the case of the spray approach, the solution is transported as a mist in the evaporation zone where precursors and solvent are rapidly volatilized due to the high surface area of the mist droplets. Separate precursor solutions can be used for the growth of multicomponent oxides, but more usually a number of precursors are dissolved in an appropriate molar ratio in a single solution.

MOCVD deposition was used by Wernberg et al for the preparation of thin films of epitaxial lithium niobate using heterometallic alkoxide precursors such as $LiNb(OEt)_6$ [135] or $LiNb(OEt)_5(dpm)$ (dpm = dipivaloylmethanate) [12]. These compounds were prepared from the reaction in inert atmosphere of niobium ethoxide with LiOEt or lithium dipivaloylmethanate, respectively, and because of their low volatility, they were engaged by a spray process. The epitaxial films obtained at ~450°C via this technique from a toluene solution of $LiNb(OEt)_6$ showed significantly improved crystalline properties and morphology compared to more conventional MOCVD routes. Although they are actually epitaxially oriented, the films obtained between 490 and 590°C (second zone) via the $LiNb(OEt)_5(dpm)$ precursor show rough surface morphology which made them unsuitable for desired device applications. More soluble and volatile modified heterometallic alkoxides were investigated and considered as single-source precursor for the preparation of $LiNbO_3$ thin films [136]. The $[LiNb(OR)_6]_2$ (R = CH_2CMe_3 and OCH_2SiMe_3) molecular compounds and an alcohol adduct, $[LiNb(OCH_2CMe_3)_6(Me_3CCH_2OH)_2]$ were synthesized and the crystal structure of the $[LiNb(OCH_2SiMe_3)_6]_2$ complex was solved. The $[LiNb(OCH_2CMe_3)_6]_2$ was engaged in a spray MOCVD process but the authors failed to deposit significant amounts of $LiNbO_3$ using the same parameters as those used for $LiNb(OEt)_6$. Other zone temperatures or carrier-gas flow rates were also examined without success. However, the bulk decomposition of this alkoxide at 600°C under nitrogen provided crystalline $LiNbO_3$. These observations suggest that $LiNb(OEt)_6$ and $[LiNb(OCH_2CMe_3)_6]_2$ display different decomposition behaviours. Alkoxides with hydrogen atoms in β position with respect to oxygen, like $LiNb(OEt)_6$, may decompose autohydrolytically in the gas phase at elevated temperatures and atmospheric pressure. This mechanism of alkoxide degradation is initiated by a trace of H_2O and yields metal oxide. According to the authors, a possible explanation for the unsuccessful deposition of $[LiNb(OCH_2CMe_3)_6]_2$ could be the lack of autohydrolytic decomposition pathway.

The classical MOCVD process was carried out to prepare PMN using multiple molecular precursors [137,138]. $Pb(Mg_{1/3}Nb_{2/3})O_3$-$PbTiO_3$ (PMN-PT) powders were grown on $SrTiO_3(100)$ single crystals under various deposition conditions and using lead tetraethyl, $PbEt_4$, niobium ethoxide, $Nb(OEt)_5$, a solid magnesium β-diketonate, $Mg(thd)_2$ (thd = 2,2,6,6-tetramethylheptane-3,5-dionate) and titanium isopropoxide, $Ti(O^iPr)_4$ as metal precursors [137]. Epitaxial pyrochlore-free PMN-PT films were obtained under all conditions. Takeshima et al also prepared PMN-PT thin films on several types of substrates, starting from lead acetylacetonate, $Pb(acac)_2$, magnesium acetylacetonate, $Mg(acac)_2.nH_2O$, $Nb(OEt)_5$ and $Ti(O^iPr)_4$ [138]. A cold-wall-type reactor was used for the deposition to prevent the reaction of the metal sources in transit before reaching the metallized substrate, Pt(100)/MgO(100) or Pt(111)/Ti/SiO_2/Si(100). Typical conditions for PMN-PT deposition are given in Table 4. The PMN-PT thin films, with the perovskite-type structure as a main phase, were obtained in the composition of Ti/(Mg+Nb+Ti) ≥ 25 mol%. Unfortunately, the metal precursors used in these multiple sources routes are not particularly compatible from the point of view of solubility and degradation behaviour. For instance, alkoxides such as $Nb(OEt)_5$ decompose at lower temperatures than lead and magnesium β-diketonates, and are also significantly more volatile. A solution to this incompatibility is the use of the single source approach. The bimetallic alkoxide precursor $Mg[Nb(OEt)_6(EtOH)_2]_2$, illustrated in Figure 15, has recently been investigated in combination with the lead β-diketonate, $Pb(thd)_2$ in the liquid injection MOCVD from a THF solution [139,140]. In the absence of $Pb(thd)_2$, $Mg[Nb(OEt)_6(EtOH)_2]_2$ gives magnesium niobium oxide films with a Mg:Nb ratio of 1:2 at lower substrate temperatures, but magnesium-deficient films at higher substrate temperatures (600°C), indicating partial decomposition in the gas phase of the mixed alkoxide. The addition of $Pb(thd)_2$ to the system was found to influence significantly the deposition chemistry of the Mg-Nb mixed alkoxide. In contrast to growth using $Mg[Nb(OEt)_6(EtOH)_2]_2$ alone, there was no incorporation of Mg at low substrate temperatures (425-500°C), but at higher substrate temperatures (550-600°C), the Mg:Nb ratio of 1:2 required in $Pb(Mg_{1/3}Nb_{2/3})O_3$ was obtained. The mechanism by which $[Pb(thd)_2]$ influences the decomposition behaviour of

$Mg[Nb(OEt)_6(EtOH)_2]_2$ was not clearly established. This can be attributed to the formation of a thermally stable magnesium β-diketonate species via a gas phase ligand exchange reaction between $Pb(thd)_2$ and $Mg[Nb(OEt)_6(EtOH)_2]_2$.

Table 4. Typical deposition conditions of PMN-PT thin films via a multiple molecular precursors route and using a cold-wall-type reactor [138]

Source temperature	
$Pb(acac)_2$	120-130°C
$Mg(acac)_2.nH_2O$	220-230°C
$Nb(OEt)_5$	120-130°C
$Ti(O^iPr)_4$	60°C
Pressure of Ti source vessel	35-100 kPa
Ar carrier gas flow rate	
$Pb(acac)_2$	50 ccm
$Mg(acac)_2.nH_2O$	50 ccm
$Nb(OEt)_5$	30 ccm
$Ti(O^iPr)_4$	50 ccm
O_2 gas flow rate	180 ccm
Substrate temperature	680-780°C
Total pressure	670 Pa

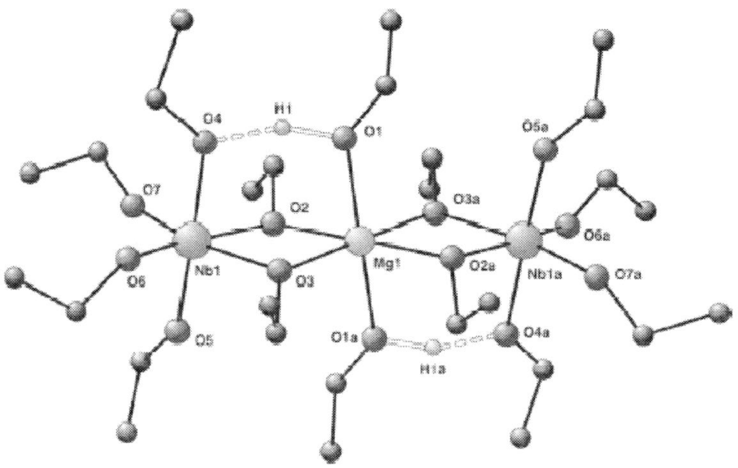

Figure 15. Crystal structure of $Mg[Nb(OEt)_6(EtOH)_2]_2$. Reproduced from Ref. [140] with permission from copyright owners.

The Sr-Nb and Sr-Ta heterometallic alkoxides, $Sr[Ta(OEt)_6]_2$ and $Sr[Nb(OEt)_6]_2$, were prepared as precursors for the MOCVD of $SrBi_2(Ta_xNb_{1-x})_2O_9$ (SBTN) films, together with $Bi(CH_3)_3$ [141-143]. The use of these double alkoxides allows to avoid the severe problem of the low vapor pressure of the most widely used source of strontium, the bis-dipivaloylmethanate compound, $Sr(dpm)_2$. Moreover, its vaporization temperature is very close to its decomposition temperature. SBTN films were prepared, from two organometallic source bottles (Nb+Ta and Bi), on a Pt-coated silicon substrate with high compositional reproducibility, the Nb/(Nb+Ta) ratio in the film being almost the same as that of the source materials. The films, deposited at 585°C, consisted of an almost single phase of $SrBi_2(Ta_xNb_{1-x})_2O_9$ [141,142]. Small amounts of the pyrochlore phase were detected. The same authors also carried out the deposition at 500°C, following by a heat treatment at 800°C in O_2 atmosphere, and came to the same results in term of crystalline composition [143]. The remanent polarization of these ferroelectric thick films however increases with the deposition temperature at an applied electric field. The principal disadvantage of the $Sr[Ta(OEt)_6]_2$ and $Sr[Nb(OEt)_6]$ mixed precursors is that they contain an unsaturated Sr centre, which makes them susceptible to attack by moisture and reduces their shelf-life, especially in solution-based liquid injection MOCVD. The insertion of donor-functionalised alkoxide ligands, such as dmae (dimethylaminoethoxide) and bis-dmap (bis(dimethylamino)propanolate), into the heterometallic alkoxide allows to increase the coordination around the central Sr atom, and render the complexes less susceptible to decomposition in solution and in the gas phase. Replacement of an ethoxo ligand in $Sr[Nb(OEt)_6]_2$ with dmae or bis-dmap leads to the complexes $Sr[Nb(OEt)_5(dmae)]_2$ and $Sr[Nb(OEt)_5(bis\text{-}dmap)]_2$ [144]. The $Sr[Nb(OEt)_5(dmae)]_2$ compound is an oil, but in the presence of a slight excess of ethanol it crystallizes as the bis-ethanol adduct, $[Sr\{Nb(OEt)_5(dmae)\}_2(EtOH)_2]$. The crystal structure of this compound is illustrated in Figure 16. The $Sr[Nb(OEt)_5(dmae)]_2$ complex is particularly suited to MOCVD applications : it was found that 1:1 mixtures of $Sr[Nb(OEt)_5(dmae)]_2$ and $Sr[Ta(OEt)_5(dmae)]_2$ form a free flowing volatile liquid which distils as a single fraction at 140°C (0.1 mmHg) without any change in composition, and thus these mixtures are promising multicomponent precursors for the liquid injection MOCVD of $SrBi_2(Ta_xNb_{1-x})_2O_9$. The deposition of $SrNb_2O_6$ from the $Sr[Nb(OEt)_5(dmae)]_2$ single source on silicon substrate was investigated [144]. The films composition was shown to be heavily dependent on the deposition temperature (300-600°C) : films grown at low temperatures were found to be Sr-rich, and progressively less Sr-rich when temperature is increased. The optimal temperature to obtain the pure $SrNb_2O_6$ phase is between 450 and 500°C. New Sr-Nb and Sr-Ta precursors, $Sr[Nb(OEt)_5(OC_2H_4OCH_3)]_2$ and $Sr[Ta(OEt)_5(OC_2H_4OCH_3)]_2$ (M = Nb or Ta), were prepared by Kadokura et al from the reaction of $M(OEt)_5$ with strontium 2-methoxyethanol, $Sr(OC_2H_4OCH_3)_2$ and totally characterized by chemical analysis, mass spectrometry, ^1H NMR and FTIR spectroscopy [145]. To investigate the behaviour of these mixed alkoxides during CVD, SBTN films on Pt-coated silicon substrates were prepared under several sets of conditions and their composition was analyzed. When deposited at around 500°C, the film displays the same composition as the starting liquid mixture of both precursors, held at 155°C and bubbled with argon.

A two-step MOCVD technique was used by Zhang et al to prepare $Sr_{1-x}Ba_xNb_2O_6$ thin films [146]. Two thin layers of single phase $SrNb_2O_6$ and $BaNb_2O_6$ were deposited alternately on a silicon substrate, and the solid solution was obtained by further high temperature annealing via interdiffusion of Sr and Ba atoms. The liquid precursors engaged in this method were double alkoxides, strontium niobium *n*-butoxide, $SrNb_2(O^nBu)_{12}$ and barium niobium *n*-butoxide, $BaNb_2(O^nBu)_{12}$, which were synthesized by alcoholysis reactions, between $Nb(OEt)_5$ and $Sr(O^nBu)_2$ or $Ba(O^nBu)_2$. The evaporation behaviour of these double alkoxides was studied. The stoichiometry control of the $SrNb_2O_6$ and $BaNb_2O_6$ films was achieved through deposition process control, according to the evaporation characteristics of the metal sources. The precursors evaporation temperature was held between 200 and 300°C, the deposition temperature of orthorhombic $SrNb_2O_6$ and $BaNb_2O_6$ was 600°C at a pressure of 1-5 Torr, and the final annealing of the films was carried out at 1000°C for 10h and provided the pure tetragonal phase of $Sr_{1-x}Ba_xNb_2O_6$. These films grown on Si(100) substrate have a strongly preferred orientation of (100) direction and the Sr/Ba molar ratio in the solid solution can be controlled by controlling the relative layer thickness. Moreover, the authors showed that the pre-crystallization of the bottom $BaNb_2O_6$ layer to its tetragonal phase can largely improve the microstructure of the Sr-Ba-Nb films.

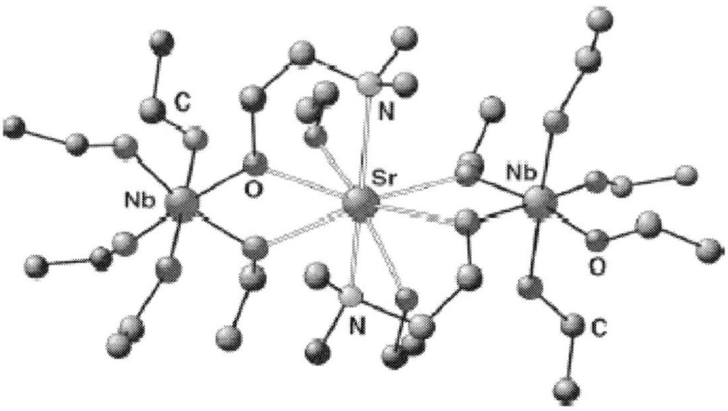

Figure 16. Crystal structure of [Sr{Nb(OEt)$_5$(dmae)}$_2$(EtOH)$_2$]. Reproduced from Ref. [161] with permission from copyright owners.

Lead niobium titanate, Pb(Nb,Ti)O_3 thin films were prepared on various substrates by MOCVD via the multiple precursors method, using the Pb(dpm)$_2$-Nb(OEt)$_5$-Ti(O^{i}Pr)$_4$ system [147]. The Nb content in the film can be controlled by changing the input gas flow rate of Nb(OEt)$_5$ while keeping an excess gas flow rate of Pb(dpm)$_2$. Polycrystalline films with (111) orientation and epitaxially grown films with *c*-axis orientation were obtained at temperatures between 400 and 620°C. The composition of the film is independent from the deposition temperature: the single phase of Pb(Nb,Ti)O_3 was deposited up to the Nb content of about 5 at% and pyrochlore oxide was co-deposited above this value. The same authors investigated the influence of deposition temperature on the microstructure of these Pb-Nb-Ti thin films [148,149].

3.4. Coprecipitation Routes

Coprecipitation techniques have been applied to the preparation of Nb-based multimetallic oxides but, because classical metal sources for this preparation mode are water-soluble salts (nitrates, chlorides,...), some modifications are generally operated to the conventional procedure. A niobium oxalate aqueous solution or the pentachloride, $NbCl_5$ in alcoholic medium are mainly used as precursors within that context, in combination with similar compounds for the other metals.

Marta et al reported the synthesis of $PbNb_2O_6$ oxide (PN) by a co-precipitation route involving a mixture of a niobium oxalate solution, prepared from the dissolution of freshly precipitated niobic acid in an excess of aqueous oxalic acid, and aqueous lead nitrate [106]. The precipitation was achieved by adding determined amounts of ammonia. The precipitate was then filtered, washed, dried at 105°C and finally subjected to thermal decomposition at 850°C for lead niobate formation. The coprecipitation mechanism was studied by potentiometric analyses and the optimal conditions for quantitative precipitation of both metals were established. The precipitated product is formed by a mixture of niobic acid and lead acetate according to the proposed equilibria in solution as follows :

$$Pb(NO_3)_2 + H_2ox + 2\ NH_3 \rightleftarrows Pbox + 2\ NH_4NO_3$$
$$2\ H_3[NbO(ox)_3] + (3{+}n)\ H_2O + 12\ NH_3 \rightleftarrows Nb_2O_5.nH_2O + 6\ (NH_4)_2ox$$

These equilibria are entirely shifted to the right when $pH > 7$. Moreover, the authors investigated the mechanism of thermal decomposition of the precipitated product by thermal analyses, IR spectroscopy and X-ray diffraction. The DTA and TG curves of the coprecipitation product exhibit endothermal events at 150°C which correspond to loss of water, and at 320 and 380°C, occurring from the decomposition of lead oxalate and the formation of lead carbonate. An exothermal event at 535°C marks the reaction between lead carbonate and niobium pentoxide with formation of $PbNb_2O_6$, and a final weak exothermal peak can be assigned to the recrystallization process.

The coprecipitation route was applied to the preparation of PMN powders, using water solutions of soluble salts such as potassium orthoniobate, K_3NbO_4, and nitrates or acetates of lead and magnesium [6]. K_3NbO_4 was synthesized from a mixture of K_2CO_3 and Nb_2O_5 which was first ball-milling with ethanol and then reacted at 740°C for 48h. Niobium aqueous species are poorly soluble in water at a pH below 7, whereas Mg and Pb hydroxides precipitate above pH 9.5 and 5.5, respectively. The principle of this process consists in mixing a basic solution, in which the niobate species are soluble, with an acidic solution, in which lead and magnesium salts are soluble. The precursor solutions were adapted to yield, upon mixing, a solution of intermediate pH which allows complete hydrolysis. The crystallization of the oxide material was followed by X-ray diffraction of the resulting solid PMN precursor calcined at different temperatures. At 350°C, the pattern displays characteristic diffraction lines of the pyrochlore phase, already associated with a small amount of the perovskite oxide. The percentage of perovskite phase progressively increases until almost pure PMN perovskite is obtained at 800°C, sometimes together with a small amount of impurities, either PbO or pyrochlore phase. This aqueous process was also applied to PFN ($Pb(Fe_{1/2}Nb_{1/2})O_3$) and PZN ($Pb(Zn_{1/3}Nb_{2/3})O_3$) oxides, replacing magnesium nitrate by iron(III) nitrate or zinc nitrate, respectively [6]. The PZN phase was obtained at 800°C but together with PbO impurities which can be eliminated through washing with acetic acid. A material containing 97% of the perovskite PFN oxide was obtained with pyrochlore structure as the impurity. PMN and PFN oxides can also be prepared by a stepwise precipitation route [3,4]. The authors engaged lead nitrate and a niobium precursor solution which can be a niobium oxalate solution [3] or a niobium peroxo solution [4], prepared from fresh niobic acid dissolved in excess aqueous oxalic acid or hydrogen peroxide, respectively. In the first case, because of the precipitation of Pb(ox) when adding $Pb(NO_3)_2$ to the Nb oxalate solution, a pre-calculated amount of lead nitrate was first added to fully eliminate the oxalate ions as Pb(ox) and with the required amounts of lead ions remaining in the precursors solution. In both cases, a two-step precipitation was carried out. A first precipitation is made with the Pb-Nb solution, and the obtained solid is then added to an appropriate amount of aqueous $Mg(NO_3)_2$ to form a slurry from which the second hydrolysis is carried out. The Pb-Mg-Nb precipitate is finally filtered, washed and calcined to provide the PMN oxide. The flow diagram for the preparation of PMN powder by a stepwise hydrolysis from the Nb peroxo solution is given in Figure 17. The PMN precursor powder begins to crystallize as pyrochlore phase near ~400°C and yields ~95% of perovskite phase after calcination at higher temperatures (800-900°C). $Pb(Fe_{1/2}Nb_{1/2})O_3$ and $Pb(Sc_{1/2}Nb_{1/2})O_3$ oxides were also prepared according to a similar method from the peroxo niobium solution, lead nitrate and iron or scandium nitrate, but from a one-step precipitation [4]. These oxides crystallize in the same conditions (800°C) as PMN but provide single perovskite phases.

Recently, strontium barium niobate, $Sr_{0.5}Ba_{0.5}Nb_2O_6$ (SBN50) were obtained by coprecipitation with $(NH_4)_2CO_3$ of $SrCl_2$ and $BaCl_2$ in a slurry of Nb_2O_5 dispersed in citric acid [7]. pH was maintained between 10 and 11 and the precipitate was filtered, washed and dried before heat-treatment between 900 and 1200°C. Powder X-ray diffraction study showed tetragonal tungsten bronze SBN50 phase formation at 1200°C.

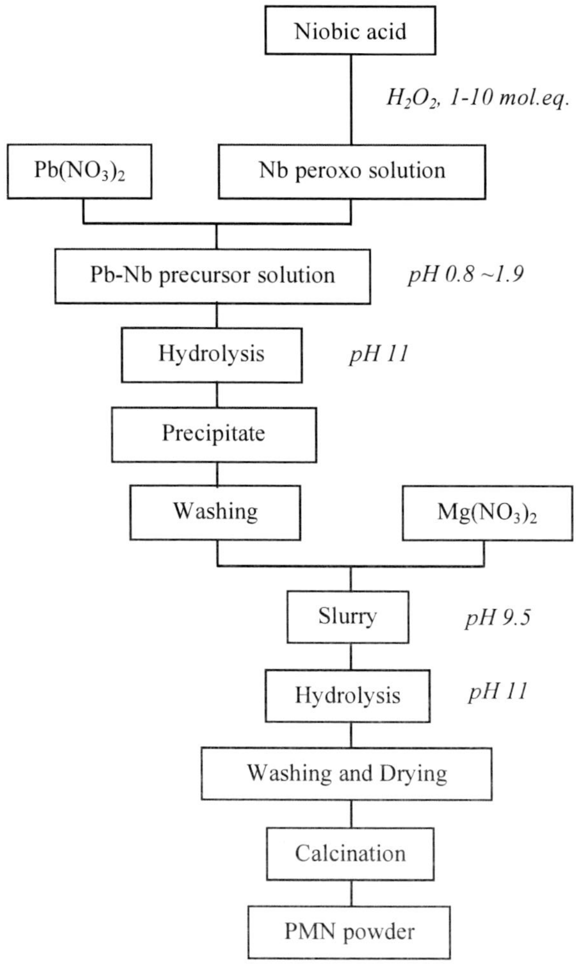

Figure 17. Flow diagram of the preparation of PMN powder from a niobium peroxo solution, by a two-step coprecipitation method [4].

Similar thermal conditions are required for the conventional solid-state method. At lower temperatures, the $BaNb_2O_6$ and $SrNb_2O_6$ oxides are also present. The particle size of powder calcined at 1200°C ranges between 250 and 300 nm. Ferroelectric barium bismuth niobate oxides can also be prepared by a coprecipitation technique as single $BaBi_2Nb_2O_9$ phase [150]. Barium chloride and bismuth nitrate are dissolved in water with a minimum amount of dilute HNO_3 to avoid precipitation of Bi ions. Nb_2O_5 is dissolved in a minimum amount of HF under heating. Both solutions are mixed and an excess quantity of HCl is added to this Ba-Bi-Nb solution to dissolve the formed barium fluoride. A mixed ammonia/ammonium oxalate aqueous solution is added until pH reaches 10 to ensure complete precipitation. After filtering, washing and drying the precipitate, the precursor is calcined at various temperatures. The pure tetragonal $BaBi_2Nb_2O_9$ phase is obtained after heat-treatment at 800°C while the ceramic method requires temperatures above 1000°C.

The synthesis of Y_3NbO_7 oxide can be achieved as powder via the coprecipitation by ammonia of metal chlorides, $NbCl_5$ and YCl_3, in an ethanolic solution [151]. When reaching pH 9, precipitate was filtered, washed and dried before being calcined between 800 and 1200°C for 2h in air or under argon. The X-ray diffraction patterns of samples show a Y_3NbO_7 single phase obtained after heating at 800°C, while at 1000°C, a YOCl phase crystallizes. By calcination at 1200°C, the undesired YOCl phase is eliminated (dissociation temperature ~ 975-1100°C) and the Y_3NbO_7 is present as single phase. A rather similar method was used by Sych et al to prepare bismuth niobate powders considering a coprecipitation from aqueous methanol solutions in the $Bi(NO_3)_3$-$NbCl_5$-CH_3OH-H_2O-NH_3 system [16]. The precipitation process of the components from the solution was studied potentiometrically. The variation of the degree of precipitation of the metals with the pH of the solution was determined and compared with that of the monometallic systems. As illustrated in Figure 18, bismuth is precipitated at pH 4.2 and niobium at 2.5 while complete precipitation occurs at pH > 3 when they are present together in the solution. Thermal analyses of the filtered solid display three events at 130, 530 and 660°C. The first one is endothermic and corresponds to the loss of water molecules, the second one is exothermic and was attributed to the crystallization of unreacted Nb_2O_5, while the last one would correspond to the crystallization process of bismuth niobate. When the coprecipitate is heated, it was shown that gradual reaction between bismuth and niobium occurs to yield $BiNbO_4$, ranging from 13% at 400°C, 92% at 550°C and 100% above 800°C. There is no mention of the structure of the obtained mixed phase in this work.

Niobium-containing catalysts are also often prepared via the coprecipitation technique. Nb-Ti oxide photocatalysts can be synthesized from chloride precursors [152]. A $NbOCl_3$ solution was prepared dissolving $NbCl_5$ in concentrated HCl and a $TiOCl_2$ solution was obtained by mixing an equal weight of $TiCl_4$ and 1 $mol.L^{-1}$ HCl. The binary oxides with different Nb:Ti molar ratios were synthesized by refluxing mixture of $NbOCl_3$ and $TiOCl_2$ solutions for 3h. The resulting solids were filtered, washed until chlorides elimination and finally dried at ambient temperature. The Nb-Ti sample showed large specific surface areas ranging from 80 to 300 $m^2.g^{-1}$, depending on the composition. X-ray diffraction revealed that the sample without niobium is constituted of pure rutile TiO_2, but when increasing the niobium content, the amount of rutile phase decreases with an increase in anatase phase. When niobium becomes dominant, i.e. Nb/(Nb+Ti) molar ratio ≥ 95%, the solids turns to amorphous. Metal (Cr, Fe, Ga and V) niobium mixed oxides were obtained with the coprecipitation method, and then used as catalysts in ODH of propane to propene [153].

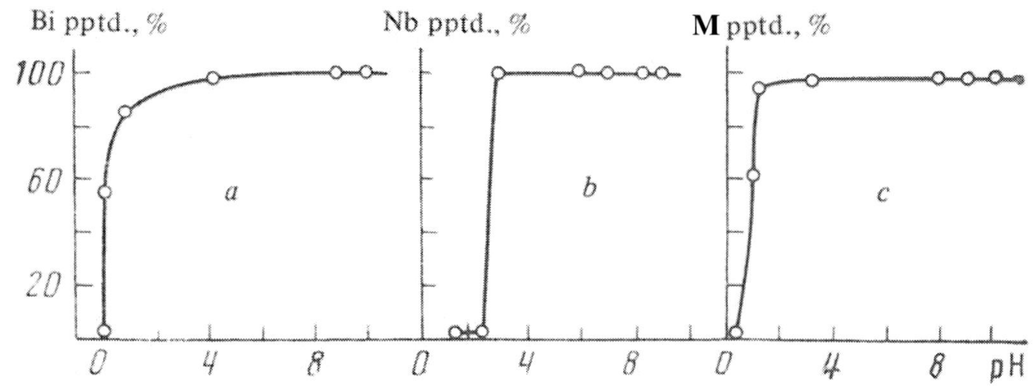

Figure 18. Variation of the degree of precipitation of the metal with pH of the solution in the systems $Bi(NO_3)_3$- CH_3OH-H_2O-NH_3 (a), $NbCl_5$-CH_3OH-H_2O-NH_3 (b) and $Bi(NO_3)_3$-$NbCl_5$-CH_3OH-H_2O-NH_3 (c). Reproduced from Ref. [16] with permission from copyright owners.

An alcoholic solution of both precursors, $NbCl_5$ and $Cr(NO_3)_3.6H_2O$, $Fe(NO_3)_3$, $Ga(NO_3)_3$ or $VO(acac)_2$, was dropped into an aqueous solution maintained at pH 7 by ammonia addition. After precipitation, the solvent was evaporated and the solids obtained were calcined either in air or in N_2 up to 700°C. In the case of the Nb-V system, $NbVO_5$ formed after thermal treatment at 500-550°C but decomposed if heated at higher temperatures into V_2O_5 and a Nb-richer phase, VNb_9O_{25}. To understand the phases formation in this system, a Raman spectroscopic study was carried out in situ at increasing temperatures on the precursor obtained by coprecipitation (Nb:V = 1:1) and compared with results obtained from a 1:1 mixture of Nb_2O_5 and V_2O_5. Rutile-type $CrNbO_4$ and monoclinic $FeNbO_4$ were obtained in almost a pure form at 650 and 500°C, respectively, and are stable up to 700°C. In the case of the Nb-Ga sample, the formation of the mixed $GaNbO_4$ phase occurred only at a minor extent, together with Nb_2O_5. Niobium pentoxide promoted with Ti, Zr, Sn, Sb, Pb or Bi was obtained by Smits et al using coprecipitation [100]. These catalysts, containing approximately 1 or 5 mol% of high-valency element oxide, were made by two different methods. In the case of V, Ti, Zr and Sn, a mixed oxalic acid solution of the two elements was slowly added to an ammonium oxalate/ammonia buffer (pH 10) at 60°C, the pH being kept around 10 by addition of ammonia. Elements of which the oxides are soluble neither in oxalic acid nor in ammonia (Sb, Pb, Bi) were incorporated by simultaneous slow addition of a nitric acid solution of the element and an oxalic acid solution of niobium to an ammonium oxalate/ammonia buffer at 60°C. In both cases, the precipitates were filtered, washed, dried at 80°C and finally, calcined at 630°C. In addition to the niobium oxalate, the precursors used in these methods were NH_4VO_3, $TiO(acac)_2$, $ZrOCl_2.8H_2O$, $SnCl_2.2H_2O$, $Sb(OAc)_3$, $Pb(NO_3)_3$ and $BiO(NO_3)$. As already mentioned in the case of similar doped niobia-based catalysts (P, V, Cr, Ge or Mo) synthesized via an oxalate method (see section 3.2.3), X-ray diffraction results indicated that *T* or *TT*-Nb_2O_5 is the major phase in the samples calcined at 630°C. Mixed oxides were actually sometimes detected in the Sb-Nb ($SbNbO_4$) or Pb-Nb ($PbNb_2O_6$) systems. According to the authors, any other phase present in the catalysts was either amorphous or present in too small quantity to be observed.

3.5. Hydro- and Glycothermal Methods

Hydrothermal syntheses of ceramic powders have interesting potential in view of increasing demand for environmentally benign materials and manufacturing methods. These routes may offer a low temperature direct route to fine oxide powders and these are now quite used for the preparation of zeolithes, aluminophosphates and mesoporous transition metal oxides. Hydrothermal reactions are performed in an autoclave or a bomb, in supercritical conditions of the water solvent (temperature > 100°C, pression ~100 MPa). The most engaged reactants in such methods are salts, hydroxides, oxides or even alkoxides of the metals. The powder generated is generally finely divided and homogeneous, but the resulting oxide is often amorphous and thus may required a further calcination step. When another reaction medium (organic solvents) is used in place of water, the method is called solvothermal process.

Next to niobium pentoxide [16,154,155], the niobium precursors most commonly engaged in hydrothermal syntheses are oxalate compounds and the materials recently prepared via the hydrothermal method correspond essentially to Mo-V-Nb-Te multicomponent oxide catalysts used in alkane selective oxidation reactions [31,32,34-36,156]. The procedures considered to obtain such materials are quite similar. It consists in treating an aqueous solution of the metal precursors, vanadyl sulfate, niobium oxalate and ammonium hexamolybdotellurate $(NH_4)_6TeMo_6O_{24}.nH_2O$, in an autoclave at 175°C for 48-72h. Different Mo/V/Nb/Te atomic ratios were engaged in these syntheses. The resulting product is then filtered, washed, dried and finally heat-treated in the 400-750°C temperature interval in nitrogen stream. Several crystalline phases can be detected in the samples, MoO_3, Mo_5TeO_{16}, $TeM_{20}O_{57}$ (M = Mo, V, Nb), $(Mo_{0.93}V_{0.07})_5O_{14}$, $Mo_3Nb_2O_{14}$, $Nb_{0.09}$ $Mo_{0.91}O_{2.80}$ and new Mo-V-Nb-Te oxides, but their nature depends on the composition of the sample and the calcination temperature. The influence of the tellurium and molybdenum precursors was also investigated by Lopez Nieto et al following the same procedure [156]. Next to the Mo(VI)-Te(VI) mixed precursors, $(NH_4)_6TeMo_6O_{24}.2H_2O$, (i) the Mo(VI)-Te(IV) compound $(NH_4)_6TeMo_6O_{24}.7H_2O$, (ii) a mixture of ammonium heptamolybdate and telluric acid, or (iii) a mixture of ammonium heptamolybdate and tellurium(IV) dioxide were also engaged. Both the characterization results of the catalysts obtained and the catalytic tests in the selective oxidation of propane to acrylic acid show important differences depending on the Mo and Te starting compounds used in the hydrothermal procedure. The most active and selective oxides were those prepared from Te(VI) compounds, while those obtained from the Te(IV) precursors presented lower activity and selectivity to acrylic acid. These observations could be related, according to the authors, to the different incorporation of tellurium, and even vanadium and niobium ions, which depends strongly on the Te-compound introduced in the starting solution. Te-doped Mo-V-Nb oxides can also be prepared by impregnation of a Mo-V-Nb material, obtained by hydrothermal synthesis, with an aqueous solution of telluric acid [35].

A solvothermal procedure was applied to the formation of niobium bimetallic oxides in glycol medium (glycothermal treatment) [157]. The reaction of niobium ethoxide or *n*-butoxide and various sources of the second metal (alkoxide, acetate or acetylacetonate), at 300°C and at the autogenous pressure of 1,4-butanediol, provides directly the crystalline $LiNbO_3$, $Zr_6Nb_2O_{17}$, $CaNb_2O_6$, $CrNbO_4$, $FeNbO_4$, $ZnNb_2O_6$ and R_3NbO_7 (R = rare earth) phases. In some cases, amorphous powders were obtained but they can be transformed at low calcination temperature into crystalline $NaNbO_3$, MNb_2O_6 (M = Mg, Sr, Ba), $MnNb_2O_6$, $Co_4Nb_2O_9$, $MNbO_4$ (M= Al, Ga), $LaNbO_4$ and Y_3NbO_7. Table 5 summarizes the starting metal precursors engaged in this method, and the nature of the product obtained directly from the glycothermal route or produced after an additional calcination step. Syntheses of Bi-Nb and Ni-Nb oxides from bismuth and nickel acetates were not successful because Bi^{3+} and Ni^{2+} ions are easily reduced by 1,4-butanediol forming Bi and Ni metals.

Table 5. Formation of Nb-M bimetallic oxides by the glycothermal method from different M precursors, and calcination of the amorphous glycothermal products [157]

Starting M precursor	Glycothermal product	Calcined sample	
		Temp. (°C) [a]	Phase [b]
Li(OAc)	$LiNbO_3$		
Na(OAc)	fluorite structure	(1000)	$NaNbO_3$, ($Na_2Nb_4O_{11}$)
K(OAc)	amorphous		
$Mg(OAc)_2$	rutile-like structure	(1000)	$MgNb_2O_6$ (orthorhombic)
$Ca(OAc)_2$	$CaNb_2O_6$		
$Sr(OAc)_2$	amorphous	700	$SrNb_2O_6$ (orthorhombic)
$Ba(C_8H_{15}O_2)_2$	amorphous	750	$BaNb_2O_6$ (orthorhombic)
$Mn(OAc)_2$	unknown	750	$MnNb_2O_6$ (orthorhombic)
$Co(OAc)_2$	amorphous	(1000)	$Co_4Nb_2O_9$
$Al(O^iPr)_3$	boehmite	750	$AlNbO_4$
$Ga(acac)_3$	amorphous	750	$GaNbO_4$
$Cr(acac)_3$	$CrNbO_4$ (rutile structure)	(1000)	$CrNbO_4$ (rutile structure)
$Fe(acac)_3$	$FeNbO_4$ (rutile structure)	800	$FeNbO_4$ (ixiolite structure)
$Zn(OAc)_2$	$ZnNb_2O_6$		
$La(acac)_3$	amorphous		$LaNbO_4$, (La_2O_3)
$Zr(acac)_4$	$Zr_6Nb_{12}O_{17}$		
$Y(acac)_3$			
Y:Nb = 3:1	amorphous	500	Y_3NbO_7
Y:Nb = 1:1	Y_3NbO_7, ($YNbO_4$)	900	Y_3NbO_7, ($YNbO_4$)
		1000	$YNbO_4$, (Y_3NbO_7)
Y:Nb = 1:3	Y_3NbO_7, ($YNbO_4$)	1000	$YNbO_4$, Nb_2O_5, (Y_3NbO_7)
$Lu(OAc)_3$	Lu_3NbO_7		

[a] phase transformation or crystallization temperature. Data in parentheses are calcination temperatures; no detailed analysis of phase transformation or crystallization was carried out by the authors.
[b] minor phases detected in the product are shown in parentheses

3.6. Other Methods

Highly oriented $LiNbO_3$ thin films on sapphire substrates were prepared using a chemically modified molecular precursor with UV sensitive chemical bonds, using ultraviolet irradiation [124]. The 1-phenyl-1,3-butanedione (pdb) was selected as a ligand for an alkoxide precursor. The complex obtained from $Nb(OEt)_5$, LiOEt and pdb in refluxing ethanol is assumed to be $LiNb(OEt)_{6-n}(pdb)_n$ in solution, with a specific UV absorption at 330 nm. The precursor films were prepared by dip-coating and then irradiated with UV light. The exposed films were heat-treated and the $LiNbO_3$ phase crystallized around 500°C. This UVpatterning of alkoxy-derived $LiNbO_3$ films offers the possibility of a feasible and low temperature process for fabricating these kinds of materials.

Ng et al reported the preparation of lead magnesium niobate from a freeze-drying process (- 40°C, ca. 200.10^{-3} mm Hg) of a Pb-Mg-Nb nitrate solution. This precursor solution is identical to that use by the same authors in a coprecipitation technique described in section 3.4 [3]. A comparison between the freeze-drying process and that of stepwise coprecipitation was made as far as the properties of the resulting oxides are concerned. Powders derived from both processes were compared in terms of their thermal decomposition, phase formation and particles characteristics. Freeze-dried PMN offered a lower formation temperature for the perovskite phase (700°C) than the coprecipitated precursor (800°C). PMN ceramics of pure perovskite phase can be obtained when the freeze-drying derived powder is sintered at 1150°C, while the pyrochlore phase persists in the PMN oxides derived from stepwise precipitation. Moreover, the particles in the obtained powders display size and morphology not significantly different between both techniques; they exhibit a rounded morphology and an average agglomerate size lower than 1 μm. Other papers reported the use of freeze-drying in less classical methods based on carboxylate modified precursor, and are thus described in section 3.2.3 [27,93,97,103,120].

Next to the combustion synthesis starting from an oxalate modified niobium precursor to prepare Sr-Ba-Nb ceramics and described in section 3.2.3 [7], a so-called combined combustion sol-gel synthesis of $LiNbO_3$ and $NaNbO_3$ was reported, engaging commercial precursors like $NbCl_5$, Li_2O or Na_2O [158]. This method is intermediate between the conventional sol-gel technique and the solid state combustion synthesis. A combustion reaction was initiated between M_2O (M = Li or Na) and $NbCl_5$ in an agate mortar in air by addition of a drop of methanol. This produced a yellow synthesis flame that moved through the material (temperature ca. 1000°C). After cooling to room temperature, more methanol was added and the mixture was stirred for 160h. A two-layered system is obtained, a methanol layer and a gel layer. The gel was isolated, washed with methanol and then dried at 150°C. Heating at 500°C for 2h yielded pure $MNbO_3$ phase in a quantitative yield. SEM analyses showed sub-micron cubic crystallites and EDX displayed no halide impurity in the samples. The overall process for the reaction can be exemplified for the formation of $LiNbO_3$ as followed:

$$5\ Li_2O + 2\ NbCl_5 \rightarrow 2\ LiNbO_3 + 8\ LiCl + Cl_2$$

The authors emphasized that the two-step process, involving combustion and gel formation, is critical in forming $MNbO_3$. The combustion process followed by annealing at 550°C yields only a mixture of Nb_2O_5 and M_2O and no mixed phase at all.

Single crystal particles of mesoporous niobium-tantalum mixed oxide with a whormhole-like structure were successfully prepared by Lee et al, by the so-called neutral templating route, using the metal pentachlorides as precursors and a block copolymer as templating agent [159]. The sample was obtained by a two-step calcination process; the first heat treatment removes the template between 400 and 450°C, forming a mesoporous structure with amorphous wall material, and the second calcination at 650°C allows crystallization while maintaining the mesoporosity. The resulting material displays thick walls and an unordered pore structure and the submicrometer mesoporous particles forming the wall were confirmed by transmission electron microscopy (TEM) to be single crystals. The pore diameter was found to increase upon crystallization, while retaining the open-pore system. As shown by experiments, this Nb-Ta oxide exhibits high thermal, hydrothermal, and mechanochemical stability. An in situ TEM study was carried out by Lu et al on the same Nb-Ta oxide in order to observe the crystallization process of the amorphous porous material [160].

4. Conclusion

From a review of the available literature, it clearly appears that the precursors methods described up to now for the preparation of multimetallic niobium-containing oxide materials are numerous. These methods, involving metal precursors, represent seducing alternatives to high-temperature solid state processes, often used as far as niobium-based oxides are concerned, and they display several advantages due to the soft chemical way adopted: (i) their use allows forming crystalline oxides under conditions (temperature, time,...) significantly milder than those employed in the ceramic method, (ii) it may provide metastable phases difficult or impossible to obtain via solid state processes, (iii) it generally results in homogeneous materials with particular morphology (specific surface areas, particle size, porosity,...). Moreover, precursors methods give easily access to specific form of multimetallic oxides, like supported materials or thin films, which are often required for many applications. In the context of the synthesis of Nb-based multimetallic oxides, multiple variants of well-known precursors routes, like the citrate or the coprecipitation routes, have been imagined to overcome some difficulties mostly occurring from the metal precursors themselves, and they have therefore generally been developed and adapted in the frame of the study of a precise oxide system. For example, the preparation process of PMN oxides via a coprecipitation route has been optimized in a stepwise process in order to limit the PbO volatilization and to form only the desired perovskite phase, while for the synthesis of the Y_3NbO_7 phase, a Pechini route, classically conducted in aqueous medium, has been carried out in methanol allowing solubilization of the niobium chloride precursor.

The method to choose when preparing Nb-containing multimetallic oxides depends on the oxide system under study and the applications aimed. The ceramic method, despite its limitations, is still now often used for such oxides. The precursors routes, even if they offer a wide range of experimental parameters to control and modified, involve metal sources which may be expensive and/or difficult to handle like the alkoxides, and sometimes they require multiple preparation steps, including the synthesis of the precursors themselves, and/or demand specific installations, i.e. an autoclave for the hydrothermal route. In general, in order to select the right way to prepare any material, a minutious analysis of the requirements, advantages/disadvantages of the accessible processes must be confronted to the chemical and physical properties necessary for the investigated materials.

Reference

[1] Costa, A. L.; Galassi, C.; Roncari, E. *J. Eur. Ceram. Soc*. 2002, 22, 2093-2100.
[2] Das, R. N.; Pramanik, P. *Nanostruct. Mater.* 1999, 11, 325-330.
[3] Ng, W. B.; Wang, J.; Ng, S. C.; Gan, L. M. *J. Mater. Chem.* 1998, 8, 2239-2244.

[4] Yoshikawa, Y.; Uchino, K. J. *Amer. Ceram. Soc.* 1996, 79, 2417-2421.
[5] Gupta, S. M.; Kulkarni, A. R. *Mater. Chem. Phys.* 1994, 39, 98-109.
[6] Guzman, G.; Aegerter, M. A.; Barboux, P.; Livage, J. *J. Mater. Sci.* 1993, 28, 6510-6515.
[7] Patro, P. K.; Kulkarni, A. R.; Harendranath, C. S. *Mater. Res. Bull.* 2003, 38, 249-259.
[8] Sakamoto, W.; Yogo, T.; Kikuta, K.; Ogiso, K. J.; Kawase, A.; Hirano, S. *J. Amer. Ceram. Soc.* 1996, 79, 2283-2288.
[9] Zhao, J. P.; Li, Y.; Wang, B.; Qiang, L. S. *Ceram. Int.* 2004, 30, 613-617.
[10] Camargo, E. R.; Kakihana, M. *Solid State Ion.* 2002, 151, 413-418.
[11] Kalinnikov, V. T.; Gromov, O. G.; Kunshina, G. B.; Kuz'min, A. P.; Lokshin, E. P.; Ivanenko, V. I. *Inorg. Mater.* 2004, 40, 411-414.
[12] Wernberg, A. A.; Gysling, H. J.; Filo, A. J.; Blanton, T. N. *Appl. Phys. Lett.* 1993, 62, 946-948.
[13] Ayyub, P.; Multani, M. S.; Palmkar, V. R.; Vijayaraghavan, R. *Phys. Rev.* B 1986, 34, 8137-8140.
[14] Castro, A.; Palem, D. *J. Mater. Chem.* 2002, 12, 2774-2780.
[15] Keve, E. T.; Skapski, A. C. *J. Chem. Soc.* A 1971, 1280-1286.
[16] Sych, A. M.; Popolitov, V. I.; Eremenko, L. A. Russ. *J. Inorg. Chem.* 1975, 20, 1442-1445.
[17] Wang, N.; Zhao, M. Y.; Yin, Z. W.; Li, W. *Mater. Lett.* 2003, 57, 4009-4013.
[18] Okubo, T.; Kakihana, M. *J. Alloy. Compd.* 1997, 256, 151-154.
[19] Thangadurai, V.; Adams, S.; Weppner, W. *Chem. Mater.* 2004, 16, 2998-3006.
[20] Amountza, A.; Riza, F.; Kasselouri, V. *Mater. Lett.* 2003, 57, 2298-2304.
[21] Tao, S.; Canales-Vazquez, J.; Irvine, J. T. S. *Chem. Mater.* 2004, 16, 2309-2316.
[22] Han, C. H.; Kim, H. J.; Chang, H.; Lee, S. K.; Park, J. D. *J. Electrochem. Soc.* 2000, 147, 2800-2804.
[23] Lee, S. K.; Chang, H.; Han, C. H.; Kim, H. J.; Jang, H. G.; Park, H. D. *J. Solid State Chem.* 2001, 156, 267-273.
[24] Bahng, J. H.; Oh, E. S.; Seo, S. H.; Kim, J. S.; Lee, M.; Park, H. L.; Lee, C. I.; Kim, G. C.; Kim, K. *J. Phys. Status Solidi A-Appl. Res.* 2002, 191, 291-295.
[25] Smits, R. H. H.; Seshan, K.; Ross, J. R. H.; van de Oetelaar, L. C. A.; Helwegen, J. H. J. M.; Anantharaman, M. R.; Brongersma, H. H. *J. Catal.* 1995, 157, 584-591.
[26] Burch, R.; Swarnakar, R. *Appl. Catal.* 1991, 70, 129-148.
[27] Burch, R.; Kieffer, R.; Ruth, K. *Top. Catal.* 1996, 3, 355-364.
[28] Ruth, K.; Kieffer, R.; Burch, R. *J. Catal.* 1998, 175, 16-26.
[29] Al Saeedi, J. N.; Guliants, V. V. *Appl. Catal.* A 2002, 237, 111-120.
[30] Holmes, S. A.; Al Saeedi, J.; Guliants, V. V.; Boolchand, P.; Georgiev, D.; Hackler, U.; Sobkow, E. *Catal. Today* 2001, 67, 403-409.
[31] Botella, P.; Nieto, J. M. L.; Solsona, B.; Mifsud, A.; Marquez, F. *J. Catal.* 2002, 209, 445-455.
[32] Holmberg, J.; Grasselli, R. K.; Andersson, A. *Top. Catal.* 2003, 23, 55-63.
[33] Holmberg, J.; Grasselli, R. K.; Andersson, A. *Appl. Catal.* A 2004, 270, 121-134.
[34] Vitry, D.; Morikawa, Y.; Dubois, J. L.; Ueda, W. *Appl. Catal.* A 2003, 251, 411-424.
[35] Lopez Nieto, J. M.; Botella, P.; Concepcion, P.; Dejoz, A.; Vazquez, M. I. *Catal. Today* 2004, 91-92, 241-245.

[36] Lopez Nieto, J. M.; Botella, P.; Vazquez, M. I.; Dejoz, A. *Chem. Comm.* 2002, 1906-1907.
[37] Zou, Z. G.; Ye, J. H.; Arakawa, H. *J. Mol. Catal. A-Chem.* 2001, 168, 289-297.
[38] Zou, Z. G.; Ye, J. H.; Arakawa, H. *Mater. Sci. Eng.* B 2001, 79, 83-85.
[39] Zou, Z. G.; Ye, J. H.; Arakawa, H. *J. Phys. Chem.* B 2002, 106, 517-520.
[40] Katou, T.; Lee, B.; Lu, D. L.; Kondo, J. N.; Hara, M.; Domen, K. *Angew. Chem. Int.* Ed. 2003, 42, 2382-2385.
[41] Kondo, J. N.; Yamashita, T.; Katou, T.; Lee, B.; Lu, D.; Hara, M.; Domen, K. *Stud. Surf. Sci. Catal.* 2002, 141, 265-272.
[42] Zou, Z. G.; Ye, J. H.; Arakawa, H. *Solid State Comm.* 2001, 119, 471-475.
[43] Zou, Z. G.; Arakawa, H.; Ye, J. H. *J. Mater. Res.* 2002, 17, 1446-1454.
[44] Gopalakrishnan, *J. Chem. Mater.* 1995, 7, 1265-1275.
[45] Gopalakrishnan, J.; Bhuvanesh, N. S. P.; Rangan, K. K. *Curr. Opin. Solid State Mater. Sci.* 1996, 1, 285-294.
[46] Rao, C. N. R. *J. Mater. Chem.* 1999, 9, 1-14.
[47] Hartmann, M.; Prakash, A. M.; Kevan, L. *Catal. Today* 2003, 78, 467-475.
[48] Nowak, I.; Ziolek, M. *Chem. Rev.* 1999, 99, 3603-3624.
[49] Ziolek, M.; Sobczak, I.; Lewandowska, A.; Nowak, I.; Decyk, P.; Renn, M.; Jankowska, B. *Catal. Today* 2001, 70, 169-181.
[50] Amini, M. M.; Sacks, M. D. *J. Amer. Ceram. Soc.* 1991, 74, 53-59.
[51] Yogo, T.; Kikuta, K.; Ito, Y.; Hirano, S. *J. Amer. Ceram. Soc.* 1995, 78, 2175-2179.
[52] Hirano, S. I.; Kato, K. *Solid State Ion.* 1989, 32-3, 765-770.
[53] Puyôo-Castaings, N.; Duboudin, F.; Ravez, J. *J. Mater. Res.* 1988, 3, 557-560.
[54] Ho, J. C.; Liu, K. S.; Lin, I. N. *J. Mater. Sci.* 1995, 30, 3936-3943.
[55] Aruga, A.; Nakamura, Y.; Aoi, T.; Watanabe, Y.; Hirayama, S.; Fujimoto, S. *J. Amer. Ceram. Soc.* 1993, 76, 2939-2941.
[56] Kato, K.; Zheng, C.; Finder, J. M.; Dey, S. K.; Torii, Y. *J. Amer. Ceram. Soc.* 1998, 81, 1869-1875.
[57] Kato, K.; Zheng, C.; Finder, J. M.; Dey, S. K.; Torii, Y. *J. Mater. Sci. -Mater. Electron.* 1998, 9, 457-464.
[58] Griesmar, P.; Papin, G.; Sanchez, C.; Livage, J. *J. Mater. Sci. Lett.* 1990, 9, 1288-1289.
[59] Yamagushi, O.; Mukaida, Y.; Shigeta, H. *Adv. Powder. Technol.* 1990, 1, 3-12.
[60] Amarilla, J. M.; Casal, B.; Ruiz-Hitzky, E. *J. Solid State Chem.* 1992, 99, 258-266.
[61] Amarilla, J. M.; Casal, B.; Ruiz-Hitzky, E. *J. Mater. Chem.* 1996, 6, 1005-1011.
[62] Amarilla, J. M.; Perez-Revenga, M. L.; Casel, B.; Ruiz-Hitzky, E. *Catal. Today* 2002, 78, 571-579.
[63] Catauro, M.; Pagliuca, C.; Lisi, L.; Ruoppolo, G. T*hermochim. Acta* 2002, 381, 65-72.
[64] Sarzi-Amade, M.; Morselli, S.; Moggi, P.; Maione, A.; Ruiz, P.; Devillers, M. *Appl. Catal.* A 2005, 284, 11-20.
[65] Barbieri, F.; Cauzzi, D.; De Smet, F.; Devillers, M.; Moggi, P.; Predieri, G.; Ruiz, P. *Catal. Today* 2000, 61, 353-360.
[66] Moggi, P.; Devillers, M.; Ruiz, P.; Predieri, G.; Cauzzi, D.; Morselli, S.; Ligabue, O. *Catal. Today* 2003, 81, 77-85.
[67] Gaikwad, S. P.; Samuel, V.; Pasricha, R.; Ravi, V. Bull. Mat. Sci. 2005, 28, 121-123.
[68] Li, Y.; Zhao, J. P.; Wang, B. *Mater. Res. Bull.* 2004, 39, 365-374.

[69] Nelis, D.; Van Werde, K.; Mondelaers, D.; Vanhoyland, G.; Van den Rul, H.; Van Bael, M. K.; Mullens, J.; Van Poucke, L. C. *J. Sol-Gel Sci. Technol.* 2003, 26, 1125-1129.
[70] Nelis, D.; Mondelaers, D.; Vanhoyland, G.; Hardy, A.; Van Werde, K.; Van den Rul, H.; Van Bael, M. K.; Mullens, J.; Van Poucke, L. C.; D'Haen, *J. Thermochim. Acta* 2005, 426, 39-48.
[71] Narendar, Y.; Messing, G. L. *Chem. Mater*. 1997, 9, 580-587.
[72] Asai, T.; Camargo, E. R.; Kakihana, M.; Osada, M. J. *Alloy. Compd.* 2000, 309, 113-117.
[73] Roth, R. S.; Waring, J. L. J. *Res. Nat. Bur. Stand.* A 1962, 66, 451-463.
[74] Bouquet, V.; Leite, E. R.; Longo, E.; Varela, J. A. *J. Eur. Ceram. Soc.* 1999, 19, 1447-1451.
[75] Bouquet, V.; Longo, E.; Leite, E. R.; Varela, J. A. *J. Mater. Res*. 1999, 14, 3115-3121.
[76] Bouquet, V.; Leite, E. R.; Longo, E.; Varela, J. A.; Viry, M. G.; Perrin, A. J. *Eur. Ceram. Soc*. 2001, 21, 1521-1524.
[77] Simoes, A. Z.; Zaghete, M. A.; Stojanovic, B. D.; Riccardi, C. S.; Ries, A.; Gonzalez, A. H.; Varela, J. A. *Mater. Lett.* 2003, 57, 2333-2339.
[78] Simoes, A. Z.; Bouquet, V.; Moura, F.; Ries, A.; Zaghete, M. A.; Varela, J. A. *Materials Science Forum* 2005, 498-499, 342-348.
[79] Nobre, M. A. L.; Longo, E.; Leite, E. R.; Varela, J. A. Mater. Lett. 1996, 28, 215-220.
[80] Simoes, A. Z.; Ries, A.; Riccardi, C. S.; Zaghete, M. A.; Varela, J. A. *Mater. Lett.* 2005, 59, 598-602.
[81] Pribosic, I.; Makovec, D.; Drofenik, M. *J. Eur. Ceram. Soc.* 2005, 25, 2713-2717.
[82] Zhong, T.; Tang, J. L.; Zhu, M. K.; Hou, Y. D.; Wang, H.; Yan, H. *J. Cryst. Growth* 2005, 285, 201-207.
[83] Spagnol, P. D.; Varela, J. A.; Bertochi, M. A. Z.; Stojanovic, B. D.; Tebcherani, S. M. *Thin Solid Films* 2002, 410, 177-182.
[84] Spagnol, P. D.; Varela, J. A.; Zaghete, M. A.; Longo, E.; Tebcherani, S. M. *Mater. Chem. Phys.* 2003, 77, 918-923.
[85] Camargo, E. R.; Kakihana, M.; Longo, E.; Leite, E. R. *J. Alloy. Compd*. 2001, 314, 140-146.
[86] Aguilar, G. G.; Costa, E. V. *Mat. Sci. For.* 2004, 455-456, 186-190.
[87] Zanetti, S. M.; Leite, E. R.; Longo, E.; Varela, J. A. *J. Mater. Res*. 1998, 13, 2932-2935.
[88] Duran, P.; Moure, C.; Villegas, M.; Tartaj, J.; Caballero, A. C.; Fernandez, J. F. *J. Am. Ceram. Soc.* 2000, 83, 1029-1032.
[89] Kakihana, M.; Szanics, J.; Tada, M. *Bull. Korean Chem. Soc.* 1999, 20, 893-896.
[90] Lanfredi, S.; Dessemond, L.; Rodrigues, A. C. M. *J. Eur. Ceram. Soc*. 2000, 20, 983-990.
[91] Nobre, M. A. L.; Lanfredi, S. *J. Phys. Chem. Solids* 2001, 62, 1999-2006.
[92] Camargo, E. R.; Longo, E.; Leite, E. R.; Kakihana, M. *Ceram. Int.* 2004, 30, 2235-2239.
[93] Heegn, H.; Trinkler, M. Fres. *J. Anal. Chem*. 1998, 361, 598-600.
[94] Narendar, Y.; Messing, G. L. *J. Amer. Ceram. Soc.* 1997, 80, 915-924.
[95] Das, R. N.; Pramanik, P. *Mater. Lett*. 2000, 46, 7-14.
[96] Das, R. N.; Ray, J. C.; Pramanik, P. J. *Mater. Res.* 2000, 15, 2273-2275.

[97] Langbein, H.; Wolki, G. *Thermochim. Acta* 1995, 264, 67-73.
[98] Resini, C.; Panizza, M.; Raccoli, F.; Fadda, M.; Carnasciali, M. M.; Busca, G.; Lopez, E. F.; Escribano, V. S. *Appl. Catal.* A 2003, 251, 29-38.
[99] Smits, R. H. H.; Seshan, K.; Leemreize, H.; Ross, J. R. H. *Catal. Today* 1993, 16, 513-523.
[100] Smits, R. H. H.; Seshan, K.; Ross, J. R. H. *ACS Symp. Ser.* 1993, 523, 380-388.
[101] Viparelli, P.; Ciambelli, P.; Volta, J. C.; Herrmann, J. M. *Appl. Catal.* A 1999, 182, 165-173.
[102] Meunier, F. C.; Yasmeen, A.; Ross, J. R. H. *Catal. Today* 1997, 37, 33-42.
[103] Bayot, D. A.; Devillers, M. M. *Inorg. Chem.* 2006, 45, 4407-4412.
[104] Desponds, O.; Keiski, R. L.; Somorjai, G. A. *Catal. Lett.* 1993, 19, 17-32.
[105] Solsona, B.; Nieto, J. M. L.; Oliver, J. M.; Gumbau, J. P. *Catal. Today* 2004, 91-92, 247-250.
[106] Marta, L.; Zaharescu, M.; Haiduc, I. *J. Therm. Anal.* 1995, 44, 431-440.
[107] Uekawa, N.; Kudo, T.; Mori, F.; Wu, Y. J.; Kakegawa, K. J. *Colloid Interface Sci.* 2003, 264, 378-384.
[108] Dey, D.; Kakihana, M. *Mater. Eng.* 2005, 28, 161-171.
[109] Belov, V.; Avdonina, L.; Mikhlin, A. *Ferroelectrics* 1992, 131, 63-66.
[110] Selezneva., K. I.; Nisel'son, L. A. *Russ. J. Inorg. Chem.* 1968, 13, 45-47.
[111] Cheng, Z. X.; Ozawa, K.; Miyazaki, A.; Kimura, H. *Chem. Lett.* 2004, 33, 1620-1621.
[112] Cheng, Z. X.; Ozawa, K.; Miyazaki, A.; Kimura, H. J. *Cryst. Growth* 2005, 275, e971-e975.
[113] Kim, M. J.; Matijevic, E. J. *Amer. Ceram. Soc.* 1994, 77, 1950-1953.
[114] Boyle, T. J.; Alam, T. M.; Dimos, D.; Moore, G. J.; Buchheit, C. D.; Al-shareef, H. N.; Mechenbier, E. R.; Bear, B. R. *Chem. Mater.* 1997, 9, 3187-3198.
[115] Nakano, H.; Yano, T.; Suyama, Y. J. *Ceram. Soc. Jpn.* 2005, 113, 59-63.
[116] Shcheglov, P. A.; Drobot, D. V.; Seisenbaeva, G. A.; Gohil, S.; Kessler, V. G. *Chem. Mater.* 2002, 14, 2378-2383.
[117] Thurston, J. H.; Whitmire, K. H. *Inorg. Chem.* 2003, 42, 2014-2023.
[118] Thurston, J. H.; Trahan, D.; Ould-Ely, T.; Whitmire, K. H. *Inorg. Chem.* 2004, 43, 3299-3305.
[119] Bayot, D.; Tinant, B.; Devillers, M. *Catal. Today* 2003, 78, 439-447.
[120] Bayot, D.; Devillers, M. *Chem. Mater.* 2004, 16, 5401-5407.
[121] Bayot, D.; Tinant, B.; Devillers, M. *Inorg. Chem.* 2005, 44, 1554-1562.
[122] Bayot, D.; Degand, M.; Devillers, M. *J. Solid State Chem.* 2005, 178, 2635-2642.
[123] Bayot, D.; Tinant, B.; Mathieu, B.; Declercq, J. P.; Devillers, M. *Eur. J. Inorg. Chem.* 2003, 737-743.
[124] Hirano, S.; Yogo, T.; Sakamoto, W.; Takeichi, Y.; Ono, S. J. *Eur. Ceram. Soc.* 2004, 24, 435-440.
[125] Boulmaaz, S.; Papiernik, R.; Hubert-Pfalzgraf, L. G. Heterobimetallic acetatoalkoxides based on niobium as precursors to multicomponent oxides. *In :* Hench, L. L.; West, J. K., editors. *Chemical processing of advanced materials*, John Wiley and Sons; 1992; 239-245.
[126] Boulmaaz, S.; Papiernik, R.; Hubert-Pfalzgraf, L. G.; Septe, B.; Vaissermann, J. J. *Mater. Chem.* 1997, 7, 2053-2061.
[127] Hubert-Pfalzgraf, L. G.; Papiernik, R.; Boulmaaz, S. *Eurogel* 1991, 93-100.

[128] Hubert-Pfalzgraf, L. G. *Polyhedron* 1994, 13, 1181-1195.
[129] Hubert-Pfalzgraf, L. G. *Inorg. Chem. Comm.* 2003, 6, 102-120.
[130] Boulmaaz, S.; Papiernik, R.; Hubertpfalzgraf, L. G.; Daran, J. C. *Eur. J. Solid State Inorg. Chem.* 1993, 30, 583-593.
[131] Hubert-Pfalzgraf, L. G. *J. Mater. Chem.* 2004, 14, 1-11.
[132] Fukui, T.; Sakurai, C.; Okuyama, M. J. *Ceram. Soc. Jpn.* 1994, 102, 393-396.
[133] da Silva, C. L. T.; Camorim, V. L. L.; Zotin, J. L.; Pereira, M. L. R. D.; Faro, A. D. *Catal. Today* 2000, 57, 209-217.
[134] Li, Y. C.; Yue, B.; Yan, S. R.; Yang, W. M.; Xie, Z. K.; Chen, Q. L.; He, H. Y. *Catal. Lett.* 2004, 95, 163-166.
[135] Wernberg, A. A.; Gysling, H. J. *Chem. Mater.* 1993, 5, 1056-1058.
[136] Goel, S. C.; Hollingsworth, J. A.; Beatty, A. M.; Robinson, K. D.; Buhro, W. E. *Polyhedron* 1998, 17, 781-790.
[137] Stemmer, S.; Bai, G. R.; Browning, N. D.; Streiffer, S. K. *J. Appl. Phys.* 2000, 87, 3526-3531.
[138] Takeshima, Y.; Shiratsuyu, K.; Takagi, H.; Tomono, K. J*pn. J. Appl. Phys. Part 1 - Regul. Pap. Short Notes Rev.* Pap. 1995, 34, 5083-5085.
[139] Davies, H. O.; Jones, A. C.; Leedham, T. J.; Wright, P. J.; Crosbie, M. J.; Lane, P. A.; Steiner, A.; Bickley, J. F. A*dv. Mater. Opt. Electron.* 2000, 10, 177-182.
[140] Jones, A. C.; Davies, H. O.; Leedham, T. J.; Wright, P. J.; Crosbie, M. J.; Steiner, A.; Bickley, J. F.; O'Brien, P.; White, A. J. P.; Williams, D. J. J. *Mater. Chem.* 2001, 11, 544-548.
[141] Funakubo, H.; Ishikawa, K.; Watanabe, T.; Mitsuya, M.; Nukaga, N. *Adv. Mater. Opt. Electron.* 2000, 10, 193-200.
[142] Mitsuya, M.; Nukaga, N.; Funakubo, H. *Jpn. J. Appl. Phys. Part 2-Lett.* 2000, 39, L822-L824.
[143] Mitsuya, M.; Ishikawa, K.; Nukaga, N.; Funakubo, H. J*pn. J. Appl. Phys. Part 2-Lett.* 2000, 39, L620-L622.
[144] Potter, R. J.; Marshall, P. A.; Roberts, J. L.; Jones, A. C.; Chalker, P. R.; Vehkamaki, M.; Ritala, M.; Leskela, M.; Williams, P. A.; Davies, H. O.; Tobin, N. L.; Smith, L. M. *Mat. Res. Soc. Symp. Proc.* 2004, 784, 97-108.
[145] Kadokura, N.; Okuhara, Y.; Mitsuya, M.; Funakubo, H. *Chem. Vapor Depos.* 2000, 6, 225-227.
[146] Zhang, R. C.; Xu, R. J. *Mater. Res.* 2000, 15, 1702-1708.
[147] Matsuzaki, T.; Funakubo, H. J. *Appl. Phys.* 1999, 86, 4559-4564.
[148] Liu, X. D.; Funakubo, H.; Noda, S.; Komiyama, H. J. *Electrochem. Soc.* 2001, 148, C227-C230.
[149] Liu, X. D.; Funakubo, H.; Noda, S.; Komiyama, H. *Chem. Vapor Depos.* 2001, 7, 253-259.
[150] Gaikwad, S. P.; Potdar, H. S.; Samuel, V.; Ravi, V. *Ceram. Int.* 2005, 31, 379-381.
[151] Kobayashi, H.; Ogino, H.; Mori, T.; Yamamura, H.; Mitamura, T. J. *Ceram. Soc. Jpn.* , *Int. Ed.* 1993, 101, 654-658.
[152] Tsai, S. J.; Cheng, S. J. *Chin. Chem. Soc.* 2001, 48, 1009-1016.
[153] Ballarini, N.; Calestani, G.; Catani, R.; Cavani, F.; Cornaro, U.; Cortelli, C.; Ferrari, M. *Stud. Surf. Sci. Catal.* 2005, 155, 81-94.

[154] Santos, I. C. M. S.; Loureiro, L. H.; Silva, M. F. P.; Cavaleiro, A. M. V. Polyhedron 2002, 21, 2009-2015.
[155] Xiaoyong, W.; Minqiang, W.; Xi, Y. *Ferroelectrics* 2001, 264, 105-110.
[156] Lopez Nieto, J. M.; Botella, P.; Solsona, B.; Oliver, J. M. Catal. Today 2003, 81, 87-94.
[157] Kominami, H.; Inoue, M.; Inui, T. *Catal. Today* 1993, 16, 309-317.
[158] Aguas, M. D.; Parkin, I. P. J. *Mater. Sci. Lett.* 2001, 20, 57-58.
[159] Lee, B.; Yamashita, T.; Lu, D. L.; Kondo, J. N.; Domen, K. *Chem. Mater*. 2002, 14, 867-875.
[160] Lu, D.; Katou, T.; Uchida, M.; Kondo, J. N.; Domen, K. *Chem. Mater*. 2005, 17, 632-637.
[161] Leedham, T. J.; Jones, A. C.; Davies, H. O.; Tobin, N. L.; Williams, P. A.; Bickley, J. F.; Steiner, A. *Integr. Ferroelectr*. 2001, 36, 111-118.

In: Progress in Solid State Chemistry Research
Editor: Ronald W. Buckley, pp. 165-204
ISBN 1-60021-313-8

Chapter 4

SYNTHESIS AND CHARACTERIZATION OF NANOSIZED MAGNETIC FERRITES AND METAL OXIDES

Shao-Yun Fu and Xian-Ming Liu
Technical Institute of Physics and Chemistry, Chinese Academy of Sciences,
Beijing 100080, P.R. China

ABSTRACT

Synthesis of nano-sized magnetic materials has become a particularly important area of research and is attracting a growing interest because of their potential applications in ferrofluids, advanced magnetic materials, catalysts, colored pigments, high-density magnetic recording media, and medical diagnostic etc. In this chapter, sol-gel method, polymer-pyrolysis method, template method, hydrothermal method and refluxing method are introduced for synthesis of nano-sized spinel ferrites (or composites) and metal oxides including $NiFe_2O_4/SiO_2$, $CoFe_2O_4/BiFeO_4$, $CoFe_2O_4$, $MnFe_2O_4$, $ZnFe_2O_4$, $NiFe_2O_4$, α-Fe_2O_3, NiO, β-MnO_2 etc, respectively. These nanosized magnetic materials prepared using the above methods are characterized using X-ray diffraction (XRD), transmission electron microscopy (TEM), electron diffraction (ED), scanning electron microscopy (SEM), electron dispersed X-ray diffraction (EDX), X-ray photoelectron spectroscopy (XPS) and vibrating sample magnetometer (VSM) measurements, respectively. Influencing parameters such as reaction conditions etc on their magnetic properties are discussed.

1. INTRODUCTION

The development of uniform nanometer sized particles has been intensively pursued because of their technological and fundamental scientific importance [1, 2]. These nanoparticulate materials often exhibit very interesting electrical, optical, magnetic, and chemical properties, which cannot be achieved by their bulk counterparts [3].

Transition metal oxides constitute one of the most fascinating classes of inorganic solids because they are capable of exhibiting a wide variety of structures and properties. In recent years, there has been renewed interest in nanoparticles of different ferrites for their enhanced ferromagnetic character [4]. Magnetic particles in the single-domain size range often exhibit superparamagnetic character and high field saturation. This has led to the generation of a vast amount of experimental data with a large variance in observed magnetic behavior [5]. A better understanding of magnetism at the nanoscale is required not only for basic physics but also for the technological use of ferromagnetic substances in data storage [6], color imaging [7], ferrofluids [8], and biomedical applications such as magnetic labeling in immunoassays [9], drug delivery [10], and hyperthermia of cancer cells [11].

Chemical methods have been widely used to produce nanostructured materials due to their straightforward nature and their potential to produce large quantities of the final product. Realizable particle sizes range in size from nanometers to micrometers, by controlling particle size during synthesis by using competition between nucleation and growth. There are several synthetic procedures for synthesizing nanostructured magnetic materials [12,13]. In the case of reverse/normal micelle emulsion method, the particle size of nanoparticles can be controlled by systematically adjusting the reaction parameters, such as time, temperature, and concentrations of reagents and stabilizing surfactants. In general, particle size increases with increasing reaction time because more monomeric species are generated, and with increasing reaction temperature because the rate of reaction is increased.

Magnetic nanoparticle and nanostructure studies combine a broad range of synthetic and investigative techniques from physics, chemistry, and materials science. In the best case, these studies provide not only information about the structural and magnetic properties of the materials but also improve understanding of the synthesis techniques. The correlation between nanostructure and magnetic properties suggests a classification of nanostructure morphologies. Development of experimental recipes for the synthesis of nanomaterials over a range of sizes, shapes, and chemical composition is an important aspect of research these days [14]. Based on our research group's work, nanosized magnetic materials are prepared using different methods, such as sol-gel method, polymer-pyrolysis method, template method and solvothermal method and magnetic properties of nanoparticles of magnetic transition metal oxides and spinel ferrites are studied in this chapter.

2. Synthesis and Characterization of $NiFe_2O_4/SiO_2$ Nanocomposites by Sol-Gel Method

Recent developments in soft magnetic materials have included what may be termed artificially structured materials in which the structure of heterogeneous materials is carefully controlled to produce the desired magnetic properties. A heterogeneous magnetic material consisting of nanocrystalline particles embedded in an amorphous magnetic matrix phase, as shown in Fig. 1, in which the magnetic properties of the particulate and matrix phases are essentially different, allows considerable scope for controlling the magnetic properties through structural changes brought about by annealing. These are multiphase materials consisting of particulate and matrix materials with different magnetic and structural phases in which the eventual properties of the entire material can be controlled by structural changes to produce a material with a matrix. Dispersion of the nanocrystals in a suitable matrix is one way for reducing particle agglomeration and this technique allows one to stabilize the particles.

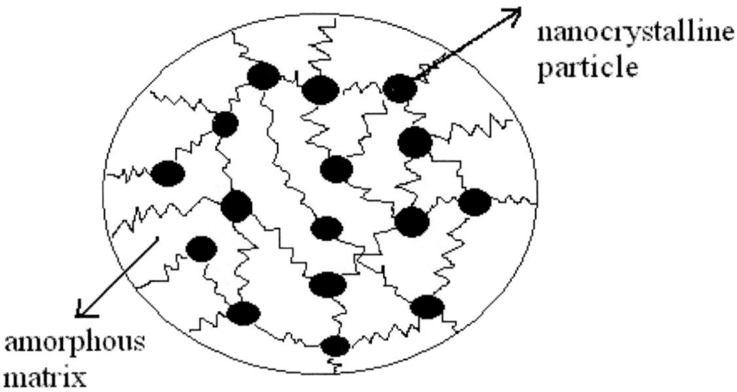

Figure 1. Schematic of an artificially nanostructured magnetic material consisting of nanocrystalline particles embedded in an amorphous matrix.

The interest in the preparation of magnetic composites has increased in the last years due to the properties presented by these materials are dependent on particle sizes and distribution in the matrix. This type of composites, if formed by nanoparticles with a high surface/volume ratio, can exhibit excellent magnetic properties compared with those of magnetic materials obtained by conventional processes. Systems such as Fe_2O_3/SiO_2 have been intensively studied over the last years [15,16], revealing magnetic behaviours different from those of bulk systems and serving as a model for the study of small particles.

$NiFe_2O_4/SiO_2$ nanocomposites are synthesized by sol-gel method using TEOS as the precursor of the silica matrix, iron and nickel nitrates as precursors of ferrites and formamide as a modifier. The magnetic properties of these materials are studied by varying the temperature of the annealing progress and the Ni ferrite to silica molar ratios [17]. Tetraethyl orthosilicate (TEOS, 98% purity), is used as the precursor for silica gel because TEOS is easily hydrolyzed and condensed in the presence of dilute HNO_3 catalyst, giving rise to homogeneous gels that are monolithic and transparent. $Fe(NO_3)_3{\cdot}9H_2O$ and $Ni(NO_3)_2{\cdot}6H_2O$ are employed as the starting iron and nickel species. The amounts of Ni and Fe nitrate are controlled for Ni/Fe molar ratio of 0.5, and the molar ratio of $NiFe_2O_4/(NiFe_2O_4+SiO_2)$ is fixed at 3:20. First, 2.18 g of $Ni(NO_3)_2{\cdot}6H_2O$ and 6.06 g of $Fe(NO_3)_3{\cdot}9H_2O$ are dissolved in 4.5 mL of distilled deionized water acidified with nitric acid as a catalyst, respectively. The concentration of HNO_3 is controlled at 0.03 M in the solution. TEOS (8.84 g), dissolved in 7 g of methanol and 2.25 g of formamide, are added to the above solution. After the mixed solution is gelatinized for 2 h at 40°C and aged for 24 h, the gel sample is obtained and dried at 40°C for three days in vacuum. Then, the gel sample is heated to 300°C in vacuum (rate of 5°C/min) followed by grinding, heating to 750, 850, 950 and 1050°C (rate of 3°C/min) in air atmosphere, and finally annealed for 3 h, respectively. In order to get more information, the analogous samples are prepared according to the above process, controlling the molar ratios of 1:20 and 1:4 for $NiFe_2O_4$/($NiFe_2O_4$+SiO_2) by varying the amounts of nitrates, but heat treatment temperature is only adjusted to 950°C.

Fig.2 exhibits the TG curve measured in air atmosphere for the dried gel 3$NiFe_2O_4$-17SiO_2 after being heated to 300°C in vacuum (rate of 5 °C/min) followed by grinding.

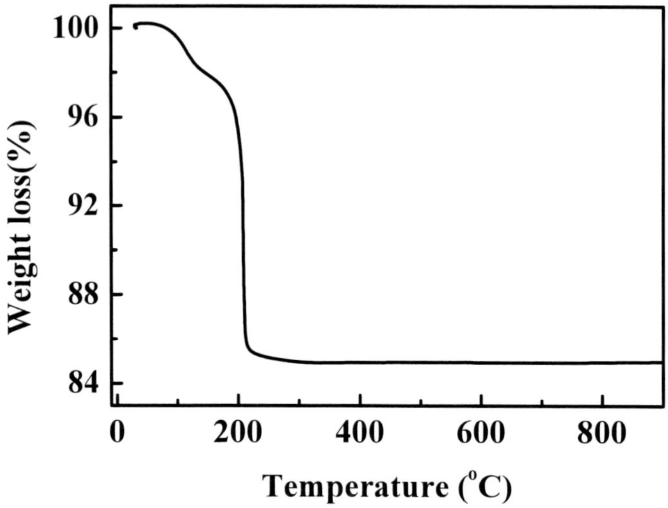

Figure 2. TG curve measured in air atmosphere for the dried gel 3$NiFe_2O_4$-17SiO_2 after being heated to 300°C in vacuum (rate of 5°C/ min). Reproduced from Ref. [17] with permission from copyright owners.

It can be seen from Fig. 2 that the weight losses of ca. 2% and 13% took place at the temperature range of 30 ~ 350°C, respectively. The first weight loss was probably due to the evaporation of the residue water and organic solvent. The weight loss over the range of 150 ~ 500°C was ascribed to the elimination of the complete decomposition of residual NO_3^- ions introduced by starting materials. Due to the dried gel sample dried for three days at 40°C and heated to 300°C in vacuum, the solvent of water and methanol had almost been evaporated, and part of the organic modifier and residual NO_3^- ions had been decomposed. There was no obvious change in weight with increase of the temperature after 400°C, which showed that $NiFe_2O_4$ nanocrystals had been formed in sequence in the silica matrix [16].

Fig.3 shows X-ray diffraction patterns of $NiFe_2O_4$/SiO_2 powders calcined at 750, 850, 950 and 1050°C, respectively. The molar ratio of $NiFe_2O_4$/SiO_2 is 3:17. According to the TG analysis and previous work [16], it can be concluded that a magnetic phase of Fe_2O_3 (hematite) is also formed together with the Ni ferrite in temperature over 75°C. The X-ray peaks of the composites are in accordance with those of a typical crystalline Ni ferrite, with spinel structure without a preferred orientation. These peaks are less intense, and present amorphous phases due to the presence of the silica matrix. The positions for these peaks coincide with the characteristic peaks for the standard spinel phase. This confirms that the cluster sizes grow into spinel nanocrystals in the silica matrix and the number of the spinel particles increase with increasing temperature. From the broadening effect of the most intense peak (311), the average grain size is determined by Scherrer formula to be ca. 15 nm for the magnetic particles annealed at 950°C. Similar broadening effect and relatively weak peak intensities have been found in nanocrystalline ferrites annealed at other temperatures.

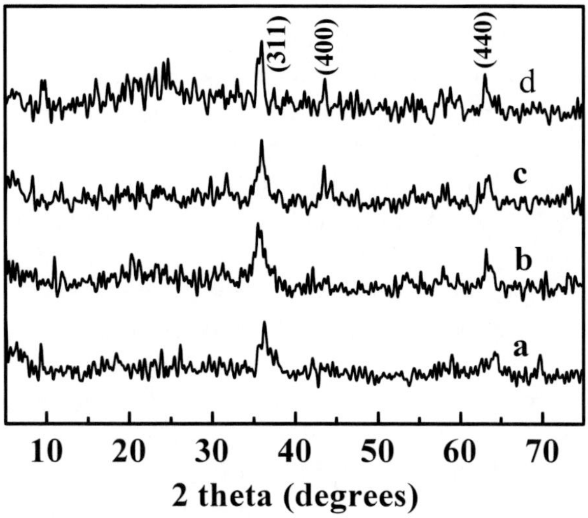

Figure 3. XRD patterns of the powders calcined at 750(a), 850(b), 950(c) and 1050°C (d). Reproduced from Ref. [17] with permission from copyright owners.

Above 750°C, spinel $NiFe_2O_4$ nanocrystals are dispersed in a silica matrix. During the formation of $NiFe_2O_4$ nanocrystals, no traces of quartz, cristobalite, or intermediate products, e.g. Fe_2SiO_4 or Ni_2SiO_4 are formed even at temperatures as high as 1050°C. Further, rearrangement of the network for the amorphous silica matrix occurs. These results show that, in these nanocomposite systems, the transition metals did not participate in the direct reaction with the silica matrix and that the crystallization of Ni ferrite is accompanied with a rearrangement process of the network for amorphous silica matrix, which is supported by the IR analysis.

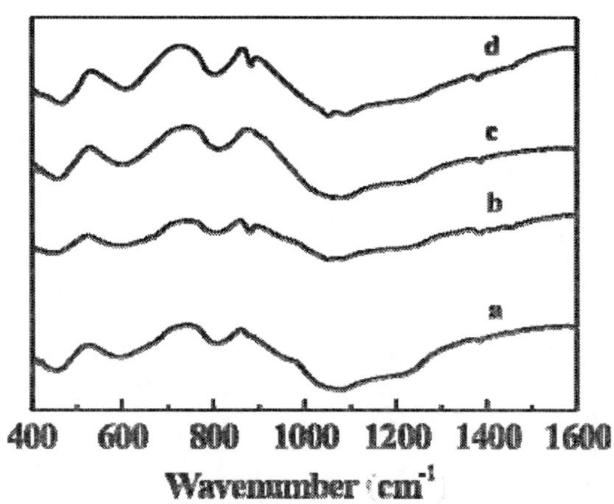

Figure 4. IR spectra of the powders calcined at 750 (a), 850 (b), 950 (c) and 1050°C (d). Reproduced from Ref. [17] with permission from copyright owners.

Fig. 4 shows IR spectra of the $NiFe_2O_4/SiO_2$ powders that are obtained by heating the dried gel at successively high temperatures. In Fig. 4, the characteristic absorptions for the silica network located at ca.1086, 798, and 461 cm^{-1} are assigned separately as follows: the broad high-intensity band at ca. 1080 cm^{-1} along with the accompanied shoulder at ca. 1200 cm^{-1} is due to the asymmetric LO and TO stretching bonds -Si-O-Si- of the SiO_4 tetrahedron associated with the motion of oxygen in Si-O-Si antisymmetrical stretch; the band at ca. 798 cm^{-1} is assigned to the Si-O-Si symmetric stretch, while the sharp band at ca. 461 cm^{-1} corresponds to the Si-O-Si or O-Si-O bending mode. The characteristic absorption at ca. 1380 cm^{-1} associated with the nitrate species disappears, which corresponds to the complete decomposition of nitrate species. With increase of the temperature, the absorption bands for silica matrix network at ca. 1080, 800, and 460 cm^{-1} increase in intensity slightly. It is interesting to note that absorptions associated with the characteristic Fe-O stretching modes in $NiFe_2O_4$ phase are found at ca. 600 cm^{-1}, which corresponds to the production of well-crystallized $NiFe_2O_4$ nanocrystals. Most of the IR bands remain the same except that the band at ca. 600 cm^{-1} grows stronger and narrower with increase of temperature. This is probably due to a growth process for the large particles in the $NiFe_2O_4$ phase. It is essential here to emphasize that the composition of the spinel clusters is $NiFe_2O_4$, because in NiO-Fe_2O_3 systems there are two other most possible compounds such as that could crystallize in the spinel structure with a very similar cell size as that of $NiFe_2O_4$. However, the IR spectra in Fig.4 do not show the absorption band at ca. 720 cm^{-1}, suggesting that α-Fe_2O_3 and Fe_3O_4 are not formed in the $NiFe_2O_4/SiO_2$ systems. These results are in good agreement with the analysis of XRD.

TEM micrographs of $NiFe_2O_4/SiO_2$ nanocomposites with different Ni ferrite molar fractions (5%, 15%, and 25%) treated at 950°C are displayed in Fig.5. It can be observed that the nickel ferrite particles are homogeneously dispersed in the silica matrix in Fig.5. These Ni ferrite particles are highly crystalline and very small. The average size of particle diameters is estimated to be ca. 10, 15 and 25 nm, respectively. With the molar ratio increase of $NiFe_2O_4$ nanocrystals in the silica matrix the rearrangement process of the silica matrix from heat treatment apparently promotes the coalescence of $NiFe_2O_4$ clusters into larger nanocrystals.

Magnetic properties of $NiFe_2O_4$ in the silica matrix annealed at different temperature are investigated at room temperature using a vibrating sample magnetometer with an applied field -10 kOe $< H <$ 10 kOe. Plots of magnetization *vs* applied magnetic field for various samples are given in Fig.6, and no hysterisis is found. At low external field, the hysteresis loops exhibit zero coercivity and remanence. At high external field, magnetization of the four samples increases almost linearly with the given field and shows a lack of saturation at a field as high as 10 kOe. The magnetization at 10 kOe as a function of temperature is presented in the inset of Fig. 6. It is of interest to note that the magnetization increases with the increase of calcination temperature. Increase in temperature leads to a steady rise in magnetization from 750°C to 950°C, which is followed by a little increase upon the temperature up to 1050°C. This trend can be well correlated to the sizes of the Ni ferrite particles and increase of γ-Fe_2O_3 phase. With the growth of $NiFe_2O_4$ nanoparticles in crystallinity at high temperature, the magnetization increases with the increase in temperature. But the ferrite spinel structure and particle sizes are almost the same over 950°C. These results are consistent with the studies of XRD and IR discussed earlier.

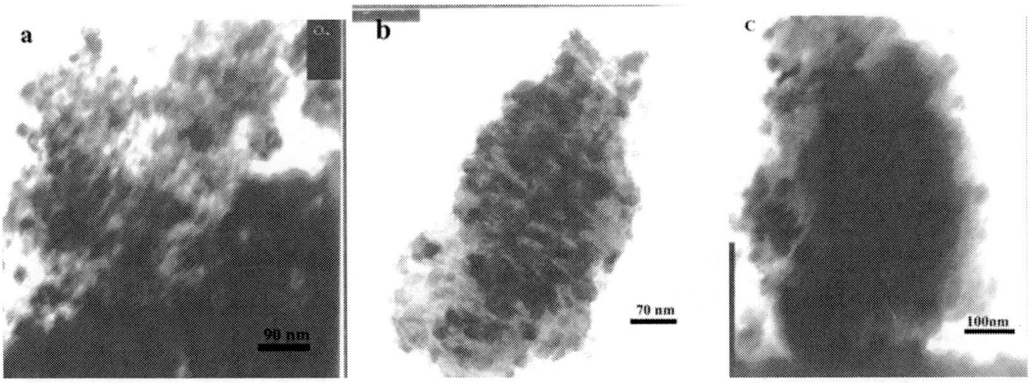

Figure 5. TEM micrographs of $NiFe_2O_4$-SiO_2 nanocomposites with different Ni ferrite molar fractions annealed at 950°C. (a) 5%; (b) 15%; (c) 25%. Reproduced from Ref. [17] with permission from copyright owners.

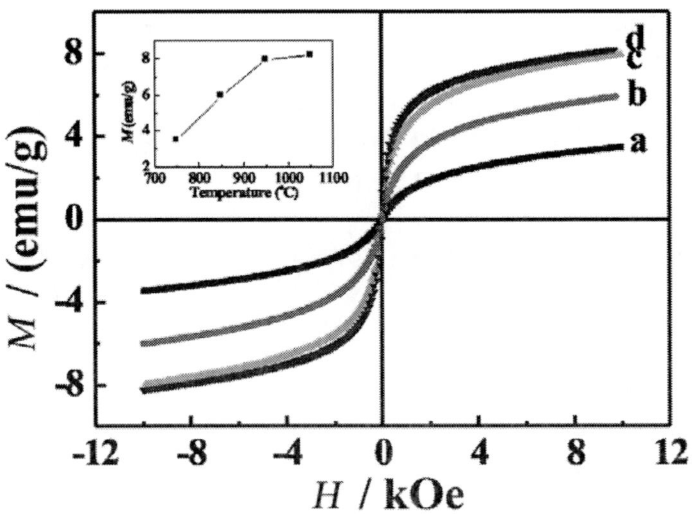

Figure 6. Plots of magnetization vs applied field for the samples annealed at different temperatures ((a) 750°C; (b) 850°C; (c) 950°C; (d) 1050°C) measured at 300 K. Inset of Fig. 6 shows the relationship between the magnetization and calcination temperature. Reproduced from Ref. [17] with permission from copyright owners.

Values of magnetization at the external field of 10 kOe for the samples annealed at 950°C are listed in Table 1. The magnetization of nanocomposites almost linearly varies with the content of Ni ferrites in the following order: M (25 wt %)$>M$ (15 wt %)$>M$ (5 wt %). For the present nanocomposite system, the magnetization with the molar fraction of 5% referring only to the magnetic nickel ferrite particles after subtracting the weight of the silica matrix is the largest. It verifies that when the particles are dispersed in an inert medium, such as $NiFe_2O_4/SiO_2$ nanocomposites, the magnetization increases even with decreasing particle size. For the pure ferrite, these results are in accordance with the literature [18]. As a result, Ni ferrite nanocomposites exhibit excellent magnetic properties.

Table 1. Magnetization data at 10 kOe for the samples measured at 300 K

$NiFe_2O_4$ molar ratio / %	Treatment temperature /°C	*Particle size / nm	Magnetization of nanocomposites / (emu/g)	Magnetization of $NiFe_2O_4$ / (emu/g)
5	950	10	3.79	22.20
15	950	15	7.95	19.48
25	950	25	11.66	20.60

*The particle sizes are estimated from Figure 5

In summary, $NiFe_2O_4/SiO_2$ nanocomposites can be well prepared by sol-gel method using tetraethylorthosilicate (TEOS) as precursor of the silica matrix, metallic nitrates as precursors of ferrites and formamide as a modifier. Ni ferrite particles are homogeneously dispersed in the silica matrix. The magnetic properties are dependent on the annealing temperature and the concentration of Ni ferrite. These results show that the magnetization of the samples annealed at different temperature increases with the increase of calcination temperature, and the magnetization of nanocomposites almost linearly varied with the content of Ni ferrites.

3. Synthesis and Characterization of Novel $CoFe_2O_4$-$BiFeO_3$ Nanocomposites by Sol-Gel Method

The synthesis of materials that exhibit simultaneous ferromagnetic and ferroelectric characteristics has recently attracted a great deal of attention not only from the viewpoint of solid-state physics but also because of their potential application in practical electronic devices [19]. The combination of both ferromagnetic and ferroelectric materials in a single material is expected to produce new properties, such as magneto-electric [20], magneto-optic [21] and other new coupling properties, due to the occurrence of coupling effects between the magnetization and electric polarization. However, there are only a few naturally occurring materials that can exhibit both spontaneous magnetization and electric polarization (or exhibiting both ferromagnetic and ferroelectric properties), and the choice of single-phase materials exhibiting coexistence of strong magnetism is limited. Thus, such interesting materials have been chemically synthesized. $CoFe_2O_4$—$BaTiO_3$ nanocomposites exhibit a strong coupling of the order parameters through the heteroepitaxy of the two lattices [19d]. A self-assembled composite thin film of $CoFe_2O_4$-PZT exhibits good magnetic properties [22]. Bismuth ferrite, $BiFeO_3$, has been considered as an attractive candidate for composite partner in a magnetoelectric material, since it is ferroelectric (Curie temperature T_c about 1123 K) and has a rhombohedrally distorted perovskite structure with $a = 3.96$ Å and belonging to the space group $R3c$ [23].

Recently, the sol-gel procedure has been employed for the preparation of $xCoFe_2O_4/(1-x)BiFeO_3$ (x=0.1, 0.2, 0.3, 0.4) spinel-perovskite nanocomposites [24]. The nanocomposites are prepared by thermal decomposition of mixed metal polymeric precursor gels. The polymeric precursor solution is prepared using the Pechini method, which has been used for synthesizing poly-cationic powders. The process is based on metallic citrate polymerization using ethylene glycol. At first, the citric acid is added into deionized water with constant agitation and kept at 60–70°C. Then, stoichiometric ferric nitrate, cobalt nitrate and bismuth nitrate are added in turn. Appropriate molar portions of metal nitrates are fixed at Co:Bi:Fe ratio of 1:9:11, 1:4:6, 3:7:13 and 1:1.5:3.5, respectively. After the mixing of the salts to their total dissolution, ethylene glycol is added into the solution with a proportion (in mass) of citric acid/ethylene glycol ratio of 60:40. The temperature is increased up to 90–110°C, in order to promote the sterification. At the end of the reaction, a colored polymeric gel (resin) is formed. The gels are dried overnight at 100°C in order to remove excess water. During the process of drying, the gel swell into the fluffy mass that is eventually broken into brittle flakes. The gel is heat-treated in air atmosphere at various temperatures ranging from 400 to 800°C for a time duration of 2 h in order to get the oxide materials.

The mechanism of mixed metal citrate precursor decomposition is quite sensitive to the formation of ultrafine single-phase ferrite materials. Fig. 7 shows representative TG plot of the mixed metal citrate precursor measured at the heating rate 10°C /min. From the TG curve, four different regions can be identified where, significant weight loss has occurred. In Fig.7, it can be observed that the weight loss of ca. 5.03% and 13.83% has occurred in the temperature range of 100-160.5°C. The weight loss is ascribed to the elimination of water and adsorbed gases. The weight loss that has occurred between 160.5 and 350.7°C can be attributed to polymeric chain decomposition, leading to CO_2 elimination. There is no obvious change in weight occurred with the increase of temperature above 350°C, which shows that the oxides had been formed.

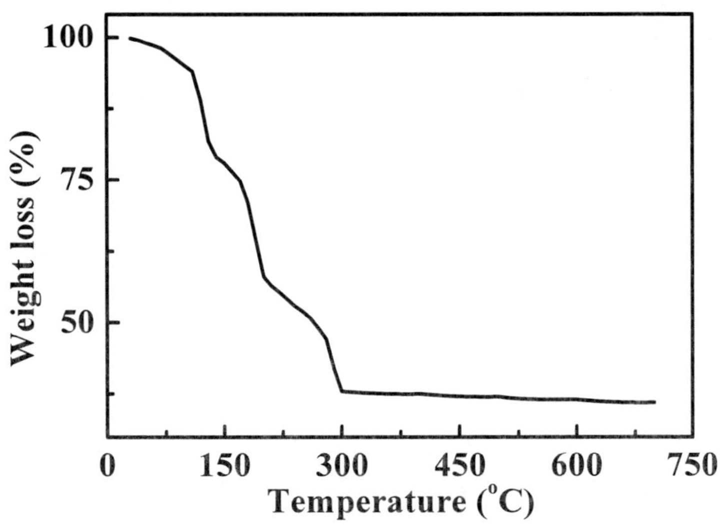

Figure 7. TG plot of mixed metal citrate precursor ($0.3CoFe_2O_4$—$0.7BiFeO_3$) . Reproduced from Ref. [24] with permission from copyright owners.

Fig. 8 shows the X-ray diffraction patterns of $0.3CoFe_2O_4$—$0.7BiFeO_3$ powders after being annealed at various temperatures ranging from 400 to 800°C for 2 h. All diffraction peaks for the materials correspond to spinel-perovskite mixed structure except for the powders annealed at 400°C shown in Fig. 8a. As the annealing temperature increases, the intensity of peaks in XRD pattern becomes stronger and sharper, indicating that there is a steady growth in crystallite size and the powders exhibit well-established broad peaks.

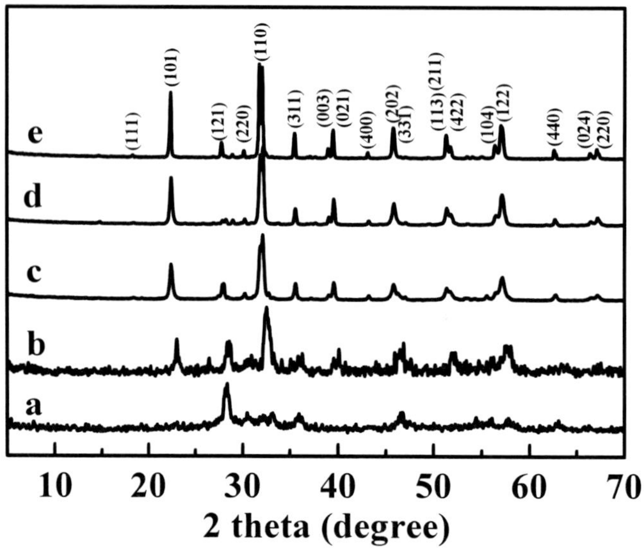

Figure 8. XRD patterns of $0.3CoFe_2O_4$—$0.7BiFeO_3$ annealed at various temperatures ranging from 400 to 800°C: a) 400°C; b) 500°C; c) 600°C; d) 700°C and e) 800°C. Reproduced from Ref. [24] with permission from copyright owners.

When the annealing temperature rises to 500°C, the intensity of these peaks become stronger and the peaks at the feature angles of $CoFe_2O_4$ and $BiFeO_3$ also appear, which shows that a composite consisting of $CoFe_2O_4$ and $BiFeO_3$ has formed after being annealed at 500°C. When the annealing temperature increases further, the intensity and sharpness of peaks are further enhanced and the d-spacing values of $CoFe_2O_4$ and $BiFeO_3$ correspond well with those of JCPDS card No. 03-0864 and 20-0169, respectively, suggesting that the powders are largely crystallized into a mixed phase of spinel-perovskite structure without traces of any impurity phases. Furthermore, the increase in intensity of X-ray diffraction shows improved crystallinity of the materials and increase in the crystallite size as a function of annealing temperature.

Fig. 9 shows the X-ray diffraction spectra of the compositions of $xCoFe_2O_4$— (1-/x)$BiFeO_3$ with x=/0.1, 0.2, 0.3 and 0.4 annealed at 700°C for 2 h. The phase analysis displays the formation of spinel-perovskite mixed structure. They exhibit strong and broadened peaks around the 2θ angles of 22.4, 32.1 and 57.1°, together with weak and broadened peaks around the 2θ angles of 27.7, 39.0, 39.5, 45.8, 51.4, 51.8, 56.4, 66.5 and 67.1°, indicating that perovskite bismuth ferrite with rhombohedral structure forms upon 2 h of annealing in the composites. This result is much identical to what reported in the literature (JCPDS card No. 20-0169). The peaks around the 2θ angles of 30.2, 35.5, 37.2, 43.2, 53.5, 57.1 and 62.7° appear, indicating that the spinel cobalt ferrites with cubic structure (JCPDS No. 03-0864) have been formed in the composites. Moreover, Fig. 9 exhibits that the intensity of $CoFe_2O_4$ peaks become stronger and broader with the increase of $CoFe_2O_4$ content in the composites. The X-ray diffraction spectra of the four compositions indicate that a mixed crystalline-spinel-perovskite phases have been formed in each case without any detectable third phase.

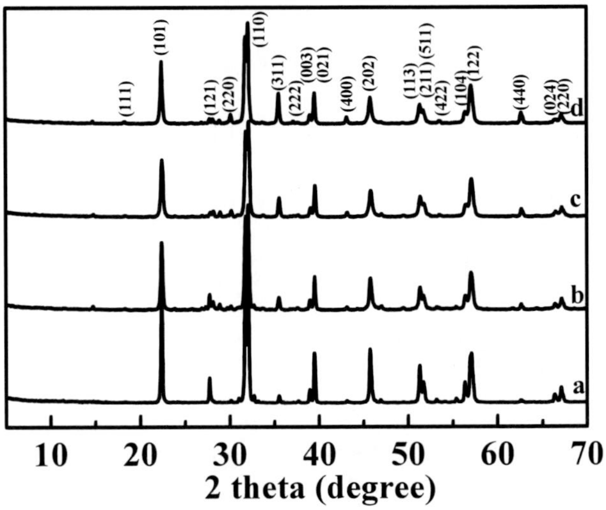

Figure 9. XRD patterns of $xCoFe_2O_4$—$(1\text{-}x)BiFeO_3$ annealed at 700°C: a) x=0.1; b) x=0.2; c) x=0.3; d) x=0.4. Reproduced from Ref. [24] with permission from copyright owners.

TEM images in Fig. 10A and 10B reveal rough, irregular but generally well-dispersed nanoparticles of $0.3CoFe_2O_4$—$0.7BiFeO_3$ spinel-perovskite powders annealed at 500 and 700℃. From the micrographs, particle sizes can be evaluated to consist of agglomerated 20-80 nm nanosized powders. The average particle sizes increases with the increase of annealing temperature. Supplementary information can be obtained from selected area electron diffraction pattern of $CoFe_2O_4$—$BiFeO_3$ spinel-perovskite powders calcined at 700 ℃. Electron diffraction (ED) patterns confirm the phases identified by XRD analysis. As a consequence of the small crystallite size, the samples (Fig. 10C) show diffuse diffraction spots and this ED pattern can be ascribed to reflection of the spinel phase ($CoFe_2O_4$) and perovskite phase ($BiFeO_3$).

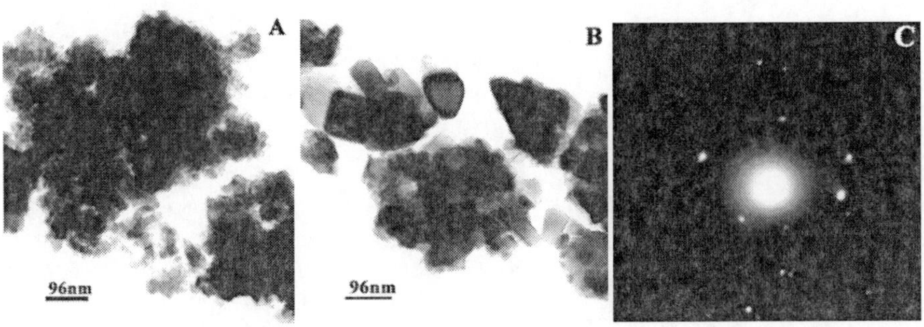

Figure 10. TEM photographs of $0.3CoFe_2O_4$—$0.7BiFeO_3$ annealed at various temperatures: A) 500 °C; B) 700 °C; C) electron diffraction pattern of the sample annealed at 700 °C. Reproduced from Ref. [24] with permission from copyright owners.

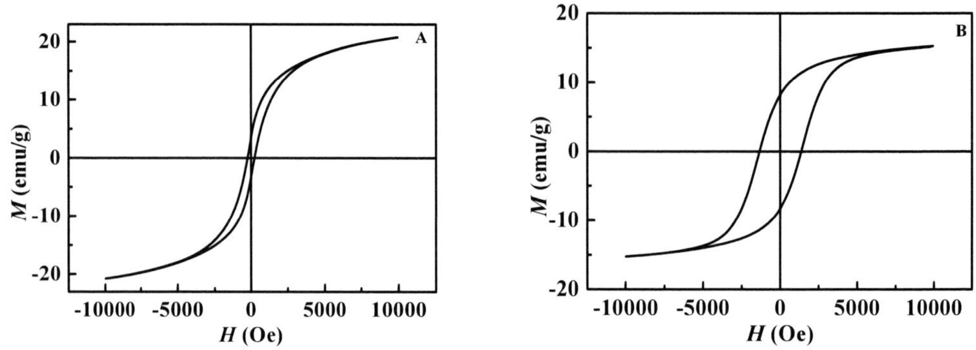

Figure 11. Hysteresis loops of the $0.3CoFe_2O_4$—$0.7BiFeO_3$ samples annealed at various temperatures measured at room temperature: A) 400 °C and B) 700 °C. Reproduced from Ref. [24] with permission from copyright owners.

Magnetic properties of $CoFe_2O_4$—$BiFeO_3$ nanocomposites annealed at various temperatures are investigated at room temperature using a vibrating sample magnetometer (VSM) with an applied field of -10 kOe≤H≤10 kOe. Fig. 11 shows the isothermal magnetic field dependence of the magnetization of the spinel-perovskite $0.3CoFe_2O_4$—$0.7BiFeO_3$ powders annealed at various temperatures. At room temperature, the samples exhibit hysteresis loop typical of magnetic behaviors, indicating that the presence of an ordered magnetic structure can exist in the mixed spinel-perovskite system. The sample annealed at 400 °C for 2 h exhibits rather low coercivity, H_c, of 229 Oe and the higher saturation magnetization, M_s, of 20.77 emu/ g, obtained by extrapolated value at 10 kOe. Low coercivity is due to disordered crystal site orientation of the ions in the spinel structure. After annealing at 700 °C, ordering takes place that in return results in lowering of magnetic moments leading to lower M_s value and higher coercivity, as shown in Fig. 11 B.

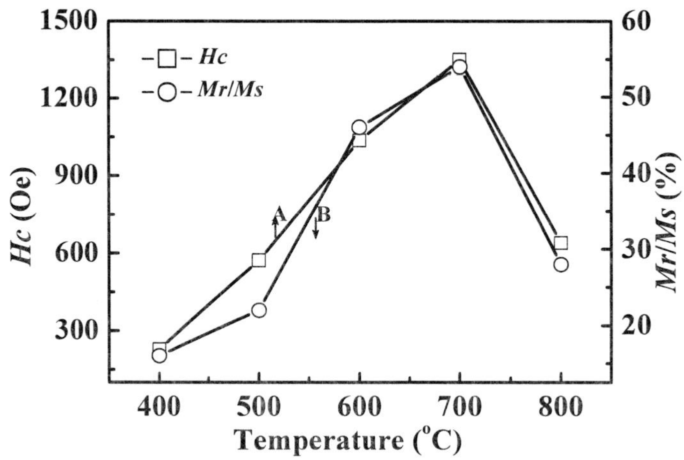

Figure 12. Plots of coercivity (H_c) (A) and squareness (M_r / M_s) (B) of $0.3CoFe_2O_4$—$0.7BiFeO_3$ nanocomposites as a function of the annealing temperature. Reproduced from Ref. [24] with permission from copyright owners.

Fig. 12 (A) and (B) show plots of coercivity (H_c) and squareness (M_r/M_s) of $CoFe_2O_4/BiFeO_3$ nanocomposites as a function of the annealing temperature ranging from 400 to 800 °C, respectively. Data are obtained for the samples annealed in 100 °C increments. The increase in coercivity is observed for the samples annealed at 400 °C and higher. This can be expected due to the rise in the size of the particles. However, annealing at 700 °C causes H_c to grow anomalously large, but decreases upon further increase in annealing temperature to 800 °C. If the increase in H_c was due only to the increase in nanocrystallite size and blocking temperature, H_c would not decrease upon annealing to higher temperature, which results in additional nanocrystal growth. One explanation for the observed excessive increase in H_c is an increase in the nanocrystallize average aspect ratio upon raising the annealing temperature from 400 to 800 °C as indicated by TEM observations. Very small increase in the aspect ratio can lead to large changes in coercivity. The decline in H_c upon higher temperature annealing can be the result of additional growth of the nanocrystals due to the increased atomic mobility at higher annealing temperatures, resulting in an overall decrease in the aspect ratio, giving rise to a "rounding" of the particles. Alternatively, one cannot rule out the possibility that magnetization pinning defects are incorporated into the nanocrystallites upon onset of significant crystallite growth at 700 °C, perhaps due to lattice mismatch between the intergrowing nanocrystallites. Annealing at higher temperature may reduce these defects and result in the decrease in H_c observed. The relationship between squareness (M_r/M_s) and the annealing temperature is shown in Fig. 12(B). The effect of annealing temperature on the squareness is similar to that of coercivity. It can be seen that the largest squareness of 54% is observed in the sample annealed at 700 °C, which indicates that the higher coercivity can be obtained under certain conditions.

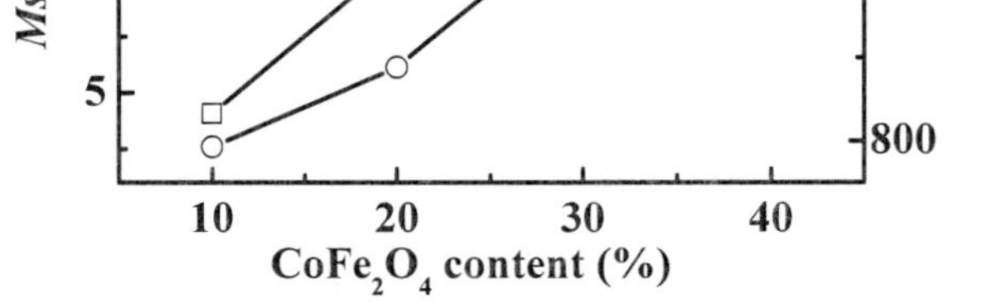

Figure 13. Plots of saturation magnetization (M_s) (A) and coercivity (H_c) (B) of $xCoFe_2O_4/(1-x)BiFeO_3$ (x=0.1, 0.2, 0.3, 0.4) nanocomposites as a function of the $CoFe_2O_4$ content. Reproduced from Ref. [24] with permission from copyright owners.

Magnetic properties of the spinel-perovskite $xCoFe_2O_4/(1-x)BiFeO_3$ (x=0.1, 0.2, 0.3, 0.4) annealed at 700 °C can be obtained at room temperature. All samples have exhibited strong magnetic characteristics. Fig. 13(A) and (B) shows the plots of saturation magnetization (M_s) and coercivity (H_c) of the nanocomposites as a function of $CoFe_2O_4$ concentration, respectively. It is found that the saturation magnetization and coercivity increase with the increase of $CoFe_2O_4$ concentration in the nanocomposites. The highest coercivity of 1847 Oe and the saturation magnetization of 23.4 emu/g are observed in the $0.4CoFe_2O_4$—$0.6BiFeO_3$ nanocomposite among the investigated compositions. The saturation magnetization and coercivity increase slightly when the $CoFe_2O_4$ concentration ranges from 10% to 20%. However, coercivity *Hc* increases sharply when the cobalt ferrite concentration is above 20%. This result may arise primarily from the magnetoelastic coupling [19d].

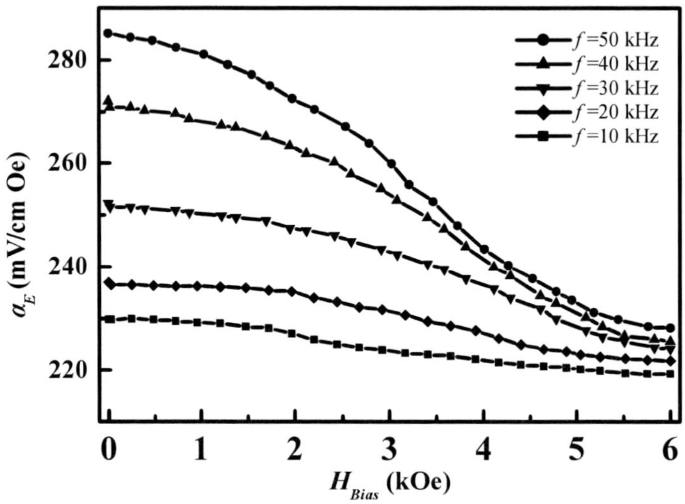

Figure 14. Variations of α_E with H_{Bias} at various magnetic frequency f for the $0.3CoFe_2O_4$—$0.7BiFeO_3$ nanocomposites. Reproduced from Ref. [24] with permission from copyright owners.

The coexistence of the ferromagnetic $CoFe_2O_4$ and ferroelectric $BiFeO_3$ phases in the present composite gives rise to a ME effect, which is characterized by the magnetoelectric voltage coefficient $\alpha_E = dE/dH$. The variations of α_E with H_{Bias} at various magnetic frequency f are shown in Fig. 14. At $H_{Bias} = 0$, the film exhibits an initial high α_E values (e.g., 230 mV/cm Oe at f=10 kHz). It can be observed that α_E is dependent on H_{Bias} with increasing H_{Bias} from zero to 5 kOe and the little variations of α_E with H_{Bias} has occurred above 5 kOe. In addition, an overall α_E increase with increasing f from 10 to 50 kHz is observed. The maximum α_E value of 285 mV/cm Oe is obtained at f=50 kHz. The reason that causes the ME effect of the present nanocomposites will now be discussed. In the magnetostrictive-ferroelectric composite system, ME coupling mainly arises from the magnetic–mechanical–electric transform through the stress-mediated transfer in the interface. The dynamic magnetoelastic coupling, which is caused by the magnetostriction of the $CoFe_2O_4$ phase, due to the domain-wall motion and domain rotation, is involved in the ME effect. Therefore, the low coercivity and high magnetostriction are advantageous for a strong ME effect [22, 25]. With respect to the variation of α_E with f, it is most likely due to the frequency dependence of the dielectric constant for the individual phases and the piezoelectric coefficient for $BiFeO_3$ [25].

In conclusion, the novel mixed spinel-perovskite structured nanocomposites of $xCoFe_2O_4$—$(1-x)BiFeO_3$ (x=0.1, 0.2, 0.3, 0.4) can be synthesized by thermal decomposition of the polymeric gel at various temperatures ranging from 400 to 800 °C. The crystallite size varies within the range of 20–80 nm. The saturation magnetization, squareness and coercivity values have been found to vary with the composition and annealing temperature due to the crystallite growth. It has been shown that optimal magnetic properties, e.g. saturation magnetization and coercivity of nagneticelectric composites can be obtained after being annealed at 700 °C. Moreover, the saturation magnetization and coercivity increase with the increase of $CoFe_2O_4$ concentration in the nanocomposites. Magnetoelectric measurements have shown that the the nanocomposites exhibit an ME effect and its ME effect strongly depends on the magnetic bias and magnetic field frequency.

4. Mass Synthesis of Spinel Ferrite Nanoparticles by Pyrolysis-Polymer Method

Large-scale applications of nanosized ferrites and tailoring of magnetic properties have prompted the development of several methods for the mass fabrication of stoichiometric and chemically pure spinel ferrite nanoparticles. Very recently, the precursor route is one of the most used ones, owing to its economics and the high degree of compositional control. In addition, the precursor synthesis requires neither extremely high processing temperature nor sophisticated processing. The above precursor method is an economical way to produce ultrafine powders. However, coarse particles may form after the decomposition process due to agglomeration. To avoid compromising the purity and properties of spinel ferrite or agglomeration, it would be desirable to prepare them from a single solid precursor which can be prepared in a pure state in which the M^{2+} and Fe^{3+} cations are uniformly distributed on an atomic level. Here, large-scale synthesis of nanocrystalline ferrites by in-situ polymerization of the mixed aqueous solution of acrylic acid in the presence of metal salts followed by a polymer-pyrolysis route is introduced. This method has the advantage that it can be readily scaled up in the form of a batch process producing highly homogeneous nanocrystalline cobalt ferrite with excellent magnetic performance [26].

In a typical experiment, nanocrystalline spinel ferrites are prepared by a polymer-pyrolysis method using polyacrylate of Me (Co, Ni, Zn or Mn) and Fe as precursor compounds. The simultaneously polymeric precursors are made by in-situ polymerization of the mixed aqueous solution of acrylic acid in the presence of metal salt and $Fe(NO_3)_3{\cdot}9H_2O$, with $(NH_4)_2S_2O_8$ as the initiator. First, ferric nitrate and target metal salt are dissolved in 10 g of acrylic acid aqueous solution (acrylic acid: H_2O = 70:30 wt %) under stirring. The molar fraction of Me/Fe is fixed at 1:2. Afterwards, a small amount (0.5 g) of 5% $(NH_4)_2S_2O_8$ aqueous solution as the initiator is added to the mixed acrylic acid solution to promote the polymerization. Under heating at 70-90 °C for 2 h, the mixed solution is dried to form the well-distributed polyacrylate salt. The Co-Fe polyacrylates are dried at 100 °C for 24 h, and calcined at different temperatures (such as 500, 700, 900 °C) for 3 h in air, then the final products are obtained after being slowly cooled to room temperature. Other polyacrylates are calcined at 500 °C for 3 h in air, then the final ferrites are obtained after being slowly cooled to room temperature.

As the first step of the ferrite synthesis, an aqueous solution of acrylic acid is mixed with the appropriate amount of Me and Fe salts. Ammonium persulfate ($(NH_4)_2S_2O_8$) is used as the initiator of polymerization reaction. This affords a composition which has the predetermined ratio of Me to Fe to be compatible with the final product ($MeFe_2O_4$) according to the following reactive formula:

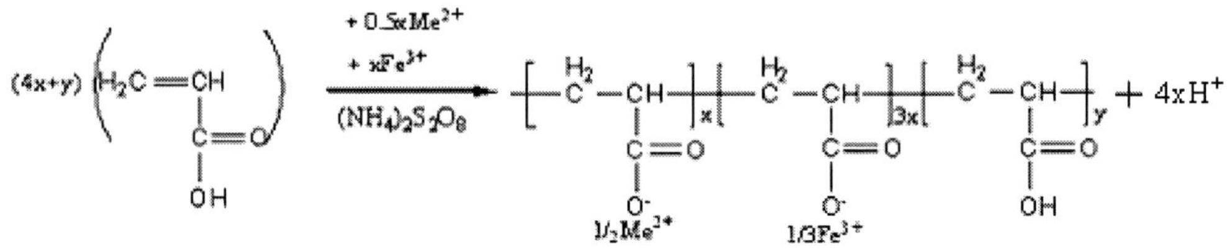

(1)

In the reactive formula, x and y represent the polymerization degrees, which are presumably distributed randomly along the polymer chain. It is preferable to chose x<<y in order to avoid the spontaneous preparation of ill-defined products. To avoid compromising the purity and properties of spinel ferrite and related materials it is desirable to prepare them from a single solid precursor which can be prepared in a pure state in which the Me^{2+} and Fe^{3+} cations are uniformly distributed on an atomic level. The co-polymeric precursor compounds are schematically represented in Fig. 15. As is shown, Me (II) and Fe (III) ions are bound by the strong ionic bonds between the metallic ions and carboxylate ions in a polymeric chain or between the polymeric chains. This uniform immobilization of metallic ions in the polymer chains favors the formation of uniformly distributed solid solution of the metallic oxides in the following pyrolysis process.

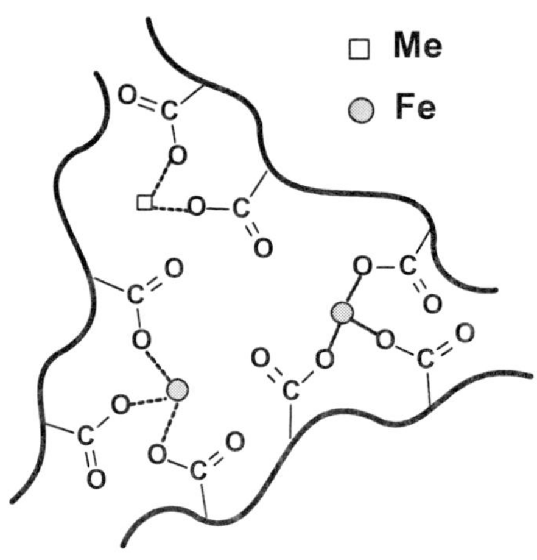

Figure 15. Schematic of polymeric chain for the copolymeric precursor of polyacrylates. Reproduced from Ref. [26] with permission from copyright owners.

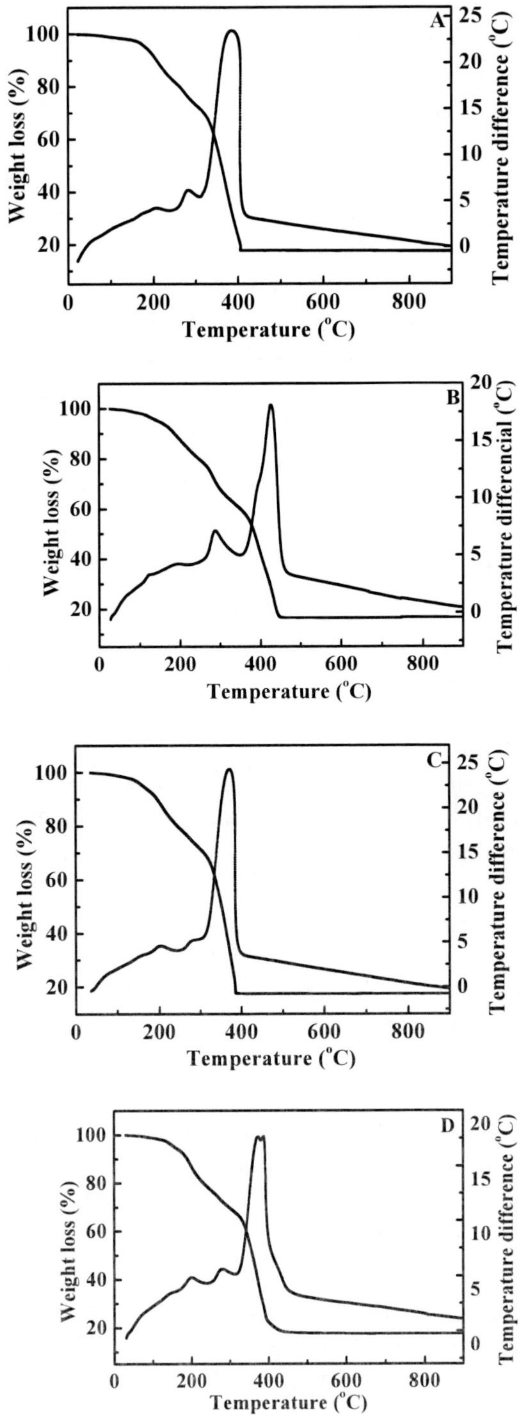

Figure 16. TG-DTA curves of metal ferrite precursors: A) $CoFe_2O_4$; B) $MnFe_2O_4$; C) $NiFe_2O_4$; D) $ZnFe_2O_4$. Reproduced from Ref. [26] with permission from copyright owners.

The copolymeric precursor compound is schematically represented in Fig. 15. As it is shown, the Me (□) and iron (○) ions are bound by the strong ionic bonds between the metallic ions and carboxylate ions in a polymeric chain or between the polymeric chains. This uniform immobilization of metallic ions in the polymer chains favors the formation of uniformly distributed solid solution of the metallic oxides in the following pyrolysis process.

To clarify the chemical reactions of the co-polymeric precursors occurring in the pyrolysis process, TG and DTA curves of the co-polymeric Me-Fe precursors are shown in Fig. 16. In the TG-DTA curves of Co ferrite precursors (Fig. 16A), there are two weight-losing processes present at RT-305 and 305-428 °C with total weight loss of 82.3%. The first step of weight loss is ca. 28.1% of total weight, corresponding to a small exothermic peak in the DTA curve, possibly due to the decomposition of nitrate. In the second step of weight loss, the weight loss is ca. 56.3% and correspondingly, a high exothermic peak in the DTA curve is observed in the temperature range of 305-428 °C, which is likely to arise from the decomposition of Co ferrite precursors into oxides. Similarly, in the thermal data of Mn ferrite precursors (Fig. 16B), two exothermic peaks are observed to appear at 242-350 and 350-454 °C, respectively, corresponding to two steps of weight loss. The first step shows a weight loss of 39.9% and the latter has a weight loss of 43.4%. Fig. 16C shows the TG and DTA curves of Ni ferrite precursors. The main feature in the DTA curve is a big exothermic peak at 371 °C, corresponding to a weight loss reaction between 246 and 408 °C, while a small exothermic peak appears at 278 °C. The total weight loss is 82.2%, possibly due to the decomposition of polyacrylates to form oxides. The TG and DTA curves of the copolymeric Zn-Fe precursor are shown in Fig. 16D. The total weight loss is 82.3% and it occurs in two steps up to 446 °C. In the first step of the weight loss (RT-316 °C), a 32.9% weight loss associated with an exothermic peak at 279 °C may be attributed to the decomposition of nitrate. In the second step (316-446 °C), the exothermic peaks at 387 °C may be related to the decomposition of Zn-Fe precursor.

The observed differences in the decomposition temperatures for different metal ferrite precursors are due to the differences in the strength of the interaction between metallic cation and the polymeric chain. It is noted that the weight of different metal ferrite precursors is almost unchanged above the decomposition temperature, indicating that the final decomposed products are ferrites.

Fig. 17 show the XRD patterns of cobalt ferrite powders prepared by the polymer-pyrolysis route at deferent temperatures. The diffraction patterns and relative intensities of all diffraction peaks match well those of JCPDS card 22-1086 for $CoFe_2O_4$. The peaks appear at around 18.5, 30.2, 35.6, 37.2, 43.0, 53.4, 57.1, and 62.5° which are well indexed to the crystal plane of spinel ferrite (111), (220), (311), (222), (400), (422), (511), and (440), respectively. The peak intensity of $CoFe_2O_4$ increases with the increase of calcination temperature, which indicates that the calcination temperature plays a role in the formation of spinel crystal structure and size. All the ferrites show a single-phase spinel ferrite with no impurities in the XRD patterns. The full-width at half-maximum (FWHM) of the peak (311) decreases with the increase of the calcination temperature. The average size of the crystallites can be estimated from the FWHM of the peak (311) using the Scherrer equation, and the results are listed in Table 2. In addition, the elemental analyses of these nanoparticles using ICP-AES showed that the molar ratios of Co^{2+}/Fe^{3+} are close to 0.5.

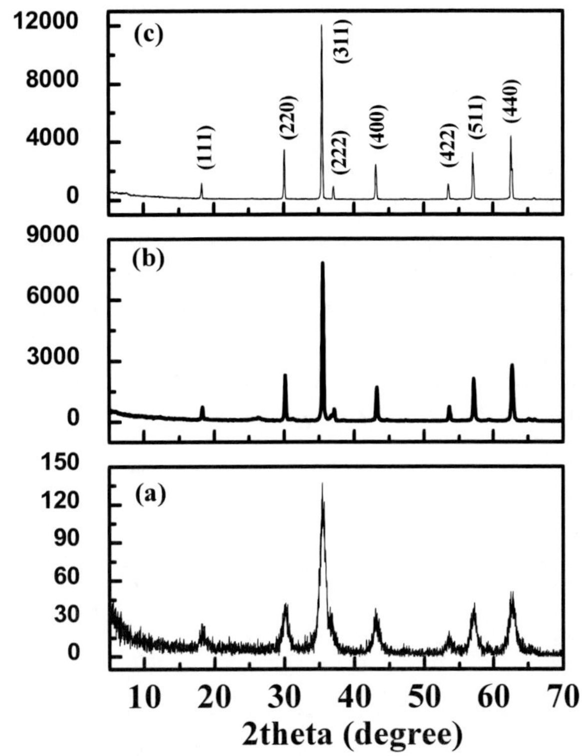

Figure 17. XRD patterns of $CoFe_2O_4$ calcined at different temperatures: 500; (b) 700; (c) 900 °C. Reproduced from Ref. [26a] with permission from copyright owners.

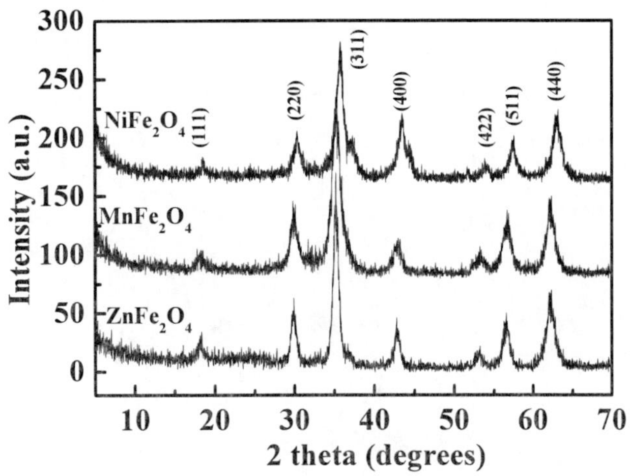

Figure 18. XRD patterns of $MeFe_2O_4$(Me=Ni, Mn, Zn) calcined at 500 °C. Reproduced from Ref. [26b] with permission from copyright owners.

Fig. 18 shows the XRD patterns of metal ferrite powders prepared by the polymer-pyrolysis route. The diffraction patterns and relative intensities of all diffraction peaks match well those of JCPDS card 10-0325 for $NiFe_2O_4$, 10-0319 for $MnFe_2O_4$ and 22-1012 for $ZnFe_2O_4$, respectively. The peaks appear at around 18.5, 30.2, 35.6, 43.0, 53.4, 57.1, and 62.5° which are well indexed to the crystal plane of spinel ferrite (111), (220), (311), (400), (422), (511), and (440), respectively. Further careful observation can find that a shoulder peak appears at 37.2° corresponding to the crystal plane (222) of spinel ferrite except $ZnFe_2O_4$. All the ferrites show a single-phase spinel ferrite with no impurities in the XRD patterns. The crystallite size can be evaluated from the major diffraction peak (311) using the well-known Scherrer's formula. The size of nanocrystalline $NiFe_2O_4$, $MnFe_2O_4$ and $ZnFe_2O_4$ are found to be about 21, 9 and 19 nm, respectively. In addition, the elemental analyses of these nanoparticles using ICP-AES show that the molar ratios of Me^{2+}/Fe^{3+} are close to 0.5.

Table 2. Properties of ferrite nanoparticles derived from XRD, TEM and magnetic measurements

Ferrites	Temperature(℃)	d_{XRD} (nm)	d_{TEM} (nm)	M_s (emu/g)	M_r (emu/g)	H_c (Oe)
$CoFe_2O_4$	500	22	25	77.4	34.8	505
$CoFe_2O_4$	700	74	80	69.3	35.6	1445
$CoFe_2O_4$	900	118	120	70.8	17.7	453
$NiFe_2O_4$	500	21	25	52.2	2.0	121
$MnFe_2O_4$	500	9	8	/	/	/
$ZnFe_2O_4$	500	19	22	/	/	/

The detailed microstructures of the products are characterized by transmission electron microscopy (TEM). Fig. 19a-c show the bright field TEM images of $CoFe_2O_4$ powders. It can be seen that the particles have nanometer-scale morphology at relative low calcined temperature and are clearly dispersive apart from each other. And their average particle sizes are 25, 80 and 120 nm for $CoFe_2O_4$ calcined at 500, 700 and 900 °C, respectively. Moreover, the solid materials obtained show a very narrow particle size distribution for cobalt ferrite.

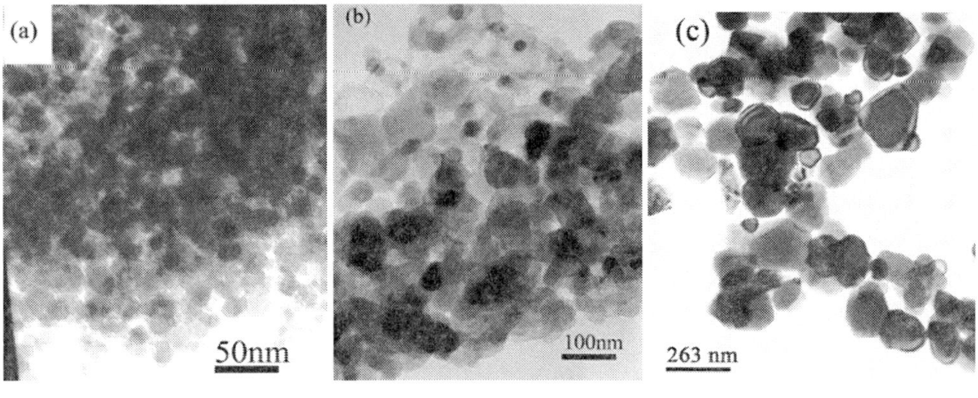

Figure 19. TEM photographs of $CoFe_2O_4$ calcined at different temperatures: (a) 500; (b) 700; (c) 900°C. Reproduced from Ref. [26a] with permission from copyright owners.

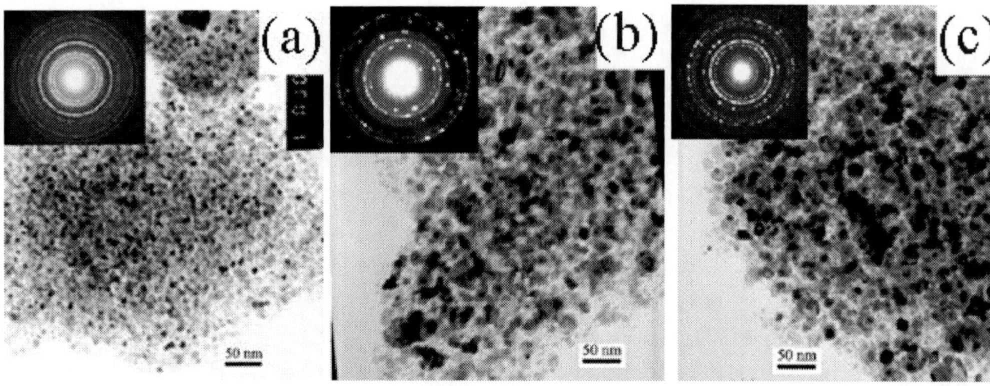

Figure 20. TEM images and ED patterns of ferrites: (a) $MnFe_2O_4$; (b) $NiFe_2O_4$; (c) $ZnFe_2O_4$. Reproduced from Ref. [26b] with permission from copyright owners.

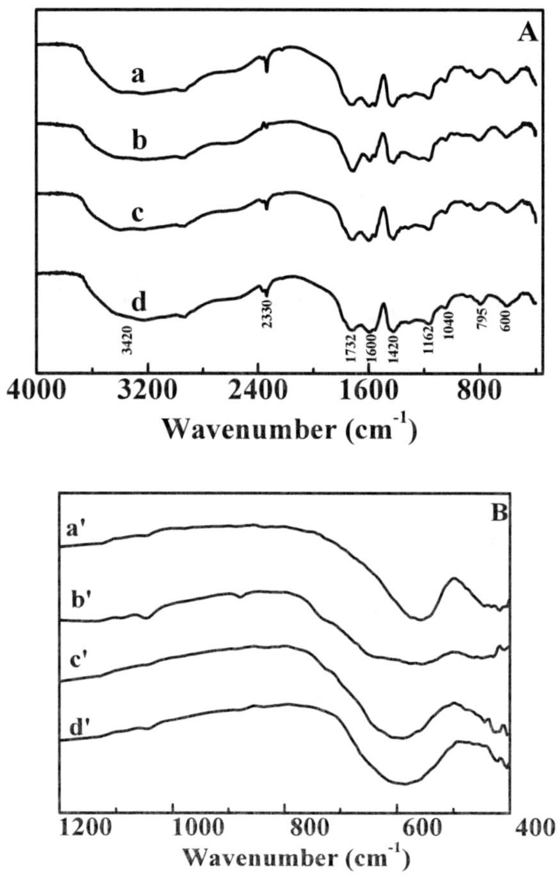

Figure 21. IR spectra of the samples: A: (a) Zn ferrite precursors; (b) Mn ferrite precursors; (c) Ni ferrite precursors; (d) Co ferrite precursors; B: (a′) $ZnFe_2O_4$; (b′) $MnFe_2O_4$; (c′) $NiFe_2O_4$; (d′) $CoFe_2O_4$. Reproduced from Ref. [26] with permission from copyright owners.

Fig. 20a-c show the bright field TEM images of $MnFe_2O_4$, $NiFe_2O_4$ and $ZnFe_2O_4$ ferrite powders. It can be seen that the particles have nanometer-scale morphology and are dispersive apart from each other. Moreover, the average size of $MnFe_2O_4$ is the smallest. The SAED pattern shown in the up-left of each of pictures should be indexed to further confirm that the structure is spinel-type.

This method used in this work is to synthesize the copolymeric precursor of the mixed metallic ions and then to pyrolyze the precursor into the required oxide structure. The IR spectra of the precursor compound of copolymerized Me-Fe polyacrylates are shown in Fig. 21A. It is found that the C=O stretching vibration (~1702 cm^{-1}) of α and β- unsaturated carboxylic acid, the C=C stretching vibration (~1640 cm^{-1}) and the =C-H vibrations (~1045 and ~984 cm^{-1}) of acrylic acid, characteristic of the vinylidene group in the acrylic monomers, are absent, suggesting that the polymerization of the acrylates took place in the acrylates of Me and Fe. Furthermore, the spectra show the appearance of the symmetric (~1618 cm^{-1}) and asymmetric (~1451 cm^{-1}) stretching bands of carboxylate salts. Compared with the standard spectrum of acrylic acid, these two bands are positively shifted, implying an associating interaction between the metallic ions and carboxylate ions. In addition, the IR bands for H_2O (~3428 cm^{-1}) and NO_3^- vibration (~1384 and ~839 cm^{-1}) are also observable due to the residual H_2O and NO_3^-, while characteristic bands of other anions with small quantity are overlapped. Fig. 21B exhibits IR spectra of spinel ferrites obtained by pyrolysis of Me-Fe polyacrylates at 500 °C. It can be seen from Fig. 21B that the absorption bands around 588 and 441 cm^{-1} are assigned to Fe–O stretching vibration mode, indicating that the pyrolysis products are inorganic metal oxide species. These results are in good agreement with the analysis of XRD that the oxides consist of spinel ferrites.

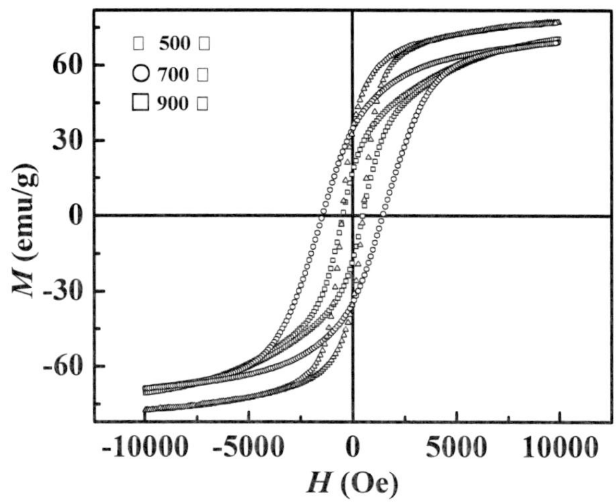

Figure 22. Hysteresis loops of the as-calcined $CoFe_2O_4$ at room temperature. Reproduced from Ref. [26a] with permission from copyright owners.

The hysteresis loops of $CoFe_2O_4$ nanoparticles are shown in Fig. 22. At room temperature, the samples exhibit hysteresis loop typical of magnetic behaviors, indicating that the presence of an ordered magnetic structure can exist in the mixed spinel system. Magnetization data are listed in Table 2. The magnetizations of the samples are much higher and tend to be saturated at a high field. After annealing at 700 or 900 °C, ordering takes place that has resulted in lowering of magnetic moments leading to lower M_s value and higher coercivity. The increase in coercivity is observed for the sample b annealed at 700 °C. This can be expected due to the rise in the size of the particles. However, annealing at 700 °C causes H_c to grow anomalously large, but decreases upon further increase in annealing temperature to 900 °C. The increase in H_c is not only due to the increase in nanocrystallite size. Otherwise, H_c should not decrease upon annealing at higher temperature, resulting in additional nanocrystal growth. One explanation for the observed excessive increase in H_c is an increase in the nanocrystallize average aspect ratio upon raising the annealing temperature as indicated by TEM observations. Very small increase in the aspect ratio can lead to large changes in coercivity. The decline in H_c upon higher temperature annealing could be the result of additional growth of the nanocrystals due to the increased atomic mobility at higher annealing temperatures, resulting in an overall decrease in the aspect ratio, giving rise to a "rounding" of the particles. Alternatively, one cannot rule out the possibility that magnetization pinning defects are incorporated into the nanocrystallites upon onset of significant crystallite growth at 700 °C, perhaps due to lattice mismatch between the intergrowing nanocrystallites. Annealing at higher temperature reduces these defects and result in the decrease in H_c observed. The effect of annealing temperature on the remanence (M_r) is similar to that of coercivity. It can be seen that the highest remanence (M_r) is observed in the sample annealed at 700 °C, indicating that the highest coercivity can be obtained under certain conditions.

Fig. 23 shows the hysteresis of $MeFe_2O_4$(Me=Ni, Mn, Zn) nanoparticles calcined at 500°C. At low external field, the hysteresis loops of Mn, Ni and Zn ferrites exhibit very small coercivity and remanence. The saturation magnetization is evaluated by extrapolated value of 10 kOe. The saturation magnetization depends on the divalent metal ions and ranges from 48.6 emu/g for $MnFe_2O_4$ and 50.6 emu/g for $NiFe_2O_4$ at the applied magnetic field of 10 kOe. However, it can also be seen from Fig. 23 that the magnetization of zinc ferrites increases almost linearly with the given field and shows a lack of saturation at a field as high as 10 kOe.

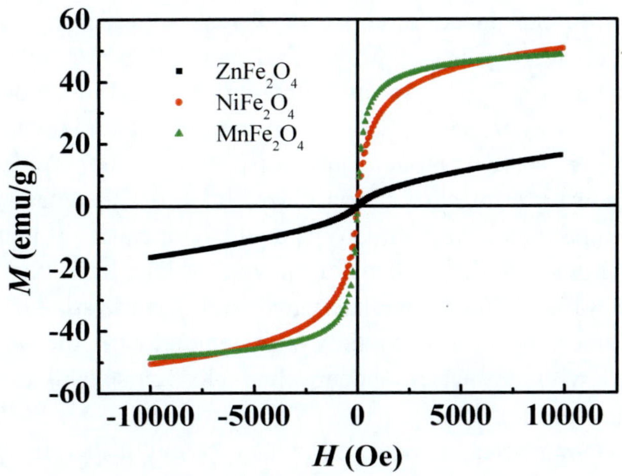

Figure 23. Hysteresis loops for $MeFe_2O_4$(Me=Ni, Mn, Zn) nanoparticles calcined at 500 °C. Reproduced from Ref. [26b] with permission from copyright owners.

The polymer-pyrolysis route possesses several advantages as a precursor to form spinel ferrites that lead to enhanced saturation magnetization. Use of a molecular precursor with cations randomly distributed with no long order facilitates the synthesis of a homogeneous spinel phase. Furthermore, the spinel is produced from a single solid precursor rather than a mixture means that the calcination process requires a much shorter time and lower temperature, leading to a lower chance of side reactions to occur.

In summary, a simple and novel route has been introduced for producing highly homogeneous nanocrystalline cobalt ferrites by the pyrolysis of Co-Fe copolymeric precursors in situ polymerized by reaction of the metal salt and acrylic acid. TG-DTA measurements show that the pyrolysis process can occur at a low temperature to form spinel ferrites. The results from the XRD and TEM measurements reveal that the ferrite powders have good crystallinity with the spinel structure.

5. SYNTHESIS AND CHARACTERIZATION OF SHUTTLE-LIKE A-FE_2O_3 NANOPARTICLES BY SUPERMOLECULAR TEMPLATE METHOD

Hematite has a rhombohedrally centered hexagonal structure of the corundum type with a close-packed oxygen lattice in which two-thirds of the octahedral sites are occupied by Fe (□) ions. Additionally, bulk α-Fe_2O_3 is an antiferromagnetic material with the Néel temperature (950 K) and the Morin transition temperature (263 K). The unusual magnetic behavior of α-Fe_2O_3 has recently been studied extensively [27]. As stimulated by both the promising applications of iron oxide and the novel properties of nano-scaled materials, synthesis of α-Fe_2O_3 with nanometer dimension is the subject of considerable topical importance. Nanostructured hematite [28], such as shuttle-like, nanowire, nanorod and nanotube has been reported. These artificial structures can be manipulated to achieve tailored materials for exploration of physical phenomena as the magnetic properties of this class of materials should differ from those of both individual nanoparticles and bulk materials. Herein, supermolecular template method has been introduced for preparing nanostructured iron oxides.

In a typical experiment, 1 g PEG, 0.36 g CTAB and 1.0 g $FeCl_3 \cdot 6H_2O$ are dissolved in 100 ml de-ionized water in a flask at room temperature. A yellow color is observed in the solution after strongly stirring for half an hour. A 30-min ultrasound treatment is necessary to ensure that all the reagents are homogeneously dispersed in the solution. The solution is stirred at ambient temperature for several hours, which is advantageous to the interaction of ferric ions within the supermolecular template. Then the temperature is increased to 90-100°C and 20 ml of 28% ammonia solution is added drop wise into the ferric and polymer-surfactant solution. The solution turns yellow immediately and is allowed to stir for an additional 2 h at this temperature. After being cooled to the room temperature, the yellow precipitate is collected and washed with hot water and absolute alcohol several times to remove the remaining CTAB, PEG and other impurities. It is then dried in air 80°C for 8 h. Finally, shuttle-like α-Fe_2O_3 nanoparticles are obtained by calcination of the precipitate at 500 °C for 4 h. For the purpose of comparison, spherical α-Fe_2O_3 nanoparticles without the supermolecular templates are also prepared under the same conditions.

A schematic diagram of the proposed growth mechanism is shown in Fig. 24.The formation of shuttle-like α-Fe_2O_3 nanoparticles involves four primary steps: formation of supermolecular template, nucleation of inner FeOOH nanostructures, washing shuttle-like β-FeOOH nanoparticles and calcination of shuttle-like β-FeOOH to obtain shuttle-like α-Fe_2O_3 nanocrystalline particles. Initially, $FeCl_3{\cdot}6H_2O$, CTAB and PEG are dissolved in deionized water and result in a clear and homogeneous solution. Since there exist interactions between water-soluble polymer and cationic surfactant [18], the surfactants associating with the polymer chains form micellar aggregates at a rather low concentration.

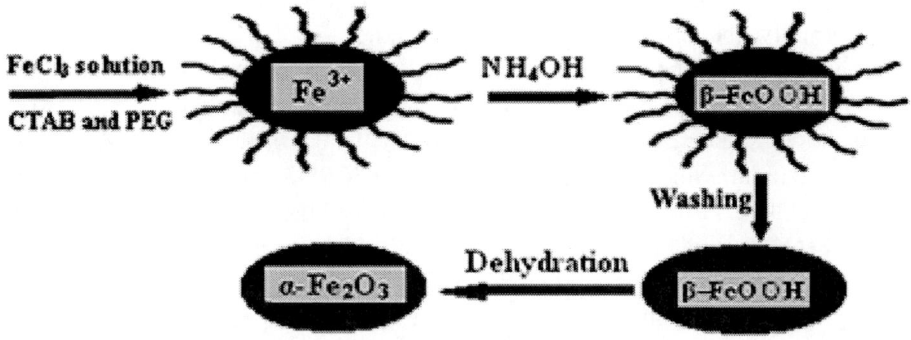

Figure 24. Schematic of the supermolecular-template preparation route of shuttle-like α-Fe_2O_3 . Reproduced from Ref. [28h] with permission from copyright owners.

Thus the supermolecular aggregates form in the solution with the coils of PEG chains as the core, around which are the rod-shaped CTAB micelles. Ferric ions react with the supermolecular template under ambient condition. As the solution is heated, ammonia is added and the FeOOH nanostructures nucleate under the control of this template. Then, shuttle-like β-FeOOH nanoparticles are collected and washed to remove the remaining CTAB, PEG and other impurities. Finally, shuttle-like α-Fe_2O_3 nanoparticles can be obtained by calcination of shuttle-like β-FeOOH nanoparticles at 500 °C for 4 h.

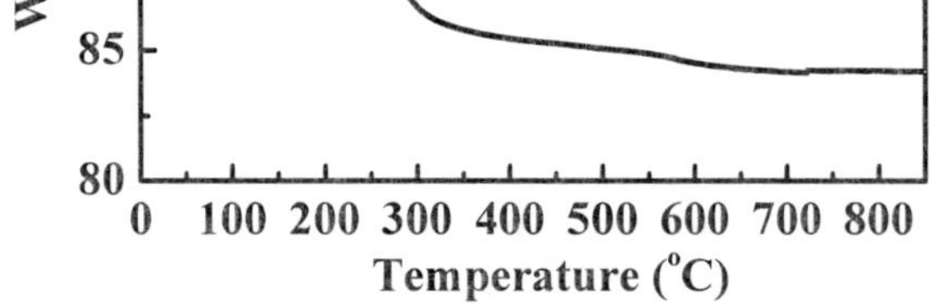

Figure 25. TG curve of shuttle-like β-FeOOH. Reproduced from Ref. [28h] with permission from copyright owners.

The dehydration process of synthetic β-FeOOH is carried out from 30 to 900 °C. The mass decrease of the samples (Fig. 25) is derived from the dehydration of nanostructured β-FeOOH powders. During the whole dehydration process, the total weight loss is 15.30% which is larger than the theoretical value (10.1%), indicating ~5.2% excess hydroxyl units are present in the synthetic β-FeOOH. The powders exhibit thermogravimetric transitions that are likely due to the loss of physisorbed and structural water. The initial weight loss from 30 to 120 °C is attributed to the loss of surface adsorbed water and ethanol. The weight loss in the range of 120—325 °C is due to the removal of the crystalline water molecules, which maintains normally the channels of akaganeite, and the dehydroxylation of the sample akaganeite. After 325 °C, the weight loss continues but gradually slowes at 400 °C and almost ceases at 500 °C. As a consequence, the stable residue can reasonably be ascribed to α-Fe_2O_3.

Fig. 26 shows XRD patterns of the samples. The peaks of the as-prepared FeOOH (Fig. 26A) are in agreement with the theoretical data of the JCPDS card no.34-1266, indicating the presence of β-FeOOH. However, some impure peaks attributed to α-Fe_2O_3 appear in the XRD pattern of β-FeOOH, indicating that a very small amount of α-Fe_2O_3 impurities may be yielded during the drying process. When β-FeOOH powders are calcined at 500 °C for 2 or 4 h (Fig. 26B), all peaks are in agreement with the theoretical data of hematite (JCPDS card no.33-0664). It indicates that β-FeOOH powders have completely transformed into hematite. At the moment, the color of samples changes from yellow to red and the XRD peaks become narrow and sharp. When the thermal transformation is nearly completed and the crystallinity of hematite is improved at 500 °C, the intensity of the (104) peak is larger than that of (110) peak, which corresponds to the standard XRD pattern of hematite. In addition, the intensity of the peaks increases with the increase of the time.

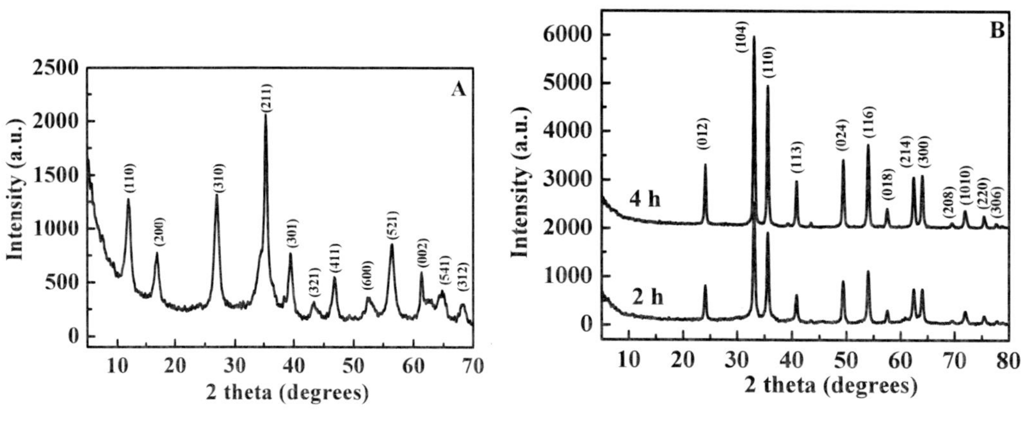

Figure 26. XRD patterns of the samples: A) shuttle-like β-FeOOH; B) shuttle-like α-Fe_2O_3 calcined at 500 °C for 2 and 4 h. Reproduced from Ref. [28h] with permission from copyright owners.

Fig. 27 shows the morphology of the samples. The morphology of β-FeOOH appears shuttle-like, while α-Fe_2O_3 calcined at 500 °C for 4 h also contains largely shuttle-like nanostructures with an average length of 120 nm and diameter of about 50 nm in the middle part. The results indicate that the existence of the polymer-surfactant can effectively affect the size and shape of ferric oxide nanocrystallites with certain aspect ratio. To further reveal the fine structure of the nanostructured α-Fe_2O_3 shown in Fig. 27C, ED analysis is also carried out. It can be concluded that the shuttle-like materials consist of polycrystals of ferric oxide. It is interesting that some mesopores are developed inside the shuttle-like β-FeOOH nanocrystals (Fig. 27A). When the temperature has reached to 500 °C, the coalescence of the pores and the recrystallization process are still in progress. The size of hematite decreases a little but the shape of hematite almost remains the same as that of the shuttle-like β-FeOOH.

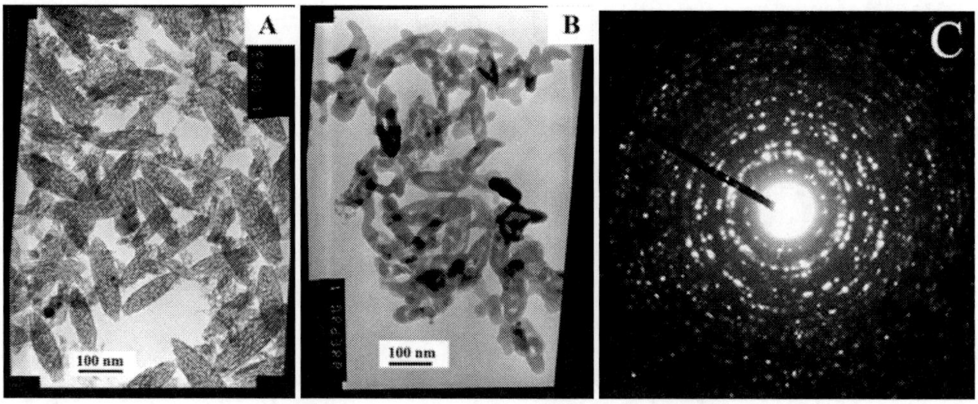

Figure 27. TEM images and ED pattern of the samples: A) shuttle-like β-FeOOH; B) shuttle-like α-Fe_2O_3 calcined at 500 □ for 4 h; C) ED pattern of shuttle-like α-Fe_2O_3 calcined at 500 □ for 4 h. Reproduced from Ref. [28h] with permission from copyright owners.

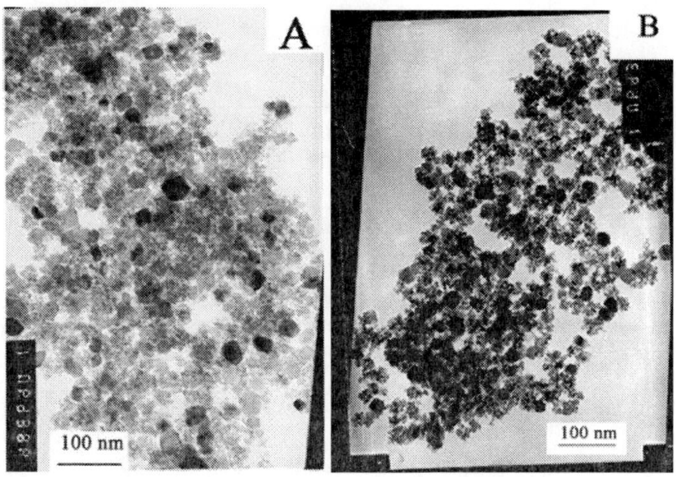

Figure 28. TEM images of the samples: A) spherical β-FeOOH; B) spherical α-Fe_2O_3 calcined at 500 □ for 4 h. Reproduced from Ref. [28h] with permission from copyright owners.

To understand well the effect of the polymer-surfactant on the formation feature of shuttle-like nanoparticles, a synthetic experiment without polymer-surfactant is carried out. It is observed that the spherical nanoparticles are formed; no shuttle-like materials that appear in the experiment with polymer-surfactant are observed. Fig. 28 shows the images of the spherical samples. The spherical α-Fe_2O_3 nanoparticles are 20-50 nm in diameter. The dipole magnetic interactions of these nanoparticles in the nanostructures are quite strong. As a rule, the samples must be sonicated continuously for more than half an hour to prepare a sample for TEM observation; however, the nanostructures have not been destroyed, indicating that the polycrystalline nanostructures are rather stable as a result of the soft-template process.

IR spectra of the samples in the range of 4000—400 cm^{-1} are shown in Fig. 29. There is a distinguished adsorption band at 3430 and 1650 cm^{-1}, except three absorbance bands located at 667, 588 and 441 cm^{-1} for β-FeOOH (Fig. 29 A). The bands at 3430 and 1650 cm^{-1} are attributed to the stretching vibration and bending vibration adsorption of OH^-, while three absorbance bands at 667, 588 and 441 cm^{-1} are assigned to the absorptions associated with the characteristic Fe-O stretching modes, suggesting that there is only β-FeOOH single phase existing in the system [22]. It gives further evidence of formation of α-Fe_2O_3 phase from Fig. 29B. After β-FeOOH was calcined at 500 °C for 2 h, the bands at 540 and 462 cm^{-1} belong to characteristic bands of polycrystal α-Fe_2O_3 in the IR spectrum [23]. And the intensity of these two peaks was in the order of 540 > 462 cm^{-1}. It indicates that α-Fe_2O_3 phase has been formed.

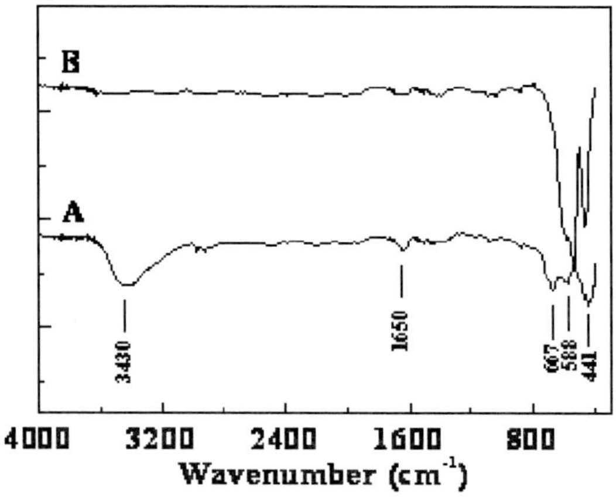

Figure 29. IR spectra of the samples: A) shuttle-like β-FeOOH; B) shuttle-like α-Fe_2O_3 calcined at 500 □for 4 h. Reproduced from Ref. [28h] with permission from copyright owners.

XPS is used to examine the oxidation state of the iron in the products, because core electron lines of ferric ions can be detected in XPS [29]. The results of Fe $2p_{3/2}$ and $2p_{1/2}$ electron binding energies are shown in Fig. 30, which are very similar to the values recorded in the literature [29]. The binding energies at 711.5 and 725.5 eV are the characteristic doublet from Fe $2p_{3/2}$ and Fe $2p_{1/2}$ core-level electrons, respectively. No other iron signal is detected in the XPS spectra.

Magnetic hysteresis measurements for α-Fe_2O_3 are carried out in applied magnetic field at room temperature, with the field sweeping from -10 to 10 kOe. The hysteresis loops of the samples (see Fig. 31) do not reach saturation up to the maximum applied magnetic field. As shown in Fig. 31, the morphology affects the magnetic properties of α-Fe_2O_3 nanoparticles. The magnetization measurements of the two type α-Fe_2O_3 exhibit hysteretic feature with the remanent magnetization and coercivity being determined to be 0.156 emu/g and 664 Oe, 0.048 emu/g and 110 Oe, respectively, suggesting that shuttle-like α-Fe_2O_3 nanoparticles are ferromagnetic while spherical α-Fe_2O_3 nanoparticles exhibit weak ferromagnetic behaviors at room temperature. It is found that the high remanent magnetization and coercivity may be associated with the aspect ratio of α-Fe_2O_3 because shape anisotropy would exert a tremendous influence on their magnetic properties. It is well known that the magnetization of ferromagnetic materials is very sensitive to the microstructure of a particular sample. Symmetrically shaped nanoparticles, such as spheres, do not have any net shape anisotropy. However, shuttle-like nanoparticles have shape anisotropy in addition to crystalline anisotropy, which will increase the coercivity.

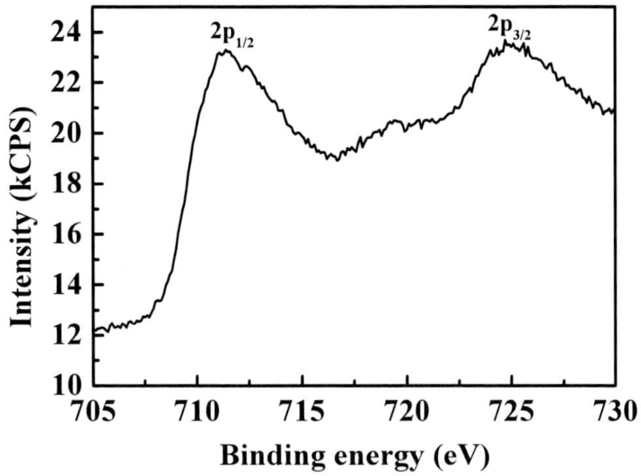

Figure 30. XPS spectrum of shuttle-like α-Fe_2O_3 calcined at 500 □ for 4 h . Reproduced from Ref. [28h] with permission from copyright owners.

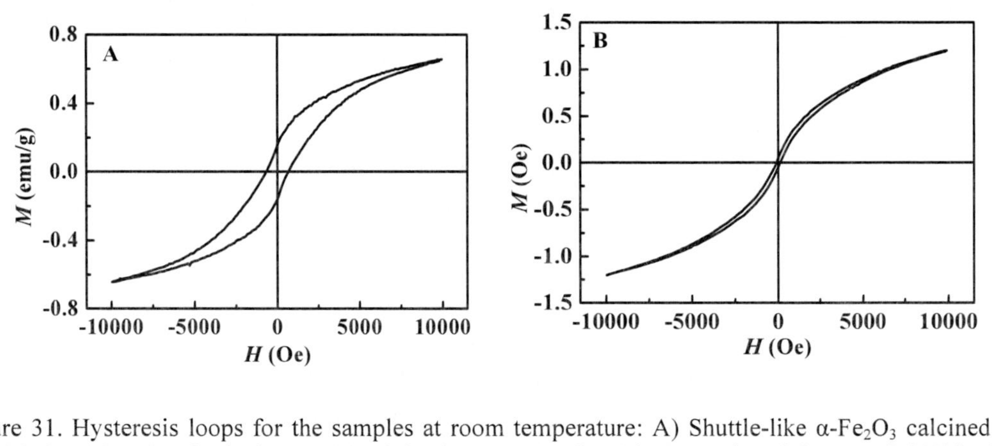

Figure 31. Hysteresis loops for the samples at room temperature: A) Shuttle-like α-Fe_2O_3 calcined at 500 °C for 4 h; B) spherical α-Fe_2O_3 calcined at 500 °C for 4 h. Reproduced from Ref. [28h] with permission from copyright owners.

This phenomenon can be explained as follows: since the shuttle-like α-Fe_2O_3 nanoparticles are polycrystalline which consist of large numbers of smaller subparticles inside them. The number of the subparticles increases and changes from the multidomain to single domain, which can result in the enhancement in coercivity. In addition, it is generally accepted that the change on the particle surface can result in the variation in magnetic properties. In this case, IR of β-FeOOH (Fig. 29A) clearly shows absorptions due to organic species. For example, around 2950 cm^{-1}, weak C—H absorption can be observed. Thus, it is presumed that the organic residues on the surface of α-Fe_2O_3 nanoparticles may result in the spin pinning owing to the increase of the magnetic surface anisotropy, and then influence the magnetic behaviors.

In conclusion, a new soft-template route has been introduced for the preparation of α-Fe_2O_3 nanostructures. The magnetization measurements of shuttle-like and spherical α-Fe_2O_3 exhibit hysteretic feature with the remanent magnetization and coercivity of 0.156 emu/g and 664 Oe, 0.048 emu/g and 110 Oe, respectively, suggesting that shuttle-like α-Fe_2O_3 nanoparticles show ferromagnetic behaviors while spherical α-Fe_2O_3 nanoparticles exhibit weak ferromagnetic behaviors at room temperature. It can be concluded that the supermolecular template significantly influences the size and shape of the α-Fe_2O_3 nanoparticles, and thus on their magnetic properties.

6. Synthesis and Characterization of Spherical NiO Nanoparticles by Hydrothermal Method

Nickel oxide (NiO), an important transition metal oxide with cubic lattice structure (a_0 = 4.195 Å), has been under extensive investigations for decades due to its interesting electronic structure, strongly affected by Ni 3d electrons which are localized in space but spread out over a wide energy range because of strong Coulomb repulsion between them. Magnetically, NiO is antiferromagnetic with the Néel temperature of 523 K and has the NaCl crystal structure. Recently, as an antiferromagnetic material and a charge-transfer insulator, fabrication of NiO nanocrystals has been attracting attention due to the technological applications using its electrical, magnetic, catalytic and optical properties [30]. NiO nanocrystals have been prepared by different methods [31], similar to the pyrolytic process in most occasions, by annealing tiny particles of hydroxide, nitrate, oxalate, carbonate to prepare NiO nanocrystals.

Spherical NiO nanocrystals are prepared by hydrothermal method using urea as precipitant and PEG as modifier. In a typical procedure, $Ni(NO_3)_2{\cdot}9H_2O$ (0.004 mol), $CO(NH_2)_2$ (0.016 mol) and PEG (0.4 g) are placed in a conical flask with 40 ml deionized water, stirred with a magnetic stirrer at room temperature until a transparent solution was obtained. The mixed solution is then transferred into a Teflon-lined stainless steel autoclave, sealed and maintained at 100 °C for 24 h. After the reaction is complete, the resulting light green solid product is filtered off, washed several times using distilled water and absolute ethanol, respectively, to remove the possible absorbed ions, and then dried at 80 °C in vacuum. The obtained green powders , the precursors of NiO are calcined in air at 300 and 500 °C, respectively, for 2 h to obtain NiO powders which color is absolutely dark. The calcination products are then collected for analysis.

The crystalline structure of NiO calcined at 300 °C and 500 °C is characterized using an X-ray diffractometer as shown in Fig. 32. The broad reflection peaks corresponding to pure cubic NiO appearing at 2θ =37.2, 43.2, 62.8, 75.2 and 79.4°can be readily indexed as (111), (200), (220), (311) and (222) crystal planes. Notably, when increasing calcination temperature to 500 °C, all the peaks belonging to NiO cubic phase are markedly sharpened with high intensity, suggesting that the crystallinity of NiO phase is higher for higher calcination temperature. All these diffraction peaks, including not only the peak positions but also their lattice parameters, can be perfectly indexed into the cubic crystalline structure of

NiO (JCPDS, No. 04-0835). The diameter of the crystalline is calculated from the major diffraction peak (200) using the well-known Scherrer's formula and is about 20 nm after being calcined at 300 °C while is about 50 nm after being calcined at 500 °C. Thus, calcination temperature has great influence on the crystalline structure and size of NiO.

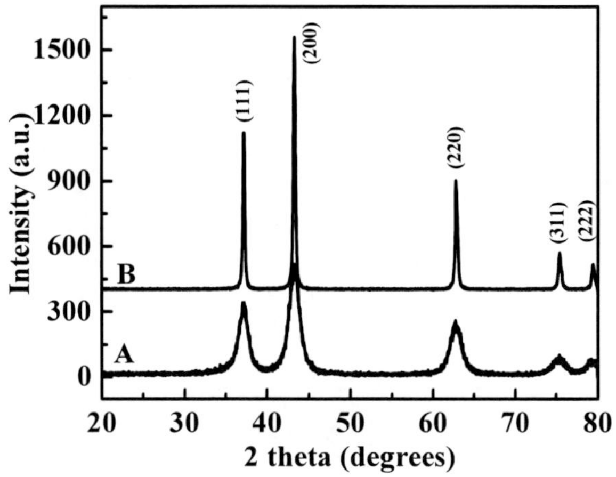

Figure 32. XRD patterns of NiO calcined at 300 (A) and 500°C (B).

The detailed morphology of the NiO nanocrystals is observed using TEM and ED pattern. Fig. 33A and B are typical TEM micrographs of NiO nanocrystals calcined at 300 and 500°C, respectively.

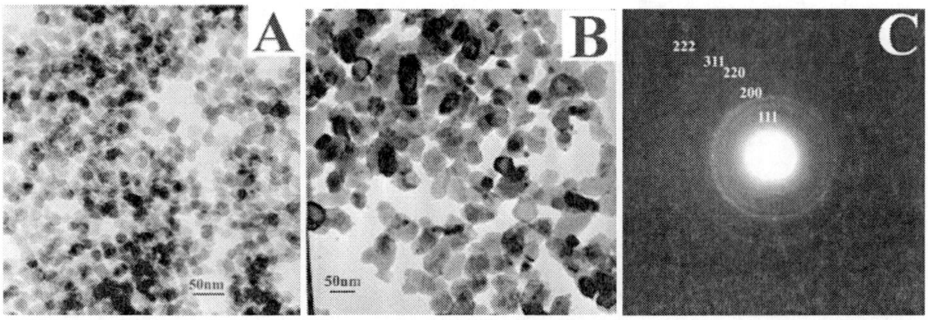

Figure 33. TEM images of NiO calcined at 300(A) and 500 °C (B); C) ED pattern of NiO calcined at 500 °C.

It can be seen that NiO nanocrystals calcined at 300 °C are uniform grains and their average size is slightly lower than 21 nm, while the average size of the NiO calcined at 500°C is about 50 nm, which are consistent with that obtained from XRD analysis. This confirms that the calcination temperature has great influence on the size of NiO. The ED pattern of NiO calcined at 500 °C shown in Fig. 33C has diffuse rings, indicating that the calcined products are in good crystalline state. From diffuse rings, the calcined products can be indexed as NiO, which is consistent with that estimated from the XRD data.

Fig. 34 exhibits IR spectra for the precursor and the as-calcined NiO nanocrystals at various temperatures. It can be seen from Fig. 34 that the broad absorptions around 3432 and 1626 cm^{-1} are assigned to the existence of water and the broad absorption around 2357 cm^{-1} is assigned to the existence of CO_2, which indicated that the powder samples prepared in air for FTIR usually have strong absorption to water and CO_2. In Fig. 34A, the wide absorption band around 1384 cm^{-1} indicates the existence of CO_3^{2-} and the peak almost disappears after calcination. The three interesting bands appearing around 1040, 825 and 680 cm^{-1} confirm the presence of the intercalated C-O species in the precursor and it disappears after calcination as shown in Fig. 34B and C. The absorption band around 440 cm^{-1} is assigned to Ni–O stretching vibration mode, indicating that the as-calcined NiO nanocrystals at 300 °C and 500°C are pure inorganic NiO species. For the bulk form NiO, the main IR absorption maximum corresponds to the long wavelength transverse optical mode and the optical photon frequency lie between 390 and 403 cm^{-1}. Due to their nanosize effects, the absorption of the as-calcined NiO is red shifted compared with that of the bulk form.

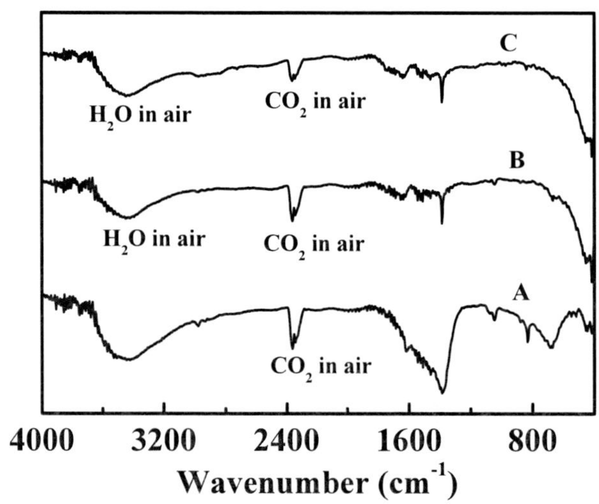

Figure 34. IR spectra of the precursor and NiO nanocrystals A) the precursor; B) 300 and C) 500 °C.

Magnetic properties at room temperature of NiO nanocrystals calcined at 300 and 500 °C are studied using a vibrating sample magnetometer with an applied field -10 kOe $\leq H \leq 10$ kOe. Hysteresis loop curves of the two samples are given in Fig. 35 and the hysteresis curves for the two samples are 'S'-shaped. At low external field, the hysteresis loops exhibit low coercivity, H_c=144.7 and 200.5 Oe, respectively, which is characteristic of weak ferromagnetism.

At high external field, the dependence of magnetization on applied field is markedly linear and the saturation magnetization is not reached especially for samples calcined at 500 °C even at the maximum applied field of 10 kOe. More importantly, it may be noted from Fig. 35A and B that the magnetization of the two samples at 10 kOe are 0.6871 emu/g and 0.1226 emu/g, respectively, which indicates that the magnetization decreases with the increase of calcination temperature. From the above discussion, it is clear that at room temperature, the magnetic properties of nanostructured nickel oxides, differ markedly from those of the corresponding bulk material which presents an antiferromagnetic behavior [30c].

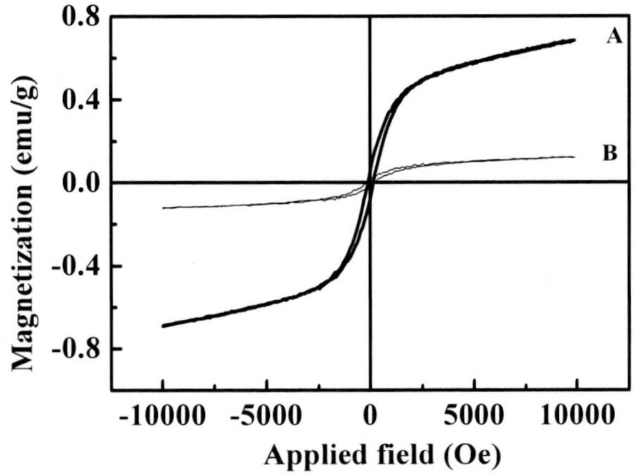

Figure 35. Hysteresis loop curves of the NiO nanocrystals measured at room temperature A) 300 and B) 500 °C.

Synthesis of spherical NiO nanocrystals has been introduced by thermal decomposition of the precursor obtained via hydrothermal method using urea as precipitant. The NiO nanocrystals are polycrystalline and the diameter is about 20 and 50 nm after being calcined at 300 and 500 °C, respectively. Magnetic hysteresis loops of NiO nanoparticles are recorded at room temperature. It is shown that the samples exhibit weak ferromagnetism with low coercivity at low external field and the magnetization curves vary linearly with applied field at high external field.

7. Synthesis and Characterization of B-MnO_2 Nanorods by Refluxing Method

Pyrolusite (β-MnO_2) is relatively pure MnO_2 with a rutile crystal structure, different from other MnO_2 structure [32]. Recently, much effort has been made for preparation of low-dimensional nanostructured β-MnO_2 because dimensionality is a crucial factor in determining the properties of nanomaterials. β-MnO_2 nanorods have been prepared generally under strict conditions [33]. Although much work has been done on nanostructural β-MnO_2 as cathodic and catalytic materials, there are few studies on nanostructural β-MnO_2 as magnetic materials.

A simple refluxing route is introduced here for preparing β-MnO_2 nanorods without any physical template and surfactant. This method is quite simple and facile, without any catalyst or any template to direct the growth of nanorods. A typical synthesis is as follows: $MnSO_4{\cdot}H_2O$ (2.00 g) and $Na_2S_2O_8$ (2.82 g) are placed in deionized water at room temperature to form a homogeneous solution completely. A certain volume of NaOH aqueous solution (the molar fraction of OH^-/Mn^{2+} was 2) is added to the above solution. The pink solution changes to brown and is stirred at room temperature for 30 min until a suspending liquid is formed. All experimental processes are under magnetic stirring. First, the suspending liquid is refluxed at 50 °C for 18 h, then refluxed at 80 °C for 5 h, finally at 100 °C for 3 h. After the reaction is complete, the resulting brownish-black solid product is filtered off, then rinsed with deionized water and absolute ethyl alcohol several times to remove ions possibly remaining in the final product, and finally dried at 80 °C in air. According to the practical and theoretical weight, the product yield can be calculated.

Wang *et al.* had proposed a wet chemical method using manganese sulfate and sodium persulfate to get the high pure battery-grade γ-MnO_2 [34]. This method is extended here for preparation of β-MnO_2 nanorods. First, the Mn^{2+} reacted with NaOH to produce amorphous $Mn(OH)_2$ at room temperature, and then the mixed system is stirred and heated according to temperature rise procedure, which leads to the formation of MnO_2 with regular shapes. The whole reaction can be represented as follows:

$$MnSO_4 + Na_2S_2O_8 + 4NaOH \rightarrow MnO_2 + 3Na_2SO_4 + 2H_2O \quad (2)$$

One-dimensional nanostructures can be obtained from the rolling of a natural or artificial lamellar structure [35]. Under certain conditions, a layer structure begins to curl, and the thus-obtained tubular structure serves as the original driving force for the growth of one-dimensional nanostructures. Among the several crystallographic forms of MnO_2, δ-MnO_2 alone has a layer structure, which is indispensable in the formation of β-MnO_2 nanorods. A schematic representation of the steps involved in the synthesis of β-MnO_2 nanorods is depicted in Fig. 36. It is suggested that the formation of β-MnO_2 nanorods by the refluxing process consist of three parts: (A) $Mn(OH)_2$ is oxidized to amorphous MnO_2 by $Na_2S_2O_8$, (B) the amorphous MnO_2 is transformed into layered structural δ-MnO_2, and (C) the β-MnO_2 nanorods is formed.

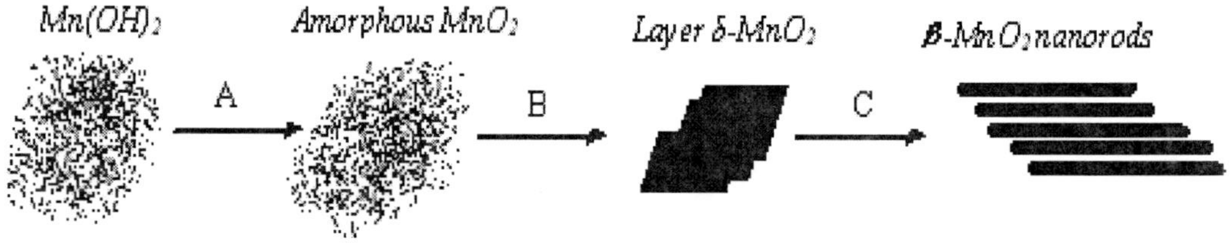

Figure 36. Schematic representation of the synthesis process of β-MnO_2 nanorods. Reproduced from Ref. [33c] with permission from copyright owners.

X-ray powder diffraction (XRD) patterns can reveal the phase and purity of the products. The typical XRD pattern of the β-MnO_2 nanorods is shown in Fig. 37. The feature peaks are assigned around the 2θ angles of 28.8°(110), 37.5°(101), 43.1°(111), 56.8°(211) and 65.0°(002). All of the reflections of the XRD pattern can be readily indexed to the pure pyrolusite (β-MnO_2) with the lattice constants a = 4.39±0.01 Å and c = 2.87±0.01 Å, which agree well with the values reported in the literature (JCPDS Card. No. 24-0735). Thus it can be determined that the crystals are of tetragonal structure. No characteristic peaks are observed for other impurities such as α-, λ-, and δ-MnO_2.

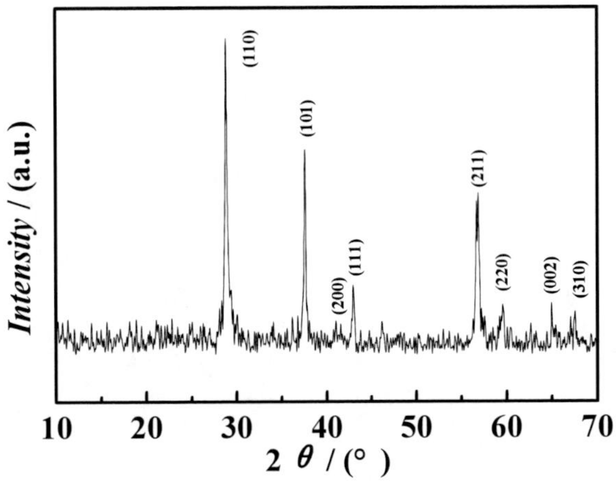

Figure 37. XRD pattern of β-MnO_2 nanorods . Reproduced from Ref. [33c] with permission from copyright owners.

The panoramic morphology of the product is obtained by SEM, in which the solid sample is mounted on conductive resin with dispersion treatment. SEM images of the products are shown in Fig. 38, indicating that all samples (Fig. 38A) have almost identical morphology and the proportion of the nanorods in the whole sample is above 95%. Careful observation (Fig. 38B) shows that nanostructured β-MnO_2 consists of nanorods with high aspect ratios.

Figure 38. SEM images of β-MnO_2 nanorods : A) Low magnification; B) high magnification . Reproduced from Ref. [33c] with permission from copyright owners.

The morphology and microstructure of the β-MnO_2 nanorods are further investigated with TEM and ED. Typical TEM images (Fig. 39A and B) show that the samples of β-MnO_2 consisted of straight and smooth nanorods with diameters in the range of 50–80 nm and lengths up to several micrometers. The ED pattern of β-MnO_2 nanorods (inset in Fig. 39B), which is taken from a single rod, indicates that these crystalline β-MnO_2 nanorods are of high quality. The yield of β-MnO_2 nanorods is above 70% of theoretical weight, which is higher than 50% by hydrothermal method.

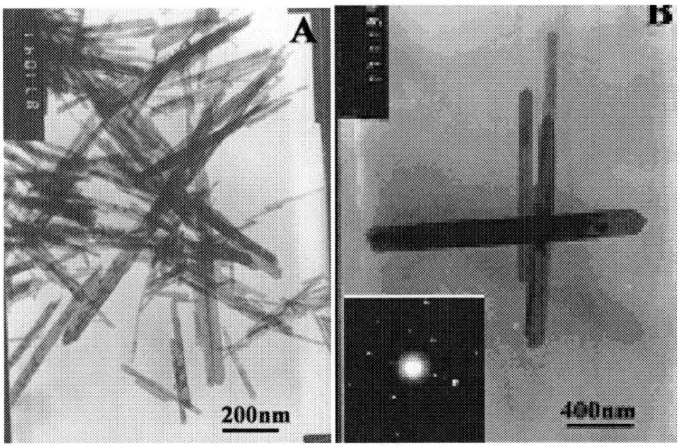

Figure 39. TEM images of β-MnO_2 nanorods (Inset in Fig. 39(B) shows ED of β-MnO_2 nanorods) . Reproduced from Ref. [33c] with permission from copyright owners.

Elemental analysis by EDX (Fig. 40A) confirms the presence of Mn and O elemental signatures. The elemental analysis show that the products have over 98% MnO_2 by mass, with a small amount of carbon ($<$2%). The presence of a small amount of C can be anticipated from the conducting resin during measurements. The ratio of Mn to O in the products is 1/2 in agreement with theoretical analysis. Fig. 40B depicts typical thermagravimetric (TG) behavior of the synthesized product in air. The slight weight loss (~0.9%) in the TG curve at 20-400 °C results from the removal of physically adsorbed water and the release of tightly bound water in the sample [23]. The sharp weight loss (~9%) at the temperature higher than 580 °C is due to the oxygen release corresponding to the formation of Mn_2O_3 [23]. The TG profile and the range of weight loss are different from those reported for α-MnO_2, λ-MnO_2, δ-MnO_2 and Na-birnessite [11,16,17,24], which suggests that the hydrous manganese oxides don't exist in the sample. Elemental and TG analysis indicated that the reaction is rather complete and the product consisted of β-MnO_2 nanorods.

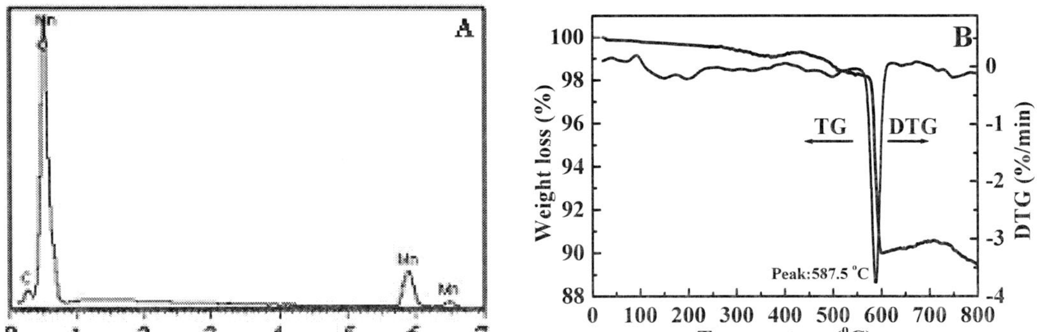

Figure 40. EDX profile (A) and TG-DTG curves (B) of β-MnO_2 nanorods . Reproduced from Ref. [33c] with permission from copyright owners.

Magnetic properties of the β-MnO_2 nanorods are investigated using a vibrating sample magnetometer with an applied field -3000 Oe≤H≤3000 Oe at room temperature. Plot of magnetization *vs* applied magnetic field for the sample is given in Fig. 41. At low external field, the hysteresis loop exhibits high coercivity (*Hc*=144 Oe), which is characteristic of ferromagnetism. At high external field, the dependence of magnetization on applied field is markedly linear and saturation magnetization is not reached even at the maximum applied field of 3000 Oe. From the viewpoint of magnetic interactions, only Mn^{4+}-O^{2-}-Mn^{4+} is in ferromagnetic coupling, while all other interactions are in antiferromagnetic coupling. Therefore, the magnetic response of nanostructured β-MnO_2 are ascribed to the ferromagnetic coupling of Mn^{4+}-O^{2-}-Mn^{4+}, because the magnetic properties are determined by interactions between the Mn ions, which in turn depend on the Mn valence distribution in the $[Mn_2]O_4$ framework [36].

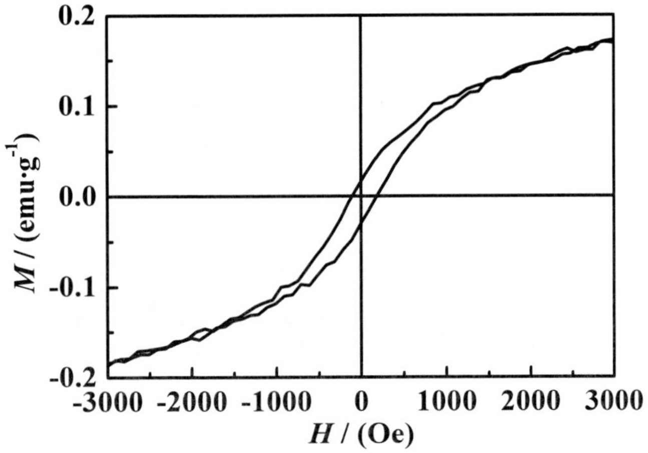

Figure 41. Plot of magnetization vs applied field for β-MnO_2 nanorods measured at room temperature . Reproduced from Ref. [33c] with permission from copyright owners.

In the end, β-MnO_2 nanorods with diameters in the range of 50–80 nm and lengths of several micrometers are synthesized by a simple refluxing route under certain conditions. This synthetic method for manganese dioxide nanorods is quite simple and facile, without any catalyst to serve as the energetically favorable site for the absorption of reactant molecules, and without any template to direct the growth of nanorods. Meanwhile, the high yield (~70%) of β-MnO_2 nanorods can be attained. Magnetic properties of β-MnO_2 nanorods prepared by the refluxing route are investigated using a vibrating sample magnetometer at room temperature, showing that the β-MnO_2 nanorods exhibit the characteristics of ferromagnetism.

8. Summary and Outlook

In this chapter, a number of chemical methods including sol-gel method, polymer-pyrolysis method, template method, hydrothermal method and refluxing method have been introduced for synthesis of nano-sized magnetic ferrites and metal oxides. It has been shown that structural and magnetic properties of these nanosized magnetic materials can be well characterized by XRD, TEM, ED, SEM, EDX, XPS and VSM, respectively.

The study of magnetic nanoparticles is clearly a fertile area of research with many unresolved scientific problems. The synthesis of magnetic nanoparticles having well-defined size, shape and properties is of great importance. The outlook of such magnetic nanoparticles is very promising because these materials will find many important industrial applications including ultrahigh-density magnetic storage media and drug-delivery systems. To improve our understanding of such systems and their applications, a collaborative multi-disciplined effort is anticipated.

REFERENCES

[1] D.C. Jiles. *Acta Mater*. 51 (2003) 5907.

[2] (a) A. P. Alivisatos, *Science*, 271 (1996) 933. (b) H. Weller, *Angew. Chem*., Int. Ed. Engl., 32 (1993) 41; (c) Clusters and Colloids, ed. G. Schmid, VCH Press, New York, 1994; (d) Nanoscale Materials in Chemistry, ed. K. J. Klabunde, Wiley-Interscience, New York, 2001; (e) *Nanoparticles and Nanostructured Films*, ed. J. H. Fendler, Wiley-VCH, Weinheim, 1998.

[3] (a) S. A. Majetich and Y. Jin, *Science,* 284 (1999) 470; (b) C. B. Murray, C. R. Kagan and M. G. Bawendi, *Science,* 270 (1995) 1335; (c) A. J. Zarur and J. Y. Ying, *Nature*, 403 (2000) 65.

[4] (a) T. Hyeon, Y. Chung, J. Park, S. S. Lee, Y. W. Kim, H. H. Park. *J. Phy. Chem*. B 106 (2002) 6831. (b) M. Moumen, M. P. Pileni. *J. Phys. Chem*. 100 (1996) 1867. (c) E. Kang, J. Park, Y. Hwang, M. Kang, J. G. Park, T. Hyeon. *J. Phys. Chem*. B 108 (2004) 13932.

[5] (a) S. Verma, P. A. Joy, Y. Khollam, B. H. S. Potdar, S. B. Deshpande. *Mater. Lett*. 58 (2004) 1092. (b) S. Sun, H. Zeng. *J. Am. Chem. Soc*. 124 (2002) 8204. (c) S. Sun, H. Zeng, D. B. Robinson, S. Raoux, P. M. Rice, S. X. Wang, G. Li. *J. Am. Chem. Soc*. 126 (2004) 273.

[6] D. E. Speliotis. *J. Magn. Magn. Mater*. 193 (1999) 29.

[7] (a) R. Semelka, T. Helmberger. *Radiology* 218 (2001) 27. (b) I. Coroiu. *J. Magn. Magn. Mater.* 201 (1999) 449. (c) C. Jung, P. Jacobs. *Magn. Reson. Imaging* 13 (1995) 661. (d) E. D. Wu, H. Tang, J. H. Jensen. *NMR Biomed.* 17 (2004) 478.

[8] R. E. Rosenweig. *Ferrohydrodynamics*, Cambridge University Press: Cambridge, U.K., 1985.

[9] (a) W. Weitschies, R. Ko¨titz, T. Bunte, L. Trahms. *Pharm. Pharmacol. Lett.* 7 (1997) 5. (b) R. Rotitz, W. Weitschies, L.Trahms, W. Brewer, W. Semmler. *J. Magn. Magn.* Mater. 94 (1999) 62.

[10] S. Goodwin, C. Peterson, C. Hoh, C. Bittner. *J. Magn. Magn. Mater*. 194 (1999) 132.

[11] R. Hiergeist, W. Andra, N. Buske, R. Hergt, I. Hilger, U. Richter, W. Kaiser. *J. Magn. Magn. Mater.* 201 (1999) 420.

[12] (a) M. Ghosh, G. Lawes, A. Gayen, G. N. Subbanna, W. M. Reiff, M. A. Subramanian, A. P. Ramirez, J. P. Zhang, R. Seshadri. *Chem. Mater*. 16 (2004) 118. (b) S. Ammar, A.

Helfen, N. Jouini, F. Fievet, I. Rosenman, F. Villain, P. Molinie, M. Danot. *J. Mater. Chem.* 11 (2001) 186.

[13] (a) S. Kale, R. D. K. Gubbala, J. Misra, *J. Magn. Magn. Mater.* 277 (2004) 350. (b) K. V. P. M.Shafi, A.Gedanken, *Chem. Mater.* 10 (1998) 3445. (c) N. Hanh, O. K. Quy, N. P. Thuy, L. D. Tung, L. *Spin. Physica* B 327 (2003) 382.

[14] C. B. Murray, C. R. Kagan, M. G. Bawendi. *Science*, 270(1995)1335.

[15] C. Savii, M. Popovici, C. Enache, et al. *Solid State Ionics*, 151(2002)219.

[16] J. B. da Silva, N. D. S. Mohallem. *J. Magn. Magn. Mater.*, 226-230(2001)1393.

[17] X. M. Liu, S. Y. Fu, C. J. Huang. *J. Magn. Magn. Mater.*, 281(2004) 234.

[18] K. V. P. M. Shafi, Y. Koltypin, A. Gedanken, et al. *J Phys Chem* B, 101(1997) 6407.

[19] (a)M. Mahesh, V.R. Palkar, K. Srinivas, S.V. Suryanarayana, *Appl. Phys. Lett.* 76 (2000) 2674; (b) T. Matsui, H. Tanaka, N. Fujimura, I.I.H. Mabuchi, K. Morii, *Appl. Phys. Lett.* 81(2002) 2764; (c) N.A. Hill, *Annu. Rev. Mater. Res.* 32 (2002) 1; (d) H. Zheng, J. Wang, S.E. Lofland, Z. Ma, L.M. Ardabili, T. Zhao, L.S. Riba, S.R. Shinde, S.B. Ogale, F. Bai, D. Viehland, Y. Jia, D.G. Schlom, M. Wuttig, A. Roytburd, R. Ramesh, *Science,* 303 (2004) 661.

[20] (a) S.V. Suryanarayana, *Bull. Mater. Sci.* 17 (1994) 1259; (b) S. S. Lopatin, I. Lopatina, I. Lisnevskana, *Ferroelectrics*, 162 (1994) 62.

[21] T. Kanai, S.I. Ohkoshi, A. Nakajima, T. Watanabe, K. Hashimoto, *Adv. Mater.* 13 (2001) 487.

[22] J.G. Wan, X.W. Wang, Y.J. Wu, M. Zeng, Y. Wang, H. Jiang, W.Q. Zhou, G.H. Wang, J.M. Liu, *Appl. Phys. Lett.* 86 (2005) 122501.

[23] C. Michel, J.M. Moreau, G.D. Achenbach, R. Gerson, W.J. James, Solid State Commun. 7 (1969) 701.

[24] X.M. Liu, S.Y. Fu, C.J. Huang, *Mater. Sci. Eng.* B 121(2005) 255.

[25] G. Srinivasan, E.T. Rasmussen, J. Gallegos, R. Srinivasan, Y.I. Bokhan, V. M. Laletin, *Phys. Rev.* B 64 (2001), 214408.

[26] (a) X.M. Liu, S.Y. Fu, H.M. Xiao, C.J. Huang, *Physica* B 370 (2005) 14; (b) X.M. Liu, G. Yang, S. Y. Fu, *Mater. Sci. Eng. C* in press.

[27] (a) R. Zboril, M. Mashlan, D. Petridis, Chem. Mater. 14 (2002) 969; (b) M. Catti, G. Valerio, R. Dovesi, Phys. Rev. B 51 (1995) 7441; (c) M.F. Hansen, C.B. Koch, S. Mørup, *Phys. Rev.* B 62 (2000) 1124.

[28] I. ul Haq, E. Matijević, J. *Colliod Interf. Sci.* 192 (1997) 104; (b) A. T. Ngo, M. P. Pileni, *J. Appl. Phys.* 92 (2002) 4649; (c) C. Pascal, J. Pascal, F. Favier, M. Elidrissi, C. Payen, Chem. Mater. 11 (1999) 141; (d) Y. Fu, J. Chen, H. Zhang, *Chem. Phys. Lett.* 350 (2001) 491; (e) Y. Fu, R. Wang, J. Xu, J. Chen, Y. Yan, A. Narlikar, H. Zhang, *Chem. Phys. Lett.* 379 (2003) 373; (f) S. Lian, E. Wang, Z. Kang, Y. Bai, L. Gao, M. Jiang, C. Hu, L. Xu, Solid State Commun. 129 (2004) 485; (g) X.G. Wen, S.H. Wang, Y. Ding, Z.L. Wang, S.H. Yang, *J. Phys. Chem.* B 109 (2005) 215; (h) X.M. Liu, S.Y. Fu, H.M.Xiao, *J. Solid State Chem.* 178 (2005) 2798.

[29] (a) Q. Guo, X.Teng, S. Rahman, H.Yang, *J. Am. Chem. Soc.* 125 (2003) 630; (b) T. Fujii, F.M.F. de Groot, G.A. Sawatzky, F.C. Voogt, T. Hibma, K. Okada, *Phys. Rev.* B, 59 (1999) 3195; (c) M. Sorescu, R.A. Brand, D. Mihaila-Tarabasanu, L. Diamandescu, *J. Appl. Phys.* 85(1999)5546.

[30] (a) W. Nolting, L. Haunert and G. Borstel: *Phys. Rev.* B Vol 46 (1992)4426; (b) O. Bengone, M. Alouani, P. Blochl and J. Hugel: *Phys. Rev.* B Vol 62 (2000) 16392; (c) F.

BØdker, M.F. Hansen, C. B. Koch and S. Mørup: *J. Magn. Magn. Mater.* Vol 221 (2000) 32; (d) X.M. Liu and X.G. Zhang: *Electrochim. Acta*, Vol 49(2004)229; (e) H. Bi, S.D. Li, Y.C. Zhang and Y.W. Du: *J. Magn. Magn. Mater.* Vol 277 (2004) 363; (f) D. Levin and J.Y. Ying: *Stud. Surf. Sci. Catal.* Vol 110 (1997) 367; (g) K. Arshak, O. Korostynska and S. Clifford : *Sens. Actuators*, A Vol 110(2004)354.

[31] (a) Y. Wang and J.J. Ke: *Mater. Res. Bull.* Vol 31(1996) 55; (b) G.J. Li, X.X. Huang, Y. Shi and J.K. Guo: Mater. Lett. Vol 51 (2001) 325; (c) N.B. Singh and A.K. Ojha: *Thermochim. Acta* Vol 378 (2001) 87; (d) V. Biju and M.A. Khadar: *Mater. Sci. Eng.* A Vol 304– 306 (2001)814. (e) X.M. Liu, S.Y. Fu, *Solid State Phenon.* In press

[32] S.A.chambers, Y. Liang, *Surf. Science*. 420 (1999)123.

[33] (a) X. Wang, Y. Li, *J. Am. Chem. Soc.* 124 (2002) 2880; (b) G. Xi, Y.Y. Peng, Y.C. Zhu, et al., *Mater.Res. Bull.* 39 (2004) 1641; (c) X.M. Liu, S.Y. Fu, C.J. Huang, *Powder Techn.* 154(2005)120.

[34] E.I. Wang, L.F. Lin, W.L. Bowden, *US Patent* 5,482,796.

[35] X. Wang, Y. Li, *Chem. Eur. J.* 9 (2003)300.

[36] C. Masquelier, M. Tabuchi, K. Ado, et al., *J. Solid State Chem.* 123(1996)255.

In: Progress in Solid State Chemistry Research
Editor: Ronald W. Buckley, pp. 205-238

ISBN 1-60021-313-8

Chapter 5

SYNTHETIC, STRUCTURAL, SPECTROSCOPIC AND MAGNETIC REVISION OF ORGANICALLY TEMPLATED IRON(III) ARSENATES

Begoña Bazán[1], Jose L. Mesa[2*], Jose L. Pizarro[1], Maria I. Arriortua[1] and Teófilo Rojo[2]

[1]Dpto. de Mineralogía-Petrología

[2]Dpto. de Química Inorgánica, Facultad de Ciencia y Tecnología, Universidad del País Vasco/EHU, Apdo. 644, E-48080 Bilbao, Spain

ABSTRACT

The mild hydrothermal synthesis is an adequate method to obtain novel microporous materials with occluding protonated diamines, which act as templating agents directing the synthesis towards a determinate product. Among these materials, iron phosphates occupy the first position because they increment a rich structural chemistry. Nevertheless, the iron arsenates are scarce. In the last years we have synthesize organically templated iron(III) arsenates with different crystalline dimensionality (1D, 2D and 3D) using the ethylenediamine, piperazine, 1,4-diazobicyclooctane diamines,... as templating agents. The synthesis was carried out in aqueous solution using as starting reagents arsenic(V) oxide, iron(III) salt, hydrofluoric acid, if the final phase contains the fluoride anions connecting the iron centers, and the diamine that regulates the pH of the initial reaction mixture. The syntheses were performed inside poly(tetrafluoroethylene) containers under autogeneous pressure at 170 °C for five days.

On this synthetic basis several iron(III) arsenates were obtained. $(C_2H_{10}N_2)[Fe(HAsO_4)_2(H_2AsO_4)].H_2O$ (1) consists of $[Fe(HAsO_4)_2(H_2AsO_4)]^{2-}$ anionic *chains*. The FeO_6 octahedra are isolated and bridged by two $HAsO_4$ and one H_2AsO_4 tetrahedra. The $(C_4H_{12}N_2)_{1.5}[Fe_3(HAsO_4)_{1.2}(HPO4)_{0.98}(AsO_4)_{0.88}(PO4)_{0.12}F_5]$ (7) *layered* phase is constructed from $[Fe_3(HAsO_4)_{1.2}(HPO4)_{0.98}(AsO_4)_{0.88}(PO4)_{0.12}F_5]^{3-}$ inorganic sheets. The layers have cavities where the piperazinium dications are placed. $(C_3H_{12}N_2)_{0.75}[Fe^{II}_{1.5}Fe^{III}_{1.5}(AsO_4)F_6]$ (12) is a three-dimensional mixed valent compound. The structure of $(C_6H_{14}N_2)[Fe_3(HAsO_4)_2(AsO_4)F_4].0.5H_2O$ (15) and the isomorphous

[*] Corresponding author: e-mail: joseluis.mesa@ehu.es; Phone: +34 946015523; Fax: +34 946013500.

mixed arsenate-phosphate compound consist of a *three-dimensional* framework formed by $[Fe_3(HAsO_4)_2(AsO_4)F_4]^{2-}$ inorganic layers interconnected by hydrogenarsenate groups.

These compounds exhibit interesting magnetic behaviors. The thermal evolution of χ_m for (1) is indicative of *antiferromagnetic* exchange couplings. The magnetic data were fitted to the equation given by Fischer and Dingle for chains of spin S= 5/2. The best fit is obtained for a value of the J-exchange parameter of J= -0.835K, where g= 2.00 was obtained from ESR spectroscopy. The magnetic susceptibility for (7) increases with decreasing temperature and reaches a maximum at 8.5K. After that, the susceptibility slightly decreases until 5K. However, in the $\chi_m T$ *vs.* T curve at low temperatures an abrupt increase is observed with a maximum at 13.4K. After this temperature the $\chi_m T$ curve decreases down to 5K. This result indicates the existence of a *ferromagnetic* behavior, which can take place due to the decompensation of the magnetic moments of the different sublattices of the FeO_6 octahedra. Compound (12) is a three-dimensional antiferromagnet with a weak hysteresis loop at low temperatures, being the remnant magnetization of 400 emu/mol and the coercive field of 1180 Gauss. In the case of compound (15) magnetic susceptibility measurements indicate the presence of antiferromagnetic interactions. In order to obtain the J exchange parameter the system was considered as a cubic network. The equation developed by Rooshbroke and Wood for a cubic antiferromagnetic system with S= 5/2 was used. Only the low temperature magnetic data, in the proximity of the maximum, can be reasonably fitted with a value of |J/K|= 0.41. The isomorphous arsenate-phosphate phase [16] is an intermediate 3D-2D antiferromagnetic system.

The more important characteristic of some organically templated iron(III) arsenates found in the literature will be studied in this revision.

1. Introduction

The concept of porous solids originates from the discovery by Crönsted [1] during the XVIIIth century, of the zeolitic properties of the mineral stilbite. Many natural zeolites were discovered later, but the importance of this family increased when chemists were able to synthesize them. The first success was due to Sainte Claire Deville in 1862 [2] and most of the synthesis required organic molecules as templates.

Zeolites are hydrated alkaline or alkaline earth aluminosilicates with the general formula $M^{n+}_{x/n}[(AlO_2)_x(SiO_2)_y]^{x-}.wH_2O$ where x indicates the number of M^{n+} cations which are necessary to compensate the negative charge of the framework, and w is the number of water molecules. From a structural point of view, zeolites exhibit a three-dimensional network of corner-sharing TO_4 tetrahedra (T= Al, Si) which delimits interconnected tunnels or cages in which water molecules and M ions are inserted. The elimination of the water molecules keeps the framework intact and the solid becomes porous.

The first breakthrough was provided by Flanigen [3], who, playing on the similarity $2Si^{4+} \leftrightarrow Al^{3+} + P^{5+}$, synthsized microporous aluminophsophates with structures related to those of zeolites. The structural studies [4] showed however a striking difference between the two families. As already mentioned, the framework of zeolites is built up exclusively from connected tetraedra, whereas in aluminosilicates and homologous gallophosphates, Al and Ga polyhedra can adopt five and six-fold coordination, which change the connectivity of the framework, and therefore the shape of the pores. The second important change was provided by Kessler et al. [5] who introduced the fluoride method in the synthesis of microporous solids. The fluoride ion is a mineralising agent that can also participate in the reaction and the structure for providing new topologies as shown by Kessled and Ferey's groups [6].

Nowadays, the research field of the microporous solid synthesized by using organic diamines as *template* agent have increased showing a great diversity in the crystal structures and in their physical properties. In this work, a revision of the more important features, from the structural, spectroscopic and magnetic point of views, of the organically templated inorganic-organic solids, based on the $(AsO_4)^{3-}$ oxoanion and the Fe^{3+} cation, is carried out.

2. Synthesis

The synthesis of these materials were performed by heating an iron(III) sal, amines and arsenic(V) oxide in appropriate proportions in water or a mixture of water and organic solvent in a teflon-lined autoclave at the temperature of 170 °C and autogeneous pressure (10-20 bars) for five days. The syntheses have so many parameters (time, temperature, relative concentration of reagents, pH, starting materials, HF acid, etc.) that it is difficult to navigate through the enormous parameter space. Subtle changes in the reaction conditions can greatly affect the course of the reactions. Several variables have to be kept constant in this exploratory study. This way iron(II,III) salts and $As_2O_5.3H_2O$ were used as the sources of iron and arsenic, respectively. The temperature, the pressure, the HF amount and the reaction time were also maintained constant in the majority of the synthetic reactions. The properties of solvent are often vital to the success or failure of the attempted syntheses. Both aqueous and non-aqueous solvent, such 1-butanol, have been used in the synthesis. Alcohols are less effective than water in hydrogen-bonding ability, and so there is a greater interaction between the framework particles and the templates. Furthermore, the butanol has higher viscosity than water, and this way, it may reduce convection currents and help produce large single crystal than polycrystalline product by sedimentation [7].

The choice of pH is also critical to the outcome of the synthesis of iron arsenates. The small ionic size of Fe(III) leads to extensive hydrolysis of its aqueous solutions. Even in strongly acidic solutions (several molar in $[H^+]$), the pale purple ion $[Fe(H_2O)_6]^{3+}$ takes on yellow coloration arising from species taking part in the slow and complex set of equlibria that are involved in the hydrolytic polymerizations process. The hydrolysis of $[Fe(H_2O)_6]^{3+}$ in acidic solution involves the main equilibrium:

$$[Fe(H_2O)_6]^{3+} \leftarrow\rightarrow [Fe(OH)(H_2O)_5]^{2+} + H^+ \quad K= 10^{-3.05}$$

A small amount of $[Fe(OH)_2(H_2O)_4]^+$ may be formed, but the second main species is the dimmer:

$$2\,[Fe(OH)(H_2O)_5]^{2+} \leftarrow\rightarrow [(H_2O)_5FeOFe(H_2O)_5]^{4+} + H_2O \quad K= 10^{-2.91}$$

As the pH is raised above 2 or 3, further condensation occurs, colloidal gel begin to form, and eventually a reddish-brown precipitate of hydrous iron oxide is formed. It is evident therefore that Fe(III) salts dissolved in water produce highly acidic solutions [8].

The mechamism of hydrothermal synthesis of microporous materials with occluding diamines is poorly understood. However, it is reasonable to propose that micropous materials nucleates within a gel particle and the organization of inorganic anions around the organic cation forms the basis of the nucleation centers from which the iron arsenate crystallites form. The synthesis or large pores may be through the use of large organic templates. Although it remains unknown how the structure are assembled at the molecular level, larger inorganic building units should also lead to large pores. One can imagine the construction of three-dimensional iron arsenates frameworks with large pores from gel precursor that already contain polynuclear iron-oxygen species. It is clear from earlier discussion that hexaquo iron(III) complexes condense to high molecular weight species as the pH increases.

This way, all the structures of the organically templated phases with occluding diamines are formed by an inorganic framework in which the concept of *Secondary Building Unit* (SBU) [1] or number of polyhedra (octahedra, tetrahedra,...) which form by interconnection and periodical repetition the structure of the organically templated phase can be defined.

The compounds studied in this revision were synthesized by using mild hydrothermal conditions using poly(tetrafluoroethylene)-lined stainless steel containers. In Table 1 the synthetic conditions used in the attainment of the discussed phases are summarized.

The limit of thermal stability of these phases is approximately 300 °C, temperature after which the crystal structure collapses giving rise to inorganic residues at temperatures of approximately 700 °C (Figure 1).

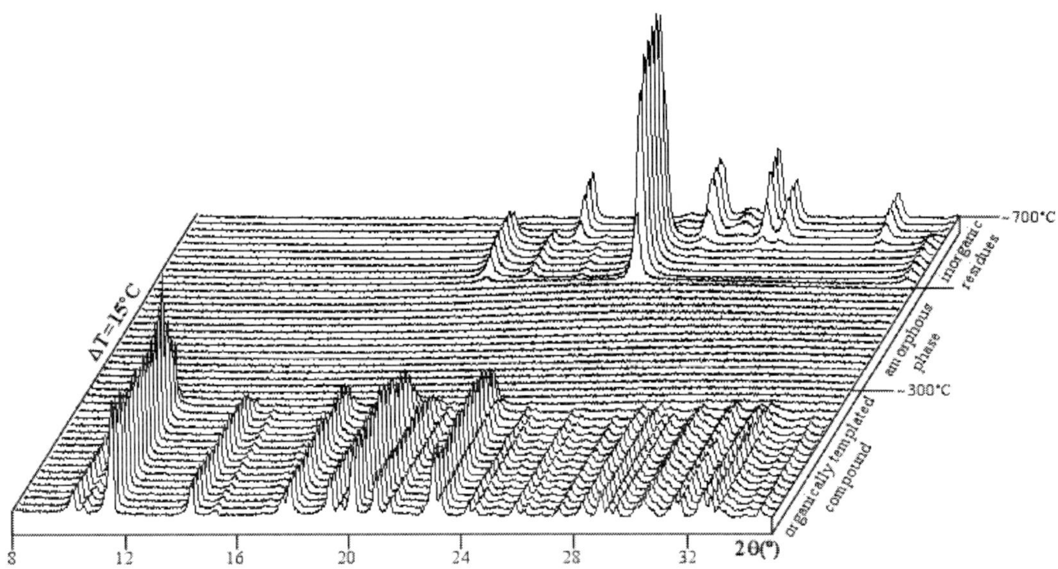

Figure 1. Schematical thermodiffractogram for the organically templated arsenates.

3. Crystal Structure

The crystal structure that exhibit these kind of compounds show diversity and range from mono-dimensional (1D) to two-dimesional (2D) and three-dimensional (3D). All the

Table 1- Summary of the synthetic conditions

		Compound	$FeCl_3$ g (mmol)	$Fe_2(SO_4)_3$ g (mmol)	FeC_2O_4 g (mmol)	$Fe(acac)_3$ g (mmol)	Fe g (mmol)	As_2O_5 g (mmol)	H_3AsO g (mmol)	H_3PO_4 mL (mmol)	HF mL (mmol)	O. M.	H_2O mL	pH/T/Days
1D	(1)	$(C_2H_{10}N_2)[Fe(HAsO_4)_2(H_2AsO_4)](H_2O)$	0.0025 (0.009)	----	----	----	----	0.0862 (0.37)	----	----	----	0.018 mL (0.269) EN	6 (*)	3/170°/5
	(2)	$\infty[(NH_2CH_2CH_2)_2NH][Fe_2F_4(HAsO_4)_2](H_2O)$	----	----	----	0.1960 (0.55)	----	0.3780 (1.64)	----	----	0.08 (2.2)	0.25 mL (2.3) DETA	2	5/180°/3
	(3)	$(C_4H_{12}N_2)[Fe(OH)(HAsO_4)(C_2O_4)](H_2O)$	----	----	0.0720 (0.4)	----	----	----	0.4542 (3.2)	----	0.02 (2.4)	0.1378 g (1.6) PIP	1.44	2/150°/3
2D	(4)	$(C_2H_{10}N_2)[Fe^{III}Fe^{II}F_2(HAsO_4)(AsO_4)]\cdot 2H_2O$	----	----	----	----	0.0279 (0.55)	----	0.2342 (1.65)	----	0.083 (10)	0.13 mL (2.0) EN	5.4	2/120°/2
	(5)	$(C_3H_{12}N_2)_{0.5}[Fe(OH)AsO_4]$	0.6757 (2.5)	----	----	----	----	----	1.0645 (7.5)	----	----	0.63 mL (7.5) DP	3 (#)	-/180°/3
	(6)	$(C_4H_{12}N_2)_{1.5}[Fe_3(HAsO_4)_2(AsO_4)F_5]$	0.2703 (1)	----	----	----	----	----	1.4194 (10)	----	0.5 (13.8)	0.8614 g (10) PIP	10	-/160°/3
	(7)	$(C_4H_{12}N_2)_{1.5}[Fe_3(HAsO_4)_{1.02}(HPO_4)_{0.98}(AsO_4)_{0.88}(PO_4)_{0.12}F_5]$	----	0.1812 (0.37)	----	----	----	0.4251 (1.85)	----	0.09 (1.85)	1 (57.5)	0.750 g (8.52) PIP	30	1.5/170°/5
	(8)	$(C_{10}H_{28}N_4)[\{FeF(OH)(HAsO_4)\}_4]$	0.1487 (0.55)	----	----	----	----	0.6248 (2.72)	----	----	0.12 (3.3)	0.441 g (2.2) AP-PIP	4 (&)	2/150°/3
3D	(9)	$(C_2H_{10}N_2)[Fe_2(AsO_4)_2F_2(H_2O)]$	----	----	----	----	0.0223 (0.4)	----	0.2839 (2)	----	0.05 (6)	0.11 mL (1.6) EN	4.32	3/120°/2
	(10)	$(C_2H_{10}N_2)[Fe_2(AsO_4)_{1.54}(PO_4)_{0.46}F_2(H_2O)]$	0.1000 (0.37)	----	----	----	----	0.4251 (1.85)	----	0.09 (1.85)	1 (57.5)	1.5 mL (22.4) EN	30	3/170°/5
	(11)	$(C_2H_{10}N_2)(NH_3)(NH_4)[Fe_3(HAsO_4)_6]$	----	----	----	----	0.0167 (0.3)	----	0.4258 (3.0)	----	0.017 (2.1)	0.08 mL (1.2) EN	3.25	-/180°/5
	(12)	$(C_3H_{12}N_2)_{0.75}[Fe^{II}_{1.5}Fe^{III}_{1.5}(AsO_4)F_6]$	----	0.1205 (0.25)	0.0887 (0.5)	----	----	0.8503 (3.7)	----	----	1 (57.5)	2 mL (23.7) DP	30	3/170°/5
	(13)	$(C_4H_{12}N_2)_2[Fe_4(HAsO_4)_6(C_2O_4)_2]$	----	----	0.099 (0.55)	----	----	0.3826 (1.66)	----	----	----	0.0478 g (0.55) PIP	2	2/150°/2
	(14)	$(C_4H_{12}N_2)[Fe^{III}_2Fe^{II}(AsO_4)_2(HAsO_4)_2]$	----	----	0.0720 (0.4)	----	----	----	0.4542 (3.2)	----	0.02 (2.4)	0.1378 g (1.6) PIP	1.44	2/180°/2.5
	(15)	$(C_6H_{14}N_2)[Fe_3(HAsO_4)_2(AsO_4)F_4]\cdot 0.5H_2O$	0.1513 (0.56)	----	----	----	----	0.8503 (3.7)	----	----	1 (57.5)	1.0108 g (9.0) DABCO	30	1.5/170°/5
	(16)	$(C_6H_{14}N_2)[Fe_3(HAsO_4)_{1.33}(HPO_4)_{0.67}(AsO_4)_{0.84}(PO_4)_{0.16}F_4]\ 0.5H_2O$	0.1513 (0.56)	----	----	----	----	0.4251 (1.85)	----	0.09 (1.85)	1 (57.5)	1.0108 g (9.0) DABCO	30	1.5/170°/5

Organic Molecules: EN: ethylenediamine; PIP: piperazine; DETA: diethylenetriamine; DP: 1,3- diaminpropane; AP-PIP: 1,4-bis(3-aminopropyl)piperazine; DABCO: 1,4-diazabicyclo[2.2.2]octane

Cosolvent: **(*)** n-butanol; 24 mL; **(#)** n-butanol; 7 mL; **(&)** n-hexanol; 2mL

Table 2- Crystallographic data

	Compound (*)	Space Group	a (Å)	b (Å)	c (Å)	α (º)	β (º)	γ (º)	V (Å^3)	Z	ρ_{calc} (g.cm^{-3})	SBU
1D	**(1)**	$P2_1$	8.705(3)	16.106(4)	4.763(1)	---	90.63(1)	---	666.7(3)	2	2.770	SBU-4
	(2)	P-1	5.5386(5)	9.5537(9)	13.6893(12)	89.780(2)	83.799(2)	76.230(2)	699.24(11)	2	2.720	SBU-4
	(3)	P-1	8.5668(3)	9.1040(3)	10.4715(4)	86.35(1)	85.34(1)	83.32(4)	618.8(1)	2	2.184	SBU-2
2D	**(4)**	P-1	5.289(1)	7.573(3)	8.888(2)	67.32(3)	82.77(3)	69.76(3)	308.2(1)	1	2.838	SBU-4
	(5)	Fdd2	9.5584(7)	41.593(3)	5.6438(4)	---	---	---	2318.45(8)	16	2.863	SBU-2
	(6)	$P2_1/c$	8.394(1)	21.992(3)	10.847(1)	---	96.188(2)	---	1990.8(4)	4	2.714	SBU-4; SBU-4
	(7)	$P2_1/c$	8.270(7)	22.028(3)	10.736(2)	---	99.79(2)	---	1927.3(2)	4	2.364	SBU-4; SBU-4
	(8)	$P2_1/n$	9.6320(3)	8.3598(3)	10.7794(1)	---	115.036(1)	---	786.42(4)	4	2.378	SBU-4
3D	**(9)**	Pbca	10.490(2)	10.090(2)	20.540(4)	---	---	---	2174.0(8)	8	3.102	SBU-2; SBU-4
	(10)	Pbca	10.457(4)	10.093(3)	20.479(1)	---	---	---	2161.4(1)	8	2.996	SBU-2; SBU-4
	(11)	C2/c	18.384(3)	8.785(1)	16.052(2)	---	111.11(1)	---	2418.5(6)	4	3.014	SBU-6
	(12)	Imam	7.566(2)	11.034(3)	13.233(4)	---	---	---	1103.9(6)	4	2.874	SBU-4
	(13)	P-1	11.1066(2)	12.9842(1)	13.4294(1)	111.345(1)	93.688(1)	93.879(1)	1706.33(4)	2	2.775	SBU-2; SBU-3
	(14)	C2/c	14.6850(3)	10.6696(1)	10.1699(3)	---	90.12(1)	---	1593.4(1)	8	3.067	SBU-7
	(15)	C2/c	18.323(1)	10.118(1)	20.623(1)	---	106.03(1)	---	3674.6(2)	8	2.840	SBU-4; SBU-4
	(16)	C2/c	18.267(1)	10.085(1)	20.436(2)	---	105.97(1)	---	3619.4(5)	8	2.749	SBU-4; SBU-4

(*) See Table 1 for the chemical formulae of the compounds

structures can be described from a very small number of SBU whose configuration is not at first glance strongly dependent of the nature of the amine chosen for the synthesis. Férey et *al.* [1] have proposed a mechanism to explain the transition from the cationic monoarsenate complexes in the initial solution to the anionic SBU in the final solid. The hypothesis proposes the existence of these SBU in the solution and implies three steps for the condensation of the final solid. First, in the solution, an oligomeric condensation of the monoarsenate complexes, during which these SBU might be formed in an identical way, despite the fact that the amines are different; the second and the third steps should correspond to the transformation of zero-charged ammonium-SBU associations (called MBU) of the solid framework from the infinite condensation of these neutral MBU, respectively [1]. The unit-cell parameters of the phases studied in this structural revision are shown in Table 2.

3.1. Mono-Dimensional Phases (1D)

3.1.1. $(C_2H_{10}N_2)[Fe(HAsO_4)_2(H_2AsO_4)]\cdot H_2O$ compound (1)

The structure of $(C_2H_{10}N_2)[Fe(HAsO_4)_2(H_2AsO_4)]\cdot H_2O$ consists of $[Fe(HAsO_4)_2(H_2AsO_4)]^{2-}$ anionic chains running along the [001] direction. The ethylenediammonium cations are displayed in the cavities of the structure delimited by three different chains and compensating its negative charge (Figure 2). The water molecules are located in the cavities formed by three arsenate anions belonging to different chains (see Figure 2).

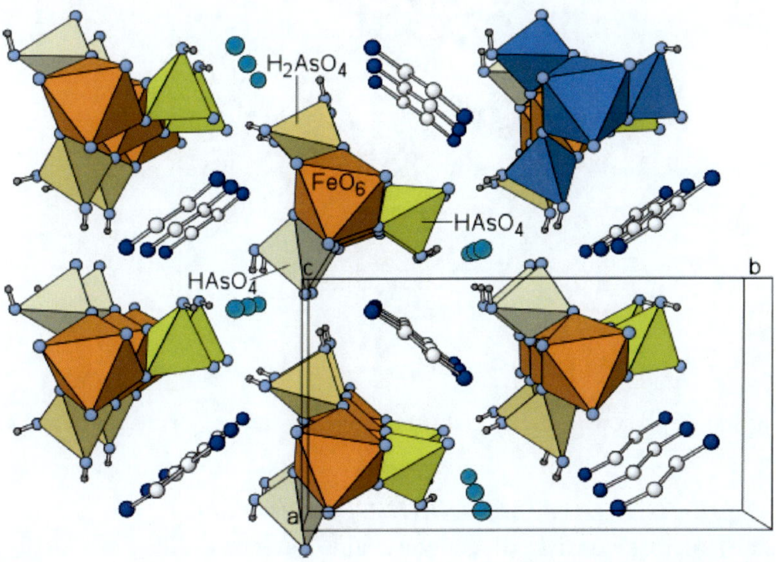

Figure 2. Polyhedral view showing the crystal structure in the form of chains of (1) and its SBU unit (blue color).

The FeO_6 octahedra are isolated and bridged by two $HAsO_4$ and one H_2AsO_4 tetrahedra. The crystal structure shows a SBU-4 unit formed by one FeO_6 octahedron, two $HAsO_4$ tetrahedra and one H_2AsO_4 tetrahedon. The union of these units along the direction [001] gives rise to the chains. This structural framework is stabilized by ionic interactions and hydrogen bonds established by the ethylenediammonium cations and the water molecules linked to the hydrogen- and dihydrogen-arsenate groups [9].

3.1.2. ∞[$(NH_3(CH_2)_2NH(CH_2)_2NH_3][Fe_2F_4(HAsO_4)_2]$ compound (2)

The crystal structure of this phase consists of $[Fe_2(HAsO_4)_2F_4]^{2-}$ anionic chains extended along the [100] direction (Figure 3). The chains are constructed by edge-sharing $[Fe_2O_6F_4]$ dimmeric SBU-4 units, linked to two $(HAsO_4)$ tetrahedra. The link between these two entities, octahedral and tetrahedral, through a common vertex gives rise to the chains. Every chain is surrounded by other six ones. The ethylelediammonium cations are disposed between these chains and establish hydrogen bonds and ionic interactions with the inorganic framework [10].

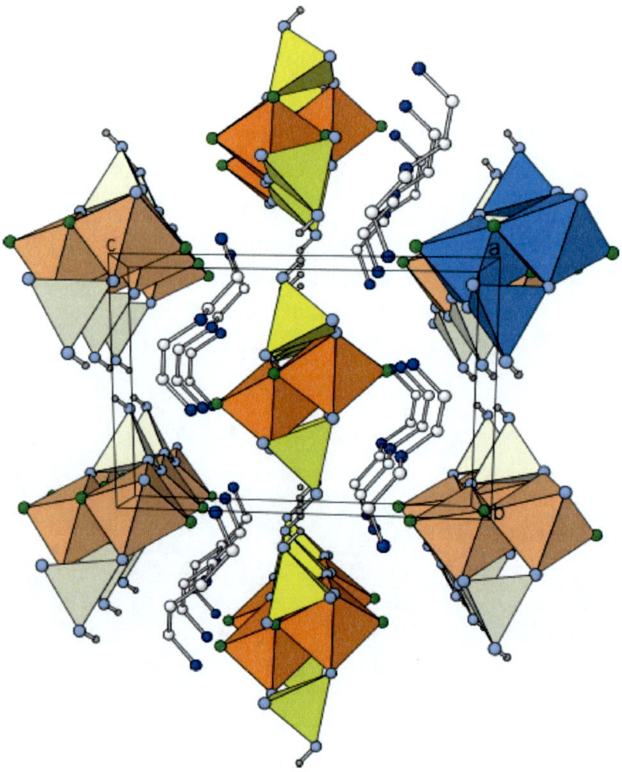

Figure 3. Polyhedral view showing the crystal structure in the form of chains of (2) and its SBU unit (blue color).

3.1.3. $(C_4H_{12}N_2)[Fe(OH)(HAsO_4)(C_2O_4)](H_2O]$ compound (3)

The crystal structure consists of double chains extentend along the (010) plane. These chains have termianal oxalate $(C_2O_4)^{2-}$ anions that establish hydrogen bonds with the hydrogenarseniate groups (Figure 4). The piperazinium cations and the water molecules, disposed in the plane (010), are located between the chains forming hydrogen bonds and establishing ionic interactions with the inorganic framework. Inside the chains there are SBU-2 units. They are constructed by an octahedron (FeO_6) and a tetrahedron $(HAsO_4)$ linked *via* common vertex. The link between these units through the vertex gives rise to the double chains [11].

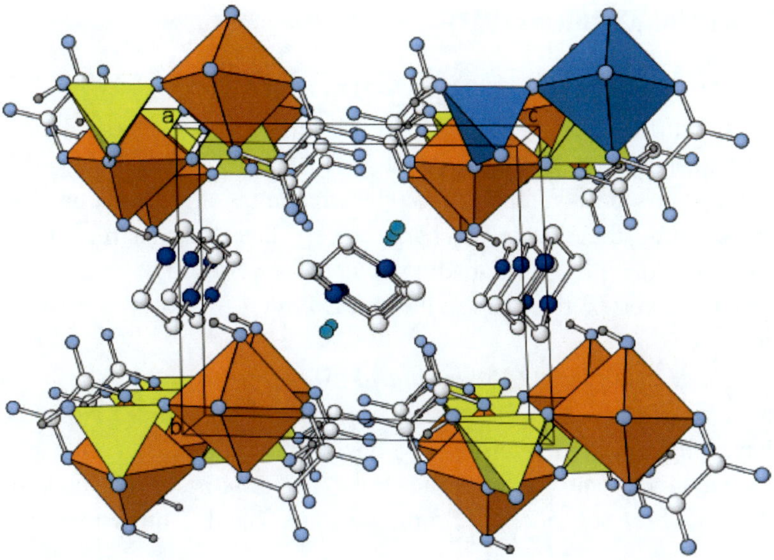

Figure 4. Polyhedral view showing the crystal structure in the form of chains of (3) and its SBU unit (blue color).

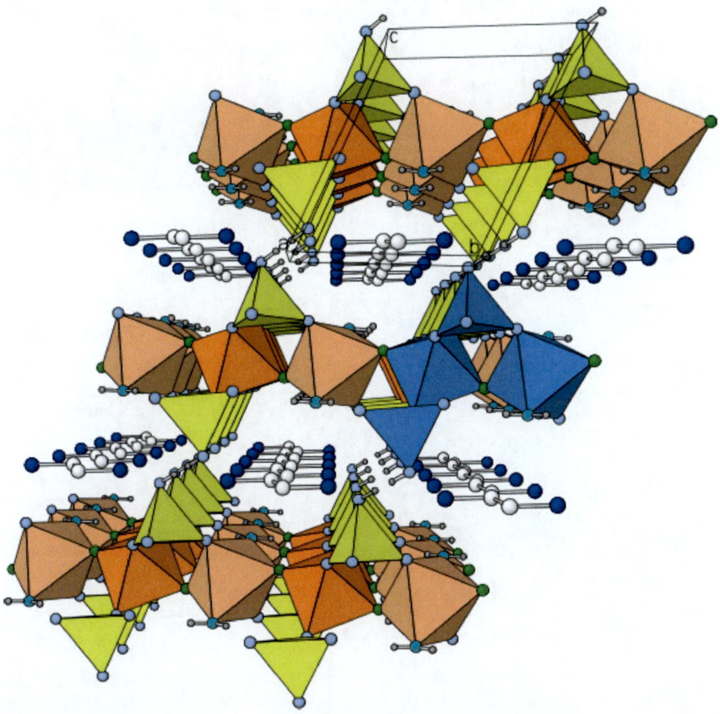

Figure 5. Polyhedral view showing the layered crystal structure of (4) and its SBU unit (blue color).

3.2. Two-Dimensional Phases (2D)

3.2.1. $(C_2H_{10}N_2)[Fe^{III}Fe^{II}(HAsO_4)(AsO_4)F_2].2H_2O$ compound (4)

The crystal structure is formed by $[Fe^{III}Fe^{II}(HAsO_4)(AsO_4)]^{2-}$ layers extended in the (001) plane (Figure 5). Inside these layers exist octahedra that form chains linked between them through the hydrogenarsenate anions. The etylenediammonium cations are located between the sheets compensating their negative charge. Every layer is constructed by SBU-4 units formed by two octahedra and two tetrahedra which share vertex. The link between the SBU-4 units through common vertex gives rise to the sheets [9b].

3.2.2. $(C_3H_{12}N_2)]_{0.5}[Fe(OH)(AsO_4)]$ compound (5)

The structure consists of $[Fe(OH)(AsO_4)]^-$ anionic chains extended along the (010) plane. The chains have zig-zag chains formed by octahedral linked by vertex along the [101] direction (Figure 6). Every sheet is constructed by SBU-2 units, an octahedron (FeO_6) and a tetrahedron (AsO_4). These secondary building units are linked to other four giving rise to the layers [12].

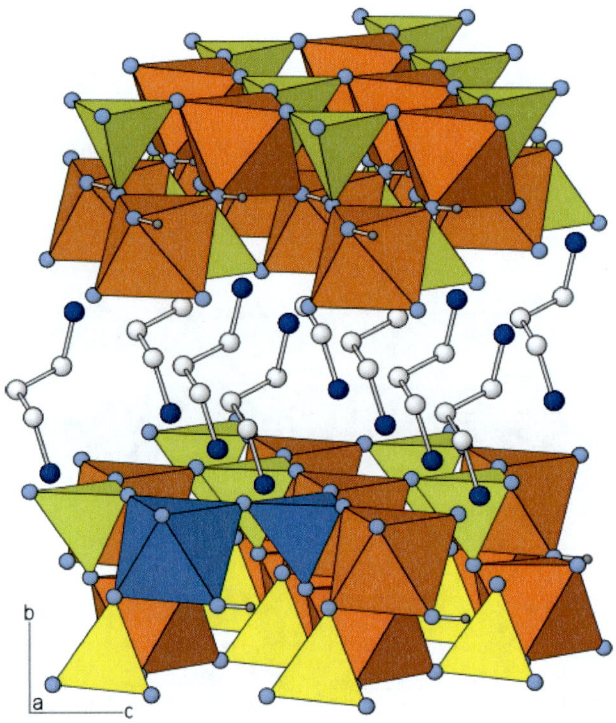

Figure 6. Polyhedral view showing the layered crystal structure of (5) and its SBU unit (blue color).

3.2.3. $(C_4H_{12}N_2)_{1.5}[Fe_3(HAsO_4)_2(AsO_4)F_5]$ compound (6) and $(C_4H_{12}N_2)_{1.5}[Fe_3(HAsO_4)_{1.02}(HPO_4)_{0.98}(AsO_4)_{0.88}(PO_4)_{0.12}F_5]$ compound (7)

The $(C_4H_{12}N_2)_{1.5}[Fe_3(HAsO_4)_2(AsO_4)F_5]$, synthesized by K-H. Lii et al. [13a], and the $(C_4H_{12}N_2)_{1.5}[Fe_3(HAsO_4)_{1.02}(HPO_4)_{0.98}(AsO_4)_{0.88}(PO_4)_{0.12}F_5]$ phase [13b] are isomorphous. We have studied the mixed arsenate-phosphate compound. $(C_4H_{12}N_2)_{1.5}[Fe_3(HAsO_4)_{1.02}(HPO_4)_{0.98}(AsO_4)_{0.88}(PO_4)_{0.12}F_5]$ shows a layered structure constructed from $[Fe_3(HAsO_4)_{1.02}(HPO_4)_{0.98}(AsO_4)_{0.88}(PO_4)_{0.12}F_5]^{3-}$ inorganic sheets extended in the (100) plane (Figure 7). The piperazinium dications are located in the interlayer space. The anionic charge of the inorganic skeleton is compensated by the organic dications, which also establish hydrogen bonds with the inorganic fragment.

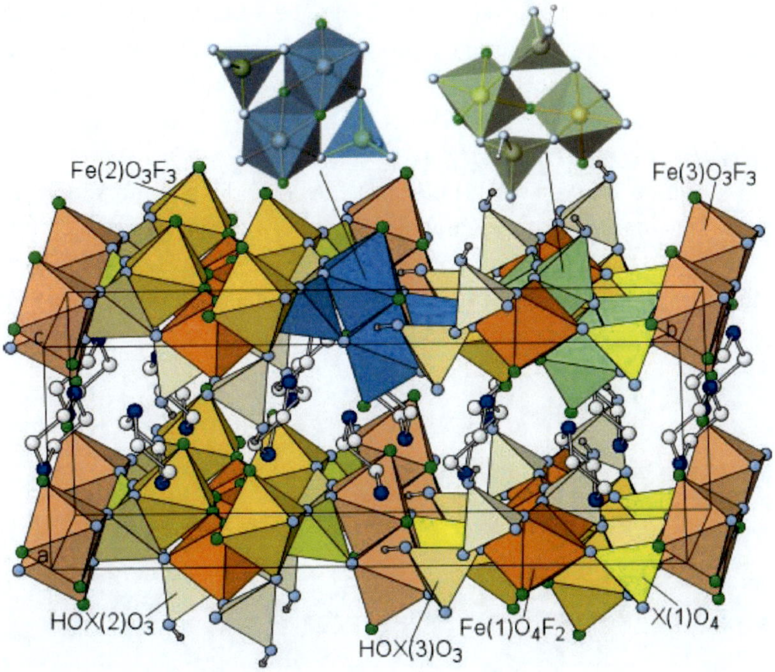

Figure 7. Polyhedral view showing the layered crystal structure of (6 and 7) and their SBU units (blue and green colors).

The layers are formed from chains built by vertex-sharing $Fe(1)O_4F_2$ and $Fe(2)O_3F_3$ octahedra, running along the *c*-axis, and the edge-sharing $Fe(3)_2O_6F_4$ dimeric entities. The chains and the $Fe(3)_2O_6F_4$ dimmeric units are connected through the $X(1)O_4$-arsenate/phosphate and $HOX(3)O_3$-hydrogenarsenate/phosphate anions. Two types of cavities can be distinguished in the structure: *i)* Cavities of six-members delimited by one $Fe(2)O_3F_3$ and two $Fe(1)O_4F_2$ octahedra, belonging to the chains, which are linked through the $X(1)O_4$ and $HOX(3)O_3$ groups to the $Fe(3)O_3F_3$ octahedron from the $Fe(3)_2O_6F_4$ dimers; *ii)* Cavities of eight-members formed by two $Fe(2)O_3F_3$ and two $Fe(3)O_3F_3$ octahedra linked in alternate way by the $X(1)O_4$ and $HOX(3)O_3$ anions. The piperazinium dications are located inside of both types of cavities, which gives rise to channels in the [100] direction by stacking of the layers.

The layer $[Fe_3(HAsO_4)_{1.02}(HPO_4)_{0.98}(AsO_4)_{0.88}(PO_4)_{0.12}F_5]^{3-}$ formed by two units SBU-4. The first, named SBU-4 (green color) is formed by two octahedral, $[Fe(1)O_4F_2]$ and $[Fe(2)O_3F_3]$, which share a vertex. The octahedra are linked to two tetrahedra, $(HOX(2)O_3)$ (X= As, $As_{0.38}/P_{0.62}$) and $(HOX(3)O_3)$ (X= As, $As_{0.64}/P_{0.36}$) every tetrahedron share two vertex with every octahedron. The link of these kinds of units gives rise to the chains of octahedra along the [001] direction. The SBU-4 is similar to that above mentioned (blue color) but in this case the octahedra form dimmers, $[Fe(3)_2O_6F_4]$ with a common egde that corresponds to two fluorine atoms. These dimmers are linked to two tetrahedra $(X(1)O_4)$ (X= As, $As_{0.88}/P_{0.12}$), every one of them share two vertex with the octahedra dimmerized. The link of these two types of SBU *via* the $(X(1)O_4)$ and $(HOX(3)O_3)$ form the sheets.

$(C_4H_{12}N_2)_{1.5}[Fe_3(HAsO_4)_{2-x}(HPO_4)_{1-y}(AsO_4)_x(PO_4)_yF_5]$ with x= 0.67, 0.88, 0.97 and y= 0.33, 0.12, 0.03, form a finite solid solution for which the $(C_4H_{12}N_2)_{1.5}[Fe_3(HAsO_4)_2(AsO_4)F_5]$ component can be obtained in the laboratory but the $(C_4H_{12}N_2)_{1.5}[Fe_3(HPO_4)_2(PO_4)F_5]$ phase can not be synthesized in similar synthetic conditions [13b].

3.2.4. $(C_{10}H_{28}N_4)[\{FeF(OH)(HAsO_4)\}_4]$ compound (8)

The crystal structure consists of $[\{FeF(OH)(HAsO_4)\}_4]^{4-}$ anionic sheets extended along the (100) plane. The organic cations are disposed between the layers compensating their negative charge (Figure 8). The sheets are formed by SBU-4 units, formed by two $[Fe_2O_8F_2]$ dimmeric octahedral linked via vertex to two hydrogenarsenate anions. Every SBU is bonded to other four ones forming the inorganic framework [14].

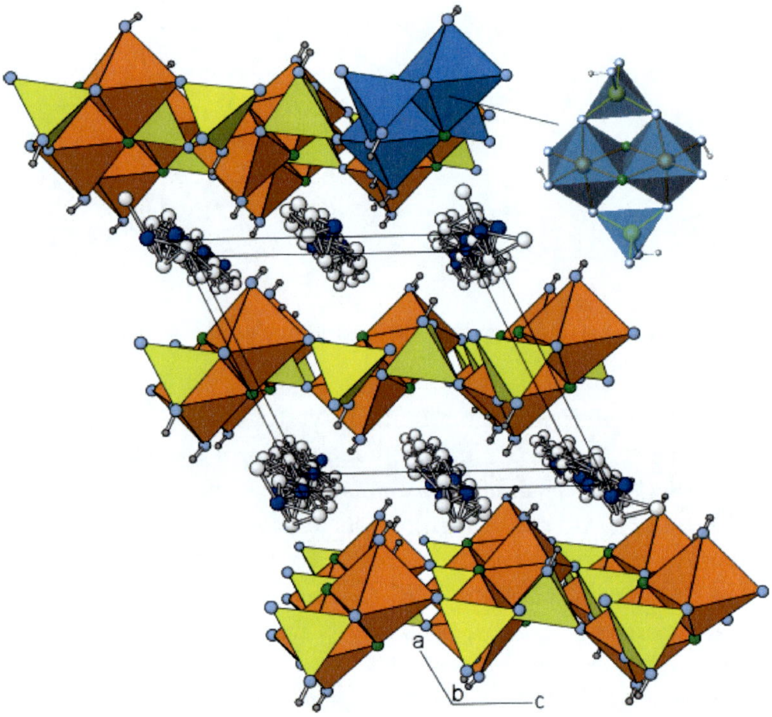

Figure 8. Polyhedral view showing the layered crystal structure of (8) and its SBU unit (blue color).

3.3. Three-Dimensional Phases (3D)

3.3.1. $(C_2H_{10}N_2)[Fe_2(AsO_4)_2(H_2O)F_2]$ compound (9) and $(C_2H_{10}N_2)[Fe_2(AsO_4)_{1.54}(PO_4)_{0.46}F_2(H_2O)]$ compound (10)

The $(C_2H_{10}N_2)[Fe_2(AsO_4)_2(H_2O)F_2]$ was synthesized by Ekambaran et *al.* [9b]. We have studied the related isomorphous $(C_2H_{10}N_2)[Fe_2(AsO_4)_{1.54}(PO_4)_{0.46}F_2(H_2O)]$ phase [15], which consists of a three-dimensional framework with channels along the *a* axis, where the etylenediammoium cations are placed (Figure 9). The organic cations stabilize the crystal structure establishing both ionic interactions and hydrogen bonds with the inorganic skeleton. Two types of puckered layers, which are stacked alternatively along the *c* axis, assemble the three dimensional network. Layer I is formed by tetrameric units constructed from $[Fe(1)_2O_8F_2]$ dimmers of $F(1),F(1)^{iii}$-edge-sharing octahedra interconnected to the $(X(1)O_4)$ tetrahedra through the $O(2)^{ii},O(3)^{i}$-oxygen atoms. Layer II contains $[Fe(2)O_4FO_w]$ octahedral and $(X(2)O_4)$ tetrahedra. This structural network has also been found in the isomorphous $(C_2H_{10}N_2)[Fe_2(AsO_4)_2F_2(H_2O)]$ [9b]. Layer I contain dimmers $[Fe(1)_2O_8F_2]$ formed by two octahedra sharing an edge, which give rise to tetrameric unit SBU-4, linked to two $X(1)O_4$ arsenate groups, through an common edge formed by two fluorine atoms. Every arsenate shares two vertex with the $[Fe(1)_2O_8F_2]$ dimmer. Layer II is formed by two SBU-2 unit constructed by an octahedron $[Fe(2)O_4FO_w]$ and one tetrahedron $(X(2)O_4)$ (X= As, $As_{0.70}/P_{0.30}$) sharing a vertex. The link of these SBU-2 units give rise to the layers [15,9b].

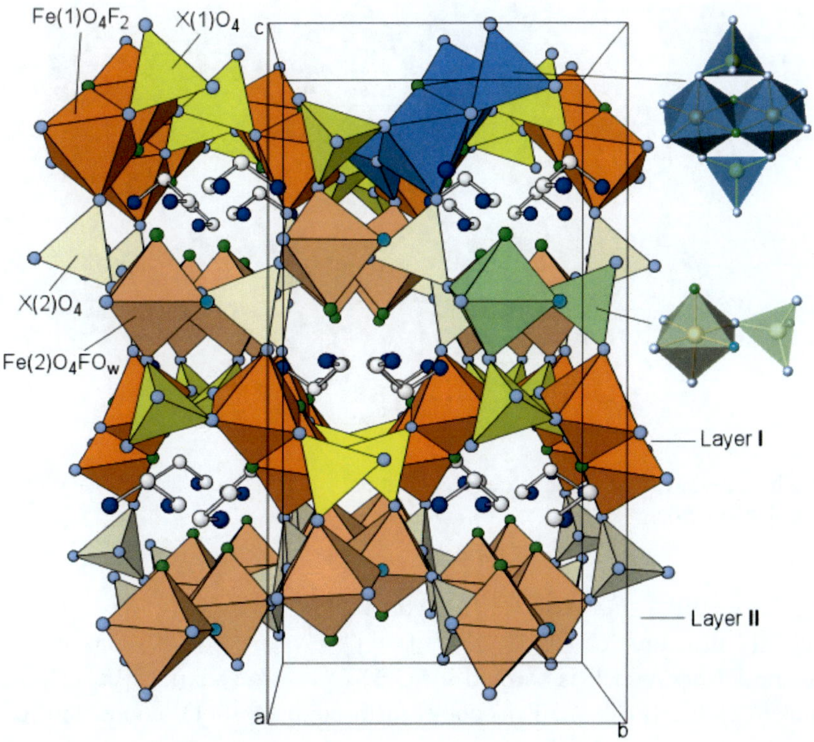

Figure 9. Polyhedral view showing the three-dimensional crystal structure of (9 and 10) and their SBU units (blue and green colors).

3.3.2. $(C_2H_{10}N_2)(NH_3)(NH_4)[Fe_3(HAsO_4)_6]$ compound (11)

The structure consists of a three-dimensional framework constructed by $[Fe_3(HAsO_4)_6]^{3-}$ anionic layers extended along the (001) plane [9b]. These sheets are connected between them through the hydrogenarsenate anions giving rise to the three-dimensional network (Figure 10). Inside the layers channels in the [1-10] and [110] direction can be distinguished, in which the etylendiammonium cations are located. The layers are formed by SBU-6 units, constructed by two octahedra $[FeO_6]$ and four tetrahedra $(HAsO_4)$. The octahedra are linked *via* vertex-shared with the hydrogenarseniate oxoanions. The link of these SBU-6 units gives rise to the layers of the three-dimensional compound.

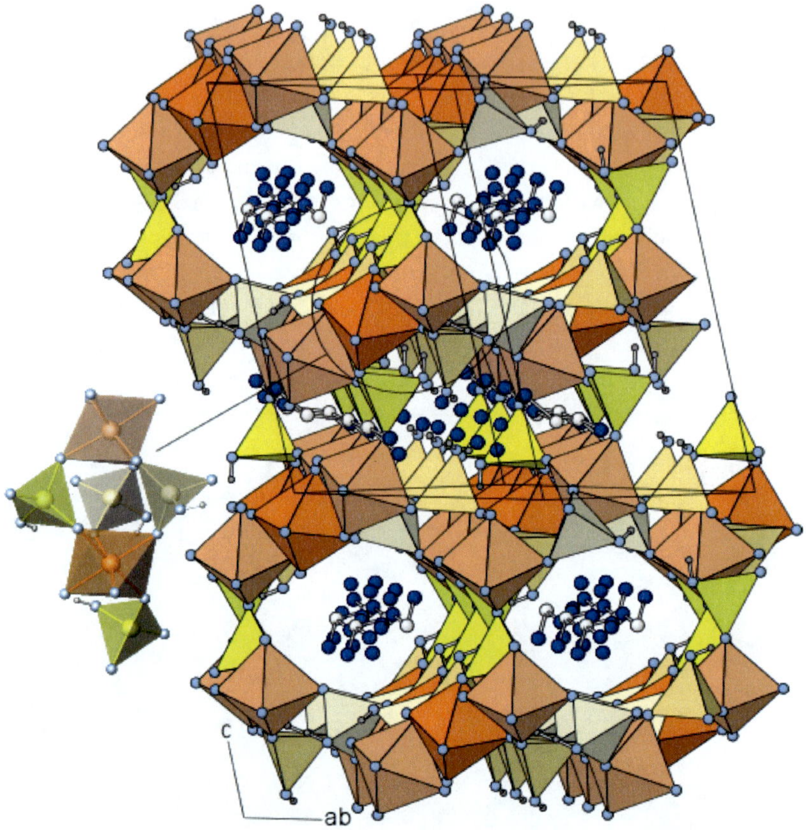

Figure 10. Polyhedral view showing the three-dimensional crystal structure of (11) and its SBU units (orange and yellow colors).

3.3.3. $(C_3H_{12}N_2)_{0.75}[Fe^{II}_{1.5}Fe^{III}_{1.5}(AsO_4)F_6]$ compound (12)

The crystal structure of $(C_3H_{12}N_2)_{0.75}[Fe^{II}_{1.5}Fe^{III}_{1.5}(AsO_4)F_6]$ consists of a three-dimensional framework which is formed by O(1)-O(1)-edge-sharing $[Fe^{II}(1)_2O_2F_8]$ dimeric octahedra linked to the $[Fe^{III}(2)O_2F_4]$ octahedra through the F(1) anions and to the (AsO_4) tetrahedra by the O(1) vertices. The latter octahedron shares the O(2) vertex with the (AsO_4) tetrahedron (Figure 11). This network yields two kinds of chains which are extended in perpendicular directions. Chain 1 is extended along the *a*-axis and is constructed from F(2)F(2)-edge-sharing $[Fe^{II}(1)_2O_2F_8]$ dimeric octahedra. Chain 2 runs along the *c*-axis and is formed by $[Fe^{III}(2)O_2F_4]$ octahedra and $(AsO_4])$ tetrahedra bonded between them through the O(2) atoms. These chains are linked by the F(1) and O(1) atoms and form cavities delimited by eight or six polyhedra along the [100] and [001] directions, respectively. The propanediammonium cations are located inside these cavities.

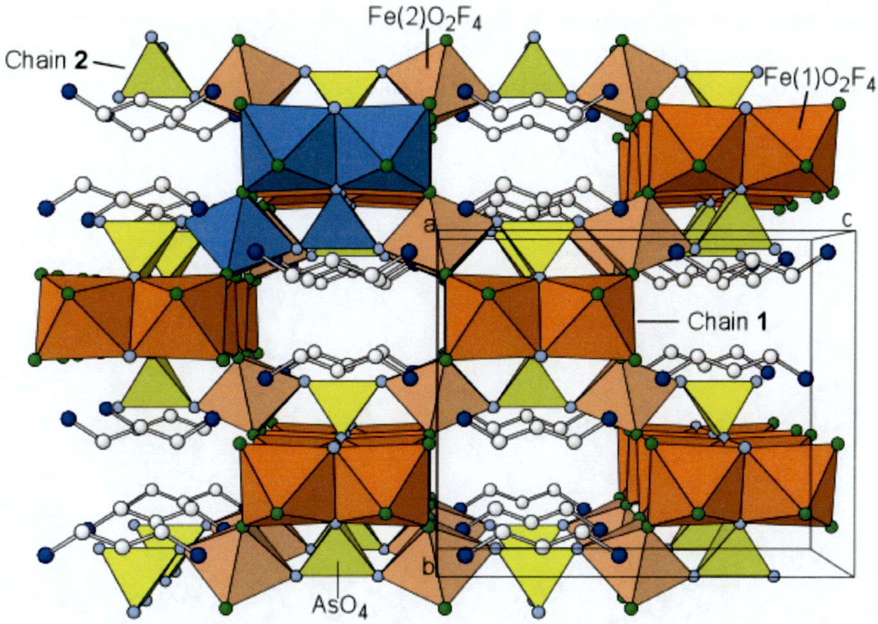

Figure 11. Polyhedral view showing the three-dimensional crystal structure of (12) and its SBU unit (blue color).

The crystal structure of this compound is formed by SBU-4 units. This unit contains one dimmer of octahedra [$Fe(1)_2F_8O_2$] sharing the edge formed by two oxygen O(1). This dimmer is simultaneously linked to the octahedron [$Fe^{III}(2)F_4O_2$], through a vertex formed by the F(1) atom and to (AsO_4) tetrahedron by the O(1) atom. Furthermore, the octahedron shares the O(2) vertex with the tetrahedron. The link of these SBU-4 units originates two kinds of chains extended along perpendicular directions [16].

3.3.4. $(C_4H_{12}N_2)_2[Fe_4(HAsO_4)_6(C_2O_4)_2]$ compound (13)

This three-dimensional compound is formed by $[Fe_4(HAsO_4)_6(C_2O_4)_2]^{4-}$ anionic layers extended along the (-10-1) plane (Figure 12). The sheets are linked through the oxalate anions to give the three-dimensional network. The layers delimit holes along the [010] direction in which the piperazinium cations are located. Every sheet is formed by two different secondary building units. The first, is a SBU-2 formed by one octahedron (FeO_6) and one tetrahedron ($HAsO_4$) sharing a common vertex. The second one, is a SBU-3 unit constructed by one octahedron and two ($HAsO_4$) tetrahedra in which every tetrahedron shares a common vertex with the adjacent octahedron. The link of these two SBU *via* shared-vertex gives rise to a layer where double chains linked between them through the hydrogenarsenate oxoanions can be distinguished [17].

3.3.5. $(C_4H_{12}N_2)[Fe^{III}{}_2Fe^{II}(HAsO_4)_2(AsO_4)_2]$ compound (14)

The crystal structure of this phase is formed by a network of [FeO_6] octahedra and (AsO_4) tetrahedra, linked *via* common vertex, giving rise to the three-dimensional framework (Figure 13). In the structure channels along the [001] direction can be distinguished, where the piperazinium cations are located. This three-dimensional phase could be also described as formed by SBU-7 units, that are constructed by three octahedra and four tetrahedra. Every three octahedra are linked *via* edge-shared giving rise to trimmeric octahedra, which are simultaneously linked through vertex-shared to the four tetrahedra. The link of these SBU-7 units gives rise to the three-dimensional structure of this phase [11].

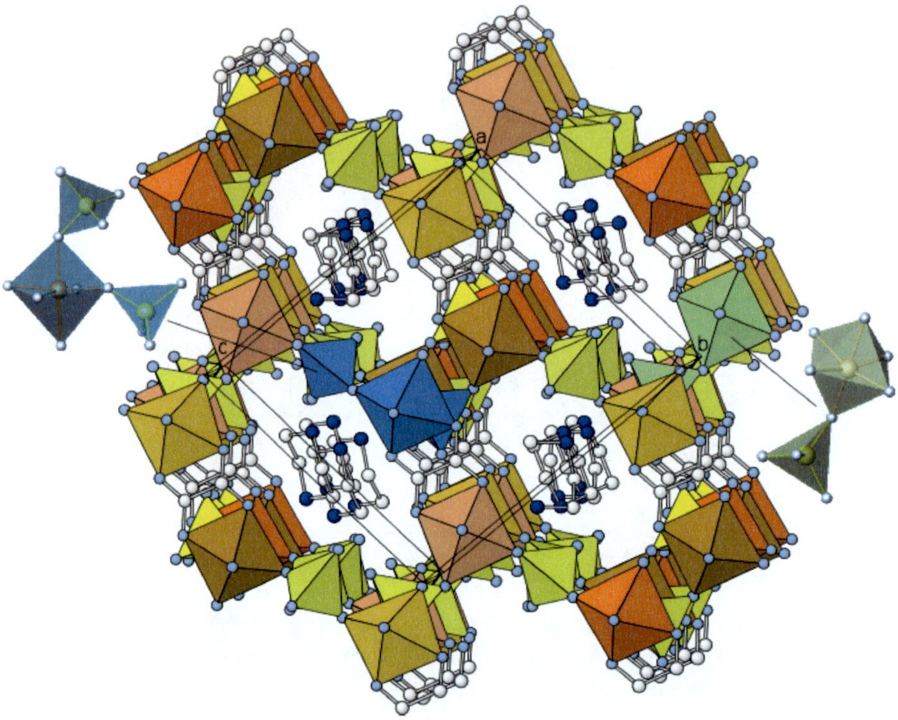

Figure 12. Polyhedral view showing the three-dimensional crystal structure of (13) and its SBU units (blue and green colors).

3.3.6. $(C_6H_{14}N_2)[Fe_3(HAsO_4)_2(AsO_4)F_4]\cdot 0.5H_2O$ compound (15) and $(C_6H_{14}N_2)[Fe_3(HAsO_4)_{1.33}(HPO_4)_{0.67}(AsO_4)_{0.84}(PO_4)_{0.16}F_4]\cdot 0.5H_2O$ (12) compound (16)

The structure of these phases consists of a three-dimensional framework formed by $[Fe_3(HAsO_4)_2(AsO_4)F_4]^{2-}$ inorganic layers extended along the (10-1) plane and interconnected by hydrogenarsenate groups [18a]. The 1,4-diazabicyclooctane dications compensate for the anionic charge of the sheets and are placed together with the water molecule in the interlayer space (Figure 14).

The layers are formed by $[Fe(1)Fe(3)O_6F_4]$ and $[Fe(2)_2O_8F_2]$ dimeric edge-sharing octahedra. The first dimmers are linked through a vertex, giving rise to chains running along the *b* axis. The chains and the $[Fe(2)_2O_8F_2]$ dimmeric units are connected through the $As(1)O_4$-arsenates and $HOAs(3)O_3$-hydrogenarsenates anions. Two types of cavities can be distinguished in the structure. The cavities of six members are delimited by two $[Fe(3)O_3F_3]$ and one $[Fe(1)O_3F_3]$ octahedra, which are linked by the $(As(1)O_4)$ and $(HOAs(3)O_3)$ groups to the $[Fe(2)O_4F_2]$ octahedra. The cavities of eight members are formed by two $[Fe(1)O_3F_3]$ and two $[Fe_2O_4F_2]$ octahedra linked in an alternate way by the $(As(1)O_4)$ and $(HOAs(3)O_3)$ anions. The 1,4-diazabicyclooctane dications are located above and below the mean plane determined by these cavities at a distance of, approximately, 6.6Å. Another organic dication appears in every side of the sheets, which is placed at a distance of, approximately, 6.4Å with respect to the above position. The layers are interconnected by the $HOAs(2)O_3$-hydrogenarsenate groups. The O(7) and O(8) oxygen atoms of this anion are linked to the $[Fe(1)O_3F_3]$ and $[Fe(3)O_3F_3]$, octahedra, respectively, whereas the O(6) atom is linked to the $Fe(2)O_4F_2$ octahedra. The water molecule is placed between two $(HOAs(2)O_3)$ anions at a distance of, approximately, 3.6Å from every hydrogenarsenate.

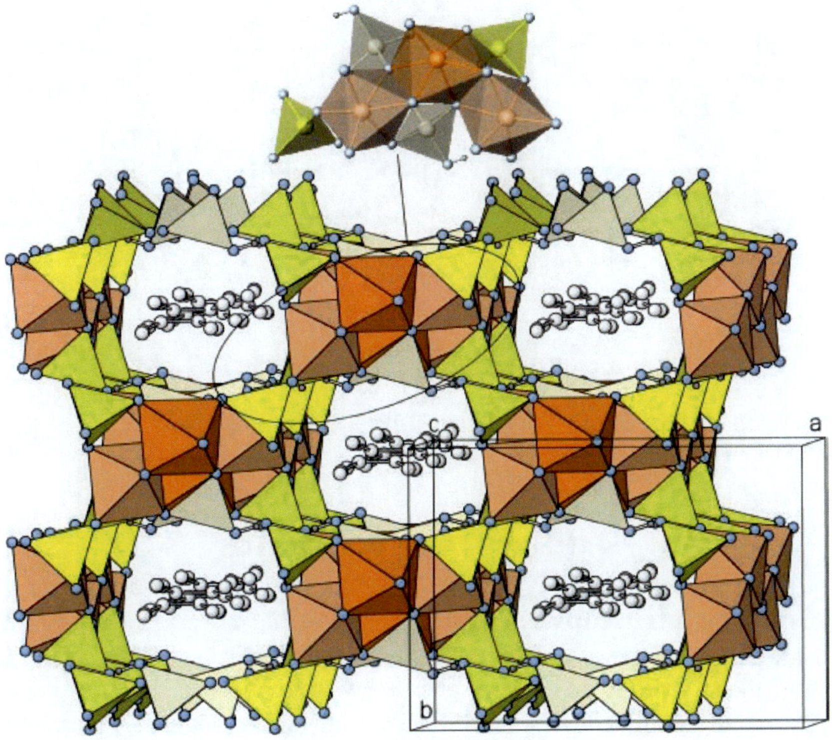

Figure 13. Polyhedral view showing the three-dimensional crystal structure of (14) and its SBU units (orange and yellow colors).

The related $(C_6H_{14}N_2)[Fe_3(HAsO_4)_{1.33}(HPO_4)_{0.67}(AsO_4)_{0.84}(PO_4)_{0.16}F_4]\cdot 0.5H_2O$ phase [18b], in which a fraction of the arsenate and hydrogenarsenate anions are substituted by the phosphate and hydrogenphosphate groups, is ismorphous with the above described arsenate compound. The layers are formed by two types of SBU units every one constructed by four polyhedra. The SBU-4 unit is formed by two octahedra $[Fe(1)O_3F_3]$ and $[Fe(3)O_3F_3]$ linked by a vertex corresponding to one fluorine atom. These octahedra are linked to two tetrahedra $(HOX(2)O_3)$ (X= As, $As_{0.66}/P_{0.34}$) and $(HOX(3)O_3)$ (X= As, $As_{0.67}/P_{0.33}$), every tetrahedron shares two vertex, formed by two oxygen atoms, with the octahedra. This SBU is linked to other equivalent *via* a common edge, formed by two fluorine atoms F(2), giving rise to chains alon the *b*-axis. The SBU-4 unit (blue color) is formed by dimmerized octahedra $[Fe(2)_2O_8F_2]$, sharing the edge formed by two fluoride anions. These dimmers are linked to two tetrahedra $(X(1)O_4)$ (X= As, As/P) through the O(2) and O(4) atoms. The link of these two SBU takes place through the $(X(1)O_4)$ (X= As, $As_{0.84}/P_{0.16}$) and $(HOX(3)O_3)$ X= As, $As_{0.67}/P_{0.33}$) tetrahedral giving rise to the layers [18b].

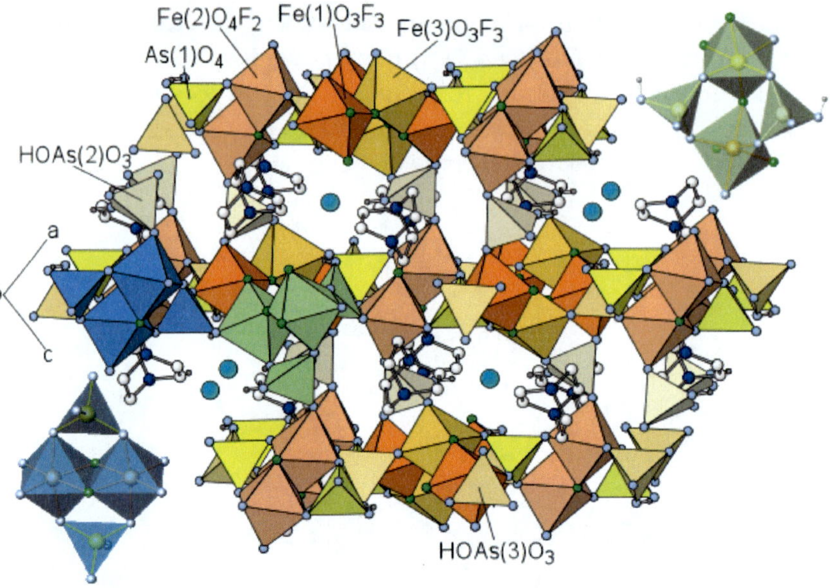

Figure 14. Polyhedral view showing the three-dimensional crystal structure of (15 and 16) and their SBU units (blue and green colors).

4. Physical Properties

4.1. UV-Visible Spectroscopy

The absorption spectra of these phases were collected by diffuse reflectance spectroscopy. In all compounds several bands with weak intensity were observed, as corresponds to a d^5 high spin iron(III) cation in slightly distorted octahedral symmetry (see Figure 15). The Dq and Racah (B and C) parameters were calculated by fitting the experimental frequencies, given in Table 3, to the energy equations derived by Tanabe and Sugano [19] and resumed in the Orgel's diagrams [20] for the Fe(III) cation.

In particular, the diffuse reflectance spectrum of the mixed valent $(C_3H_{12}N_2)_{0.75}[Fe^{II}_{1.5}Fe^{III}_{1.5}(AsO_4)F_6]$ compound (12) exhibits two bands at approximately 8000 and 9900 cm^{-1}, characteristic of the Fe(II) cations in a slightly distorted octahedral environment. This band corresponds to the spin allowed transition from the $^5T_{2g}(^5D)$ to $^5E_{2g}(^5D)$ excited level, in good agreement with the existence of a high spin d^6 cation. The splitting of this band is due to the existence of non regular $Fe^{II}O_6$ octahedra. A strong band is also observed at approximately 15500 cm^{-1}, which was assigned to the transfer charge Fe^{2+}-Fe^{3+} intervalence transition [21]. This transition can be observed when the metallic centres are close, sharing in this case a fluoride anion. The bands corresponding to the electronic transitions of the (d^5) Fe(III) cations are weak and probably obscured by the stronger ones of the Fe(II) cations.

Table 3. Summarized data of the Visible-UV measurements

	Compounds Transitions	*(1)* (cm^{-1})	*(6)* (cm^{-1})	*(7)* (cm^{-1})	*(9)* (cm^{-1})	*(15)* (cm^{-1})	*(16)* (cm^{-1})	Transitions	*(12)* (cm^{-1})
ν_1	$^6A_{1g}(^6S)\rightarrow{}^4T_{1g}(^4G)$	11850	13405	13820	12910	12535	12790		
ν_2	$^6A_{1g}(^6S)\rightarrow{}^4T_{2g}(^4G)$	17345	18585	18880	17915	18585	18835	$t_{2g}(Fe^{2+})\rightarrow t_{2g}(Fe^{3+})$	15500
ν_3	$^6A_{1g}(^6S)\rightarrow{}^4A_{1g}$, $^4E_g(^4G)$	22940	23705	23860	23370	23345	23385	$^5T_{2g}(^5D)\rightarrow{}^5E_{2g}(^5D)$	8000, 9900
ν_4	$^6A_{1g}(^6S)\rightarrow{}^4T_{2g}(^4D)$	26345	26900	26930	26385	26595	26590		
ν_5	$^6A_{1g}(^6S)\rightarrow{}^4E_g(^4D)$	------	------	------	------	------	------		
	Dq (cm^{-1})	1010	995	1005	1100	1035	1020		
	B (cm^{-1})	1135	1065	1020	1005	1080	1070	Dq (cm^{-1})	895
	C (cm^{-1})	2320	2610	2725	2665	2500	2540		
	% B reduction	98	92	88	87	94	93		

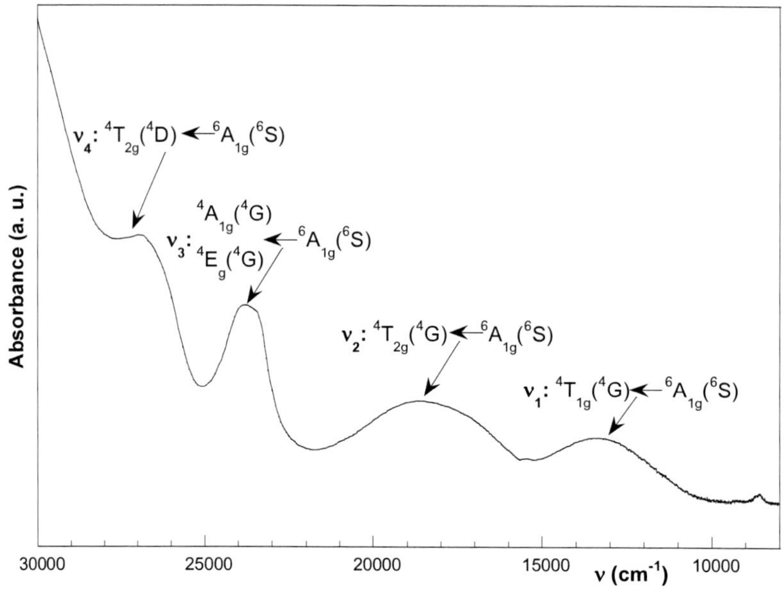

Figure 15. A general view of a diffuse reflectance spectrum for the iron(III), d^5, compounds.

4.2. Mössbauer Spectroscopy

The Mössbauer spectrum of the three-dimensional $(C_6H_{14}N_2)[Fe_3(HAsO_4)_{1.33}(HPO_4)_{0.67}(AsO_4)_{0.84}(PO_4)_{0.16}F_4]\cdot 0.5H_2O$ compound (16) recordered in the paramagnetic state at 300K is typical of high-spin ions [see Figure 16, (top)]. The spectrum exhibits two broadened lines with a weak asymmetry. Taking into account the existence of three iron(III) cations crystallographycally independent in the crystal structure of this compound, the spectrum was fitted to three doublet of Lorentzians by using the NORMOS program [22]. The values obtained of the isomer shift and the quadrupole splitting are characteristic of Fe(III) cations (see Table 4). The relative spectral areas indicate the existence of equivalent proportions for every iron(III) cation, in good agreement with the structural results.

The Mössbauer spectrum of the mixed valent three-dimensional $(C_3H_{12}N_2)_{0.75}[Fe^{II}{}_{1.5}Fe^{III}{}_{1.5}(AsO_4)F_6]$ compound (12) at 300 K is given in Figure 16. It has been simulated with the NORMOS program [22] using two doublets for iron(III) cations distributed over the positions Fe(1) and Fe(2) and one doublet for iron(II) in the Fe(1) position. The ratio of the relative spectral areas of Fe(1) and Fe(2) is near to 2:1, in good agreement with the crystallographic multiplicities for these two cations. There are two crystallographically independent positions for the irons ions. Fe(1) is placed on a binary axis with a reduction in the multiplicity of the position of 0.5, which determines the existence of eight Fe(1) in the unit cell. The Fe(2) cation is located on an inversion centre being 0.25 the reduction in the multiplicity, so, four Fe(2) appear in the unit cell. The analysis of the Fe(1)-O,F bond distances indicates that they are slightly smaller than those expected for a position completely occupied by iron(II) but slightly greater than those corresponding to iron(III). A partial substitution of Fe^{III} by Fe^{II} must therefore occur in the Fe(1) positions. The values of the isomer shift and quadrupolar splitting given in Table 4 show the characteristic values for the Fe^{II} and Fe^{III} ions.

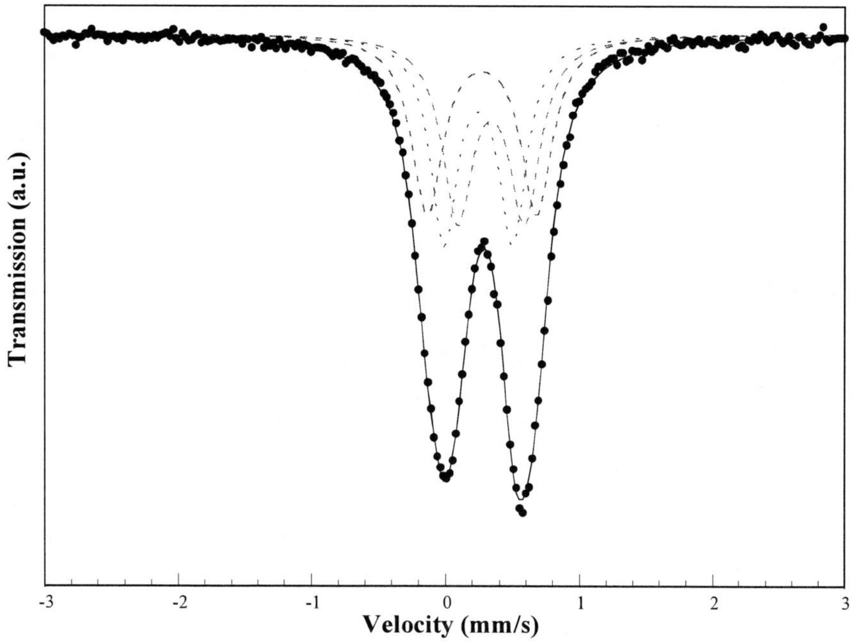

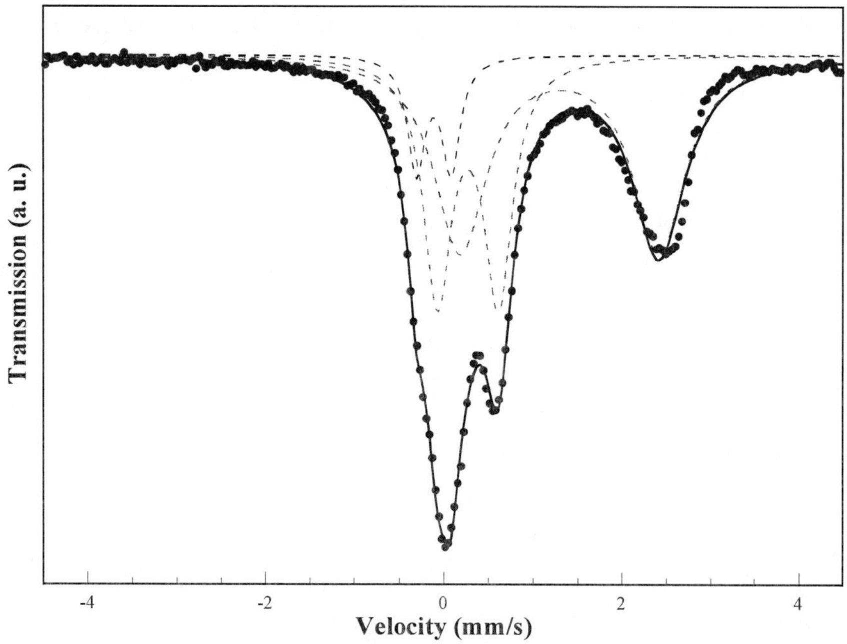

Figure 16. Mössbauer spectra at room temperature of compound (16) (top) and compound (12) (above).

Table 4. Parameters of the Mössbauer measurements for compounds (16) and (12)

	Site	ΔE ($mm.s^{-1}$)	δ ($mm.s^{-1}$)	M. R.	Atoms per unit cell		% Theoretical		% Observed	
$(C_6H_{14}N_2)[Fe_3(HAsO_4)_{1.33}$ $(HPO_4)_{0.67}(AsO_4)_{0.84}$ $(PO_4)_{0.16}F_4]$ 0.5H_2O (16)	$Fe(1)^{3+}$	0.700	0.390	1.0	8		33.3		33.3	
	$Fe(2)^{3+}$	0.400	0.360	1.0	8		33.3		33.3	
	$Fe(3)^{3+}$	0.400	0.390	1.0	8		33.3		33.3	
$(C_3H_{12}N_2)_{0.75}[Fe^{II}_{1.5}$ $Fe^{III}_{1.5}(AsO_4)F_6]$ (12)	$Fe(1)^{2+}$	2.229	1.414	0.5	8	6	66.6	49.9	64.4	53.7
	$Fe(1)^{3+}$	0.392	0.001			2		16.7		10.7
	$Fe(2)^{3+}$	0.680	0.384	0.25	4	4	33.3	33.3	35.6	35.6

4.3. Electron Spin Resonance

The X-band powdered ESR spectra of these phases containing Fe(III) cations are isotropic, from room temperature to 5 K. The signals are centred at approximately 3400 Gauss and have a "g"-value of 2.0(1), characteristic of iron(III) cations in octahedral symmetry.

For the two-dimensional $(C_4H_{12}N_2)_{1.5}[Fe_3(HAsO_4)_{1.02}(HPO_4)_{0.98}(AsO_4)_{0.88}(PO_4)_{0.12}F_5]$ compound (7) a behaviour completely different was observed. This way, the spectra remain unchanged upon cooling the sample from room temperature to approximately 90K. Below this temperature the linewidth of the spectrum broadens up to approximately 45K. When the sample is additionally cooled the linewidth of the spectrum narrows and its intensity increases vigorously. The isotropic signals observed in the 290-15K range change their line-shape below 10K and the resonance becomes less symmetrical. The resonance at 10 and 4K are centred at 244 and 216 mT, respectively, with "g"-values of 2.73(1) and 3.11(1), as a consequence of the higher internal field that shifts the signal from its usual value of 2.0(1) as expected for an iron(III) cation octahedrically coordinated (Figure 17). These features are probably due to a ferromagnetic resonance phenomenon, as was observed in other similar systems [23].

The thermal variations of the intensity and the linewidth of the signals were calculated by fitting the experimental spectra to Lorentzians curves (Figure 18). The intensity increases slightly from room temperature to approximately 50K. Below this temperature the intensity of the signals increases vigorously. The linewidth of the ESR signals slightly increases in the 300-100K temperature range, as a consequence of the dipolar homogenous broadening [24]. When the temperature is further decreased the linewidth increases vigorously due to a strong spin correlation [24], and reaches a maximum at approximately 40K. From this temperature up to approximately 20K the linewidth of the signals narrows, and they broaden again below this temperature. This later broadening is probably due to an effect of integration of the signals, caused by the large linewidth of the curves at low temperatures. The narrowing of the ESR signals, the strong increase of their intensities and the large displacement of the g-tensor from its usual value of 2.0 for the iron(III) cation, observed in this compound in the 40-15K range, can be related to the existence in this phase of a ferromagnetic resonance phenomenon as described before.

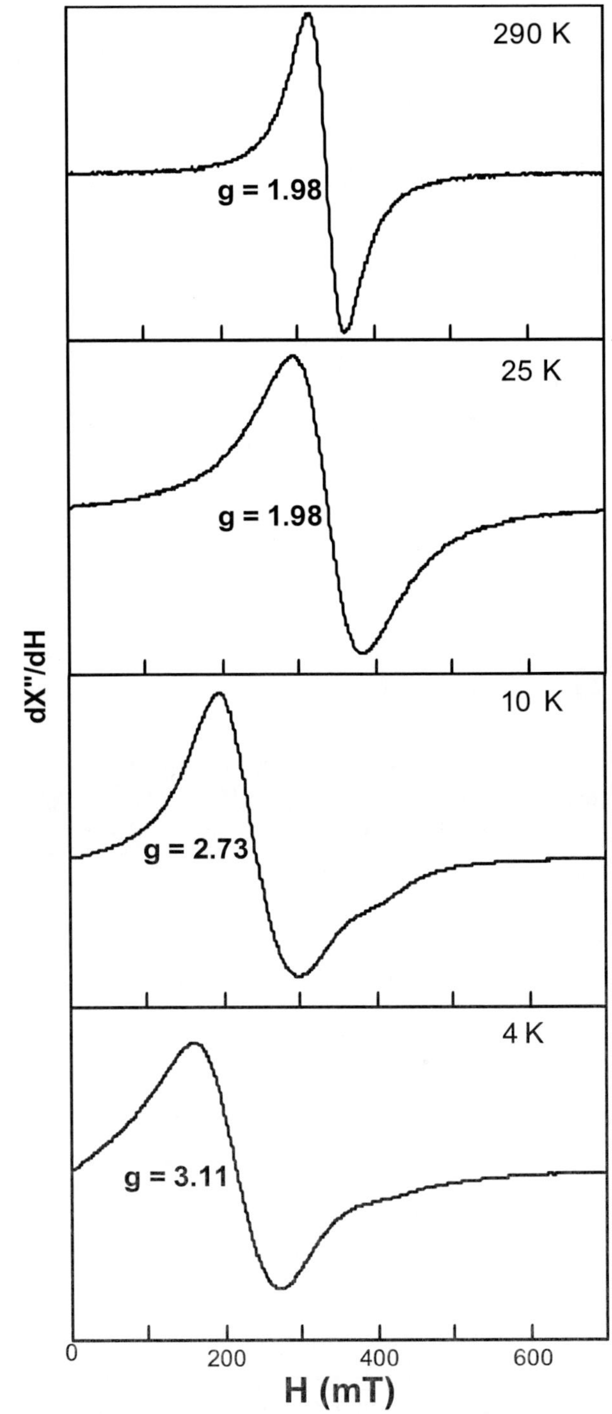

Figure 17. Thermal variation of the ESR signals for compound (7) showing a ferromagnetic resonance phenomenon.

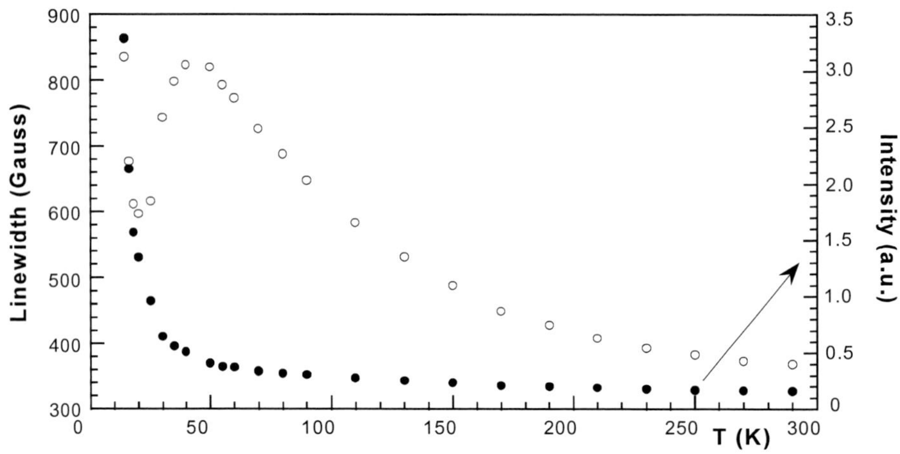

Figure 18. Thermal evolution of the linewidth and intensity of the ESR for phase (7).

4.4. Magnetic Behavior

4.4.1. Mono-Dimensional Phases (1D)

4.4.1.1. $(C_2H_{10}N_2)[Fe(HAsO_4)_2(H_2AsO_4)]\cdot H_2O$ compound (1)

Variable-temperature magnetic susceptibility measurements were carried out on a powdered sample from room temperature to 5.0K. Plot of χ_m and $\chi_m T$ *vs.* T curves are shown in Figure 19. The thermal evolution of χ_m follows the Curie-Weiss law at temperatures higher than 30K, with C_m= 4.34 cm^3K/mol and θ= -13.1K. The molar magnetic susceptibility increases from room temperature with decreasing temperature and reaches a maximum *ca.* 8K, indicating that a magnetic ordering is established at this temperature. This result together with the continuous decrease in the $\chi_m T$ *vs.* T curve, from 4.089 cm^3K/mol at room temperature up to 0.736 cm^3K/mol at 4.2K, is indicative of antiferromagnetic exchange couplings in the compound.

Considering the structural features of this compound, in which chains of $[Fe(HAsO_4)_2(H_2AsO_4)]^{2-}$ formula run along the [001] direction, the magnetic data have been fitted to the equation given by Fischer and Dingle [25,26] for chains of spin S= 5/2, Eq. (1):

$\chi = [Ng^2\beta^2S(S+1)/3kT]/[(1-n)/(1+n)]$ Eq. (1)

where $n= (T/T_0) - \coth(T_0/T)$ and $T_0= [2JS(S+1)]/k$, k is Boltzman's constant, N is Avogadro's number, and β is the Bohr magneton. The best fit (solid lines in Figure 19) is obtained for a value of the J-exchange parameter of J= -0.835K, where g= 2.0 obtained from ESR spectrum. The agreement factor, defined as $F= \Sigma i(\chi_{i,obs} - \chi_{i,calc})^2/(\chi_{i,obs})^2$, is 1.03×10^{-4}.

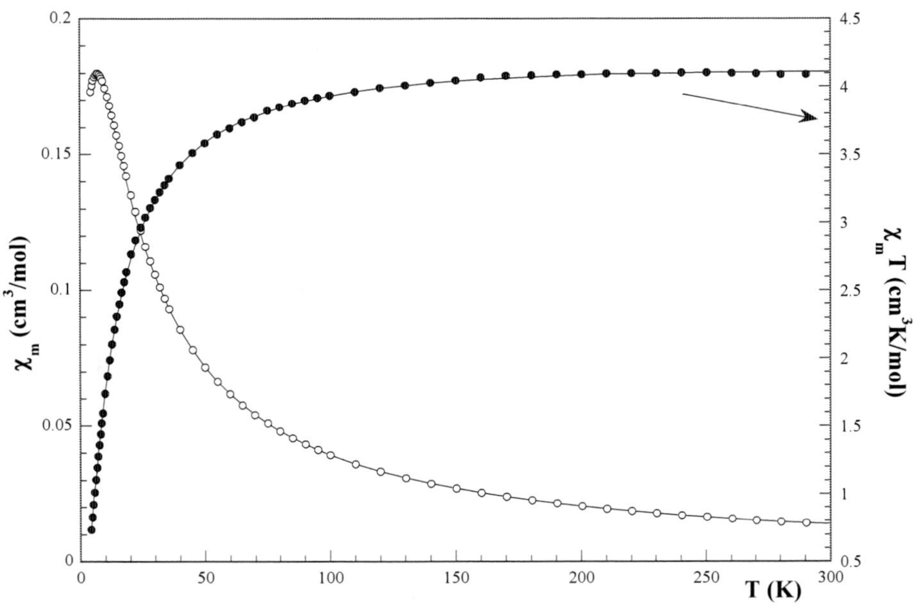

Figure 19. Thermal evolution of the χ_m and χ_mT curves for compound (1). The solid lines show the best fit to a model for lineal chains of S= 5/2.

4.4.1.2. $\infty[(NH_3(CH_2)_2NH(CH_2)_2NH_3][Fe_2F_4(HAsO_4)_2](H_2O)$ compound (2)

From the magnetic point of view this compound can be considered as a ladder-like chain. The compound shows antiferromagnetic couplings with Curie-Weiss constant of -22K and a maximum in the thermal evolution of χ_m at 14K [10].

4.4.2. Two-Dimensional Phases

4.4.2.1. $(C_2H_{10}N_2)[Fe^{III}Fe^{II}(HAsO_4)(AsO_4)F_2]\cdot 2H_2O$ compound (4)

This compound shows Curie-Weiss behavior at higher temperatures with an effective moment of 8.32 μ_B, which is somewhat higher than the calculated spin-only 7.68 μ_B for one Fe(II) and one Fe(III) in high-spin octahedral environments. The Curie-Weiss temperature is -127K and indicates strong antiferromagnetic interactions [9b].

4.4.2.2. $(C_4H_{12}N_2)_{1.5}[Fe_3(HAsO_4)_{1.02}(HPO_4)_{0.98}(AsO_4)_{0.88}(PO_4)_{0.12}F_5]$ compound (7)

Variable-temperature magnetic susceptibility measurements were performed on a powdered sample in the 300 to 5.0K temperature range, Figure 20 shows the thermal evolution of the χ_m and χ_m *vs.* T curves. χ_m increases from room temperature with decreasing temperature and reaches a maximum at 8.5K, indicating that a magnetic ordering is established at this temperature. After that, the susceptibility slightly decreases to 5K. The thermal evolution of χ_m follows the Curie-weiss law at temperature higher than 100K, with C_m= 13.52 cm^{3}K/mol and θ= -95.8K. Moreover, the χ_mT *vs.* T product (see Figure 20) decreases from 10.50 cm^{3}K/mol at room temperature down to 4.90 cm^{3}K/mol at 40.0K. However, at lower temperatures an abrupt increase is observed with a maximum at 13.4K. After this temperature the χ_mT curve again decreases down to 5K. These results indicate the existence of a ferromagnetic behavior, which can take place due to incomplete compensation of the magnetic moments of the Fe(1)O_4F_2, Fe(2)O_3F_3 and Fe(3)O_3F_3 sublattices.

In order to study the ferromagnetic contribution magnetization measurements were carried out (Figure 21). The variation of the magnetization with the magnetic field at 5K shows a small hysteresis loop in which the values of the coercive field and remnant magnetization are, approximately, 16 Gauss and 55 emu/mol, respectively (see inset in Figure 21).

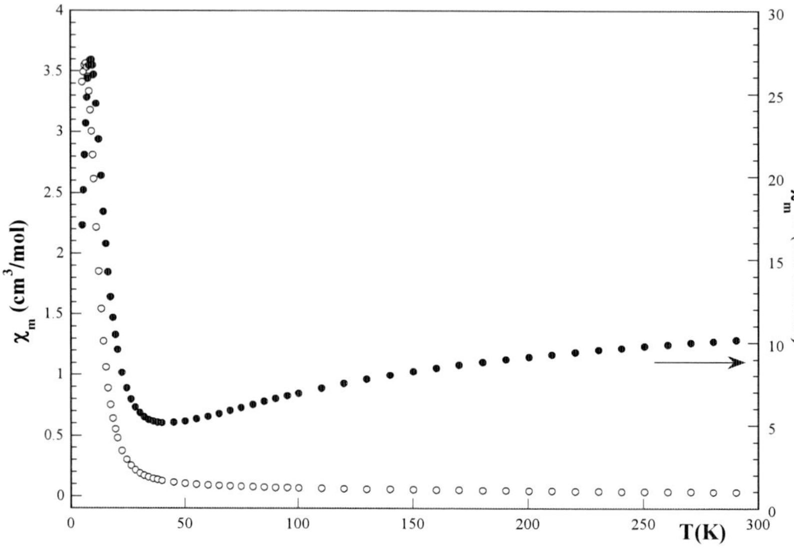

Figure 20. Thermal evolution of the χ_m and χ_mT curves for compound (7) showing its ferromagnetic behaviour.

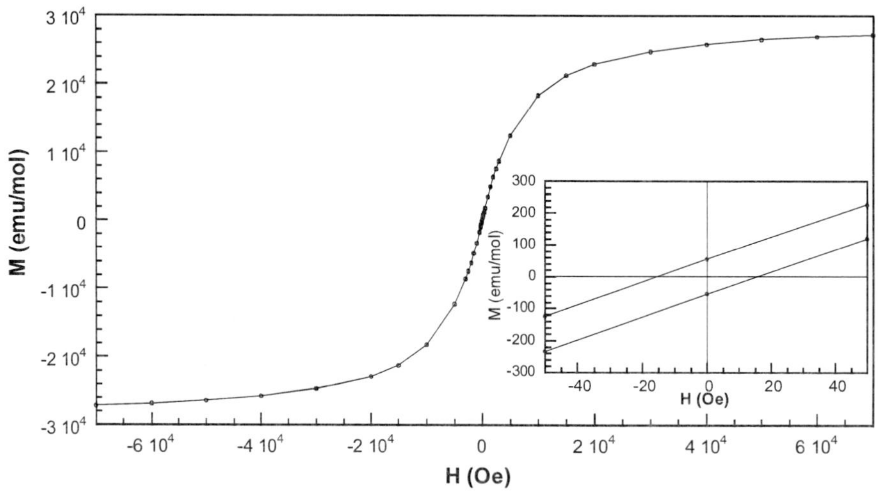

Figure 21. Hysteresis loop performed at 5K for compound (7).

Furthermore, as can be seen in Figure 21 the magnetization curve practically reaches the saturation at, approximately, 26000 emu/mol at 6 T. This value corresponds to 4.65 electrons, close to the five electrons corresponding to a high spin iron(III) cation completely magnetized. These results confirm the existence of a ferromagnetic contribution intrinsic to the sample [13b].

4.4.2.3. $(C_{10}H_{28}N_4)[\{FeF(OH)(HAsO_4)\}_4]$ compound (8)

The molar magnetic susceptibility of this compound increases with decreasing temperature and reaches a sharp maximum around 20K. If temperature is reduced further, the susceptibility decreases sharply. This behavior is characteristic of antiferromagnetic exchange couplings. The Curie constant and Curie-Weiss temperature are 16.24 cm^3K/mol and -31K, respectively. The estimated J-exchange parameter is -12K [14].

4.4.3. Three-Dimensional Phases (3D)

4.4.3.1. $(C_2H_{10}N_2)[Fe_2(AsO_4)_2(H_2O)F_2]$ compound (9)

This phase follows the Curie-Weiss law at temperatures higher than 25K with θ= -50K and μ_{eff}= 8.41 μ_B, in good agreement with two Fe(III) centres per formula unit. Below 25K the magnetization shows the beginning of antiferromagnetic ordering [9b].

4.4.3.2. $(C_2H_{10}N_2)(NH_3)(NH_4)[Fe_3(HAsO_4)_6]$ compound (11)

This compound orders ferrimagnetically below 40K. Above this temperature it is paramagnetic with θ= -53K and μ_{eff}= 10.05 μ_B, the latter in good agreement with the calculated spin-only moment for the three high-spin octahedra Fe(III) centers, 10.25 μ_B [9b].

4.4.3.3. $(C_3H_{12}N_2)_{0.75}[Fe^{II}{}_{1.5}Fe^{III}{}_{1.5}(AsO_4)F_6]$ compound (12)

Variable-temperature magnetic measurements of $(C_3H_{12}N_2)_{0.75}[Fe^{II}{}_{1.5}Fe^{III}{}_{1.5}(AsO_4)F_6]$ have been carried out on a powdered sample from 5K to room temperature after cooling in the presence (FC) and absence (ZFC) of an applied magnetic field of 1000 G. Plots of χ_m and χ_mT curves are shown in Figure 22. The thermal evolution of χ_m follows the Curie-Weiss law at temperatures higher than 100 K, with C_m= 12.40 cm^3K/mol, corresponding to the simultaneous existence of Fe(III) and Fe(II) cations in the paramagnetic state, with g-values of 2.0 and 2.2, respectively, and θ= -204.7K. The molar magnetic susceptibility increases from room temperature with decreasing temperature and reaches a maximum at ca. 6 K, indicating that a three-dimensional antiferromagnetic ordering is established at this temperature. Below 6K the magnetic susceptibility shows an abrupt decrease. The χ_mT *vs.* T curve continuously decreases from 7.37 cm^3K/mol at room temperature up to 2.0 cm^3K/mol at 22K. This result is characteristic of the existence of antiferromagnetic exchange coupling in the compound as the major magnetic interactions (Figure 22). The small maximum observed at 18K is probably due to the spin descompensation between the iron(III) and iron(II) cations present in this compound, indicating the existence of some ferromagnetic contribution around this temperature. Then, the χ_mT *vs.* T curve decreases again showing a very small second maximum which can be associated to the existence of a magnetic impurity. Finally, this curve decreases down to 1.13 cm^3K/mol at 5K.

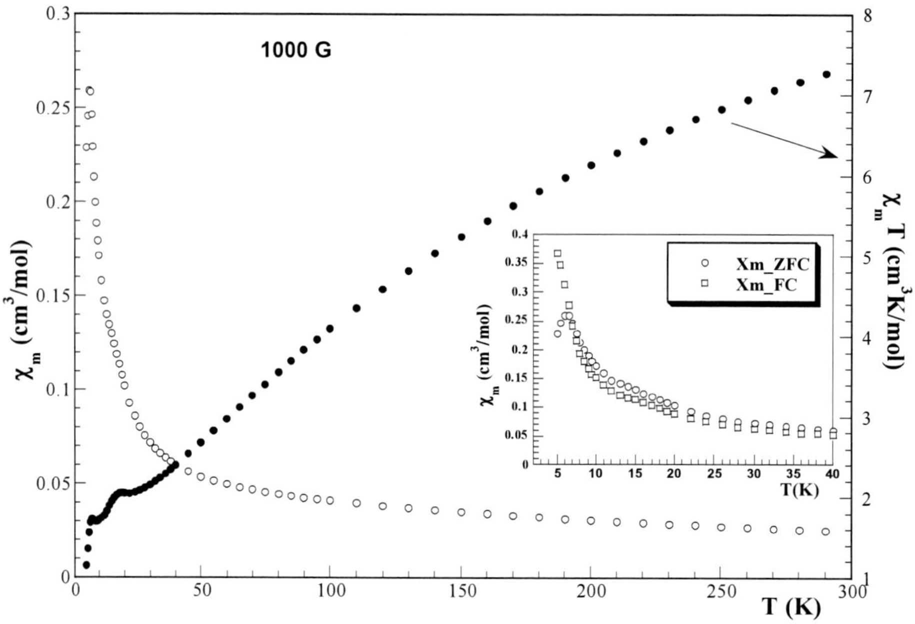

Figure 22. Thermal evolution of the χ_m and $\chi_m T$ curves for phase (12). The inset shows the irreversibility of the ZFC and FC curves at 1000 Gauss.

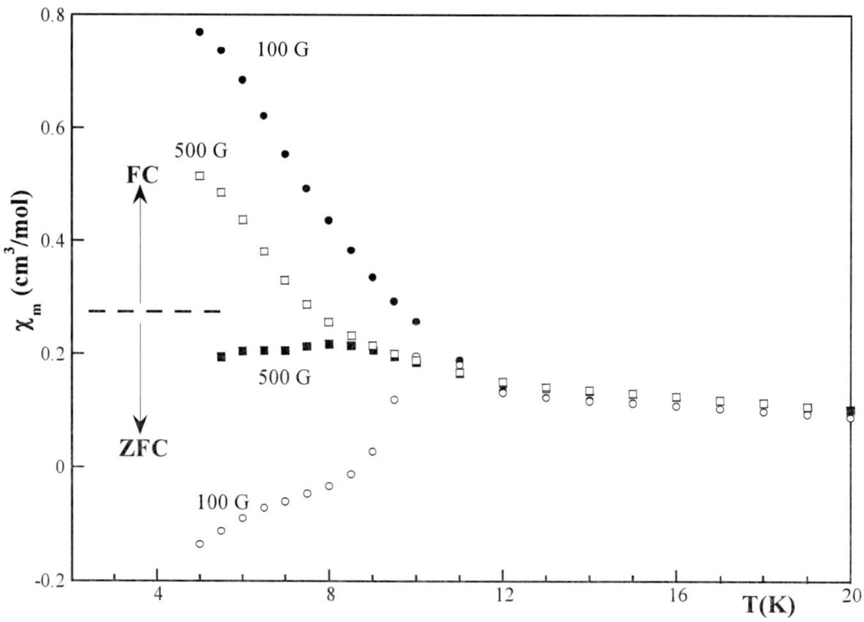

Figure 23. Separation of the χ_m and $\chi_m T$ curves at 500 and 100 gauss for compound (12).

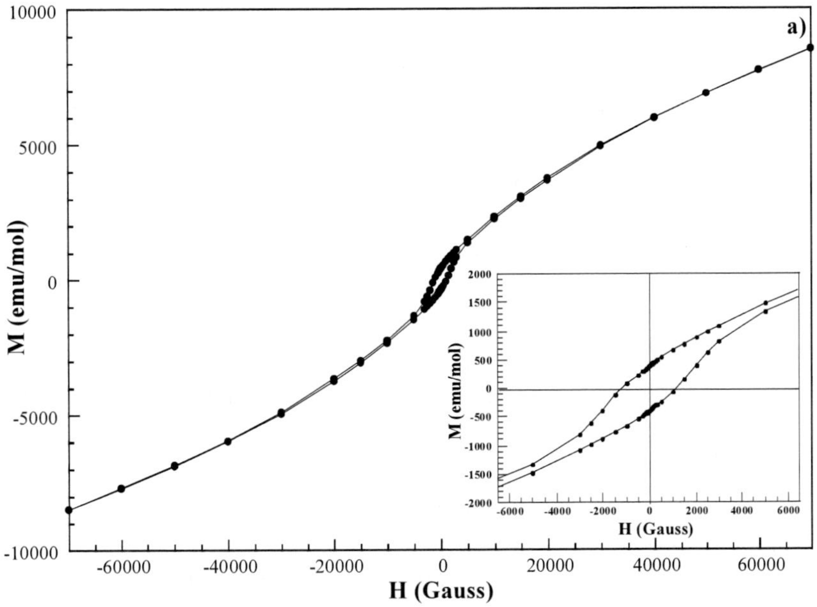

Figure 24a. Hysteresis loop at 5K for compound (12). See Figure 24b for the thermal variation of the remnant magnetization.

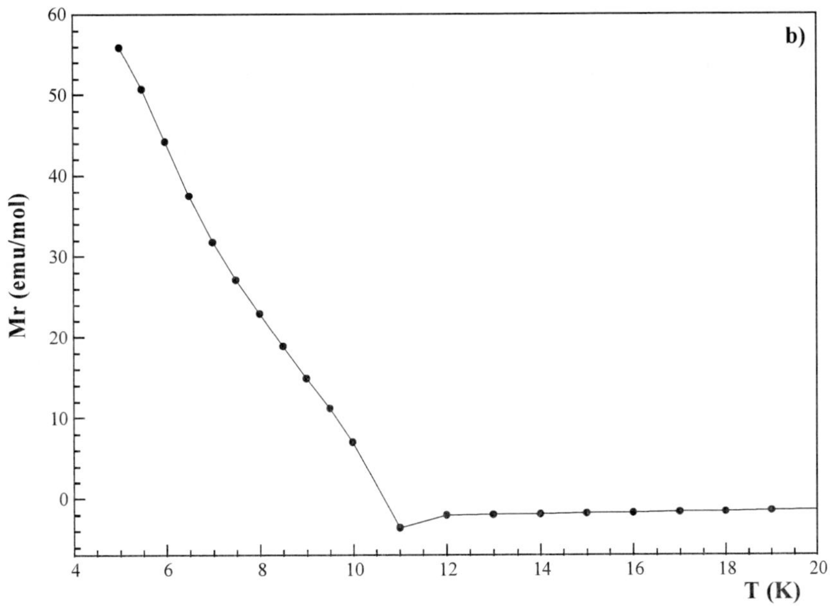

Figure 24b. Thermal variation of the remnant magnetization for compound (12).

The temperature dependence of the ZFC and FC curves shows irreversibility at low temperatures, this feature being more appreciable when the magnetic field used in the measurement decreases down to 100 and 500 Gauss (see Figure 23). These results are indicative of the existence of some ferromagnetic contribution intrinsic to the sample and predominant at low temperatures. In order to corroborate the existence of this phenomenon, magnetization measurements *vs.* magnetic field at 5.0K were performed (Figure 24a). The curves show the presence of a small hysteresis loop, with values of the remnant magnetization and coercitive field of 400 emu/mol and 1180 G. Besides, the remnant magnetization, measured after cooling the sample under a field of 100 G, disappears at approximately 12 K (Figure 24b). These results confirm the existence of some ferromagnetic contribution, probably due to the simultaneous existence of two kinds of metallic centres with different spin in this phase, iron(III), S= 5/2, and iron(II), S=1. This fact indicates the existence of a small ferromagnetic component at low temperatures, although the main interactions are antiferromagnetic in nature [16].

4.4.3.4. $(C_4H_{12}N_2)_2[Fe_4(HAsO_4)_6(C_2O_4)_2]$ compound (13)

This compound shows a strong antiferromagnetic interaction with a Néel temperature around 40K [17].

4.4.3.5. $(C_6H_{14}N_2)[Fe_3(HAsO_4)_2(AsO_4)F_4]\cdot 0.5H_2O$ compound (15)

Variable-temperature magnetic susceptibility measurements were carried out on a powdered sample in the range from 5.0 to 300K. Plots of χ_m and χT *vs.* T curves are shown in Figure 25. The molar magnetic susceptibility increases with decreasing temperature and reaches a sharp maximum at *ca.* 11.2K, indicating that a magnetic ordering is established at this temperature. After that, the susceptibility decreases continuously. This result together with the decrease of the magnetic moment from 9.60 cm^3K/mol at room temperature to 0.55 cm^3K/mol at 5.0K indicates the presence of antiferromagnetic interactions.

The three-dimensional magnetic ordering can be understood by considering the crystal structure formed by $[Fe_3(HAsO_4)_2(AsO_4)F_4]^{2-}$ inorganic layers interconnected by hydrogenarsenate groups [18a]. To fit the magnetic data to a theoretical model and to obtain the J exchange parameter, we consider the system as a simple cubic network. In this sense, the high temperature series expansion developed by Rushsbrooke and Wood for a simple cubic antiferromagnetic system with S= 5/2 was used [27]. Only the low temperature magnetic data, in the proximity of the maximum, can be reasonably fitted to a value of $|J/K|=$ 0.41K (Figure 25). From the magnetic point of view, another approximation to the problem could be the use of the molecular field theory for a three-dimensional antiferromagnetic system [28]. This way, the J exchange parameter is estimated by the following expression:

$$\chi_{max}= (N\beta^2g^2/4zJ) \qquad \text{Eq. (1)}$$

where χ_{max} is the maximum value of the magnetic susceptibility, N is Avogadro's number, β is the Bohr magneton, and z is the magnetic coordination number of a lattice site. The $|J/K|$ value calculated is 0.44, taking z= 6 and g= 2.01, as obtained from the ESR spectra. The value of the J exchange parameter obtained from both procedures is in a reasonable accordance. This fact allows us to confirm the three-dimensional character of the magnetic interactions at low temperature in this compound.

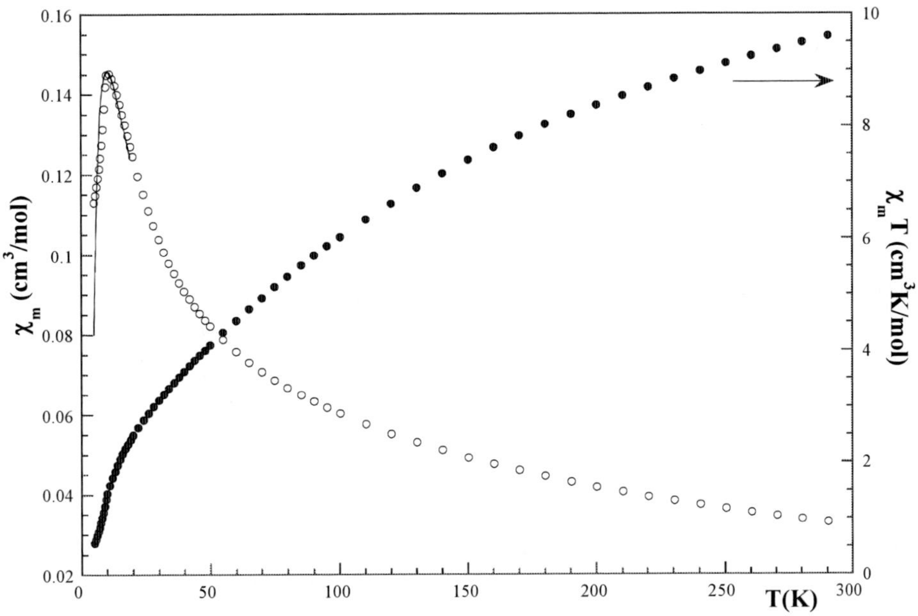

Figure 25. Thermal evolution of the χ_m and $\chi_m T$ curves for phase (15). The solid line shows the fit of the magnetic data in the proximity of the maximum obtained by using the Rushbrooke and Wood expression for an antiferromagnet 3D.

4.4.3.6. $(C_6H_{14}N_2)[Fe_3(HAsO_4)_{1.33}(HPO_4)_{0.67}(AsO_4)_{0.84}(PO_4)_{0.16}F_4]\cdot 0.5H_2O$ compound (16)

Variable-temperature magnetic susceptibility measurements were performed on a powdered sample in the range from 300 to 5.0K. The thermal evolution of χ_m and $\chi_m T$ product are shown in Figure 26. The χ_m *vs.* T curve increases from room temperature with decreasing temperature and reaches a maximum at 6.1K, indicating that a magnetic ordering is established at this temperature. After that, the susceptibility decreases to 5K. This behaviour, together with the continuous decrease of the $\chi_m T$ *vs.* T product from 9.628 cm^3K/mol at 300 K up to 0.957 cm^3K/mol at 5.0K, indicates the existence of antiferromagnetic interactions in this phase [18b]. Taking into account the three-dimensional network of this compound, as described in the structural description, we have fitted the magnetic data to the expression derived by Rushbrooke and Wood for a antiferromagnetic 3D simple cubic framework of spin S= 5/2, [27]

$$\chi = [(35N\beta 2)/(g^2/12kT)]\ [1 + 35.0\ x + 221.67\ x^2 + 608.22\ x^3 + 26049\ x^4 - 210986.5\ x^5 + 8014980.0\ x^6]^{-1} \qquad \text{Eq. (2)}$$

where x= |J|/kT. As can be seen in Figure 26, the magnetic data cannot be fitted to the above equation (2) in the entire range of temperatures studied. This result is really different from that obtained for the related three-dimensional $(C_6H_{14}N_2)[Fe_3(HAsO_4)_2(AsO_4)F_4]\cdot 0.5H_2O$ antiferromagnetic arsenate. In the later compound a three-dimensional magnetic ordering occurs below 11.2K with an estimated J-exchange parameter of -0.40K, calculated from Eq. (1). In both, the arsenate and arsenate-phosphate isomorphous compounds the intralayer magnetic interactions inside the $Fe(2)_2O_8F_2$ edge-sharing dimmeric entities are propagated through the fluoride anion with a Fe(2)-F(4)-Fe(2) bond angle of approximately 101°. The $Fe(1)Fe(3)O_6F_4$ dimmers in the chains exhibit Fe(1)-F(1)-Fe(3) and Fe(1)-F(2)-Fe(3) bond angles of approximately 101°, for both compounds. The connection between these dimmers is established with a Fe(1)-F(3)-Fe(3) angle of approximately 132°. These bond angles, together with those established by the arsenate and phosphate anions linking the iron(III) ions inside the layers and interlayer should favour the antiferromagnetic interactions, in good agreement with that observed in these compounds [29].

Despite the structural similarities between these phases they exhibit some important differences in the magnetic behaviour. This fact could be explained if we consider that, the arsenate-phosphate compound is derived from the arsenate phase by a partial substitution of the arsenate and hydrogenarsenate anions by the smaller phosphate and hydrogenphosphate groups. This chemical substitution approaches the iron(III) cations inside the $[Fe_3(HAsO_4)_{1.33}(HPO_4)_{0.67}(AsO_4)_{0.84}(PO_4)_{0.16}F_4]^{2-}$ layers, increasing the bidimensional magnetic interactions of this mixed-anion compound. So, the temperature of the three-dimensional magnetic ordering of $(C_6H_{14}N_2)[Fe_3(HAsO_4)_{1.33}(HPO_4)_{0.67}(AsO_4)_{0.84}(PO_4)_{0.16}F_4]\cdot 0.5H_2O$ must be lower than that obtained for the arsenate compound, as observed in the thermal evolution of the magnetic susceptibility. However, attempts performed to fit the magnetic data of the mixed arsenate-phosphate compound to a model for a 2D antiferromagnetic framework of spin S= 5/2, by using the Rushbrooke and Wood expression [27] [Eq. (3)] were unsuccessful (see Figure 26).

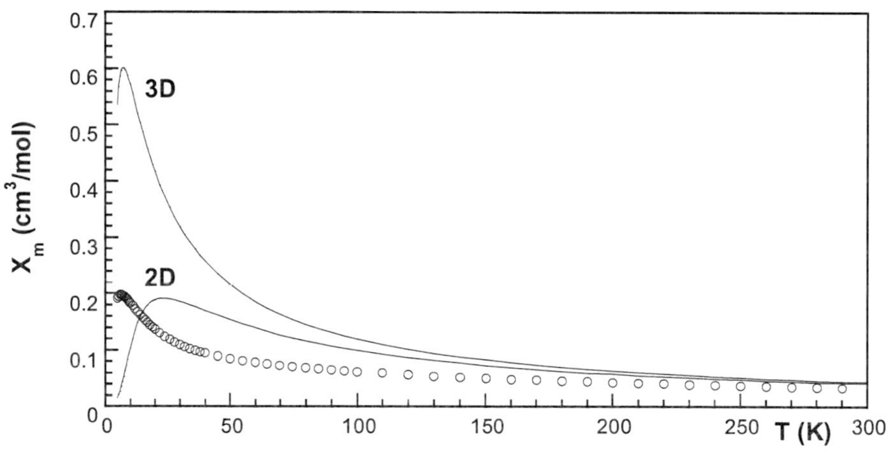

Figure 26. Thermal evolution of the the χ_m curve for phase (13) showing its magnetic intermediate character between a 2D and a 3D theoretical system.

$$\chi = [(35N\beta^2)/(g^2/12kT)]\,[1 - 23.333\,x + 147.78\,x^2 - 405.8\,x^3 + 8171.3\,x^4 - 6461.8\,x^5 - 158110\,x^6]^{-1} \qquad \text{Eq. (3)}$$

From the magnetic point of view, a reasonable explanation of this result should be to consider the $(C_6H_{14}N_2)[Fe_3(HAsO_4)_{1.33}(HPO_4)_{0.67}(AsO_4)_{0.84}(PO_4)_{0.16}F_4]\cdot 0.5H_2O$ compound as an intermediate 3D-2D antiferromagnet (Figure 26). This way, the estimation of the J-exchange parameter for this compound, by using any magnetic model, should be precluded [18b].

5. Limits and Future

The results exposed about this research field indicate the increase experimented in the synthesis, structural characterization and the study of the physical properties of the organically templated iron(III) arsenates. However, similarly to that found for the analogous phosphate compounds the mechanism of formation also show the limits for obtaining "tailor made" microporous compounds with given shape and size of the pores. If the amine is the driving forced for the synthesis of microporous compounds, it is also the major limiting factor. Its charge induces strong electrostatic interactions with the framework and makes its extraction after the synthesis difficult. Other interest of this kind of compound is the study of their physical properties (spectroscopic and magnetic) which are related with the structural type obtained in the synthesis (1D, 2D or 3D materials) and the metallic cation, in this case, the iron(III) ion or the mixed cationic valent iron(II,III) compounds. In this sense, an enormous field with other 3d-cations of the first series of transition metals (Co^{2+}, Mn^{2+}, Ni^{2+}, V^{3+},…) remains unexplored. Our efforts in the next years are directing towards this exciting research field.

Acknowledgements

This work has been financially supported by the "Ministerio de Educación y Ciencia" (MAT2004-02071) and the "Universidad del País Vasco" (UPV/EHU) (9/UPV00130.310-13700/2001 and (9/UPV00169.310-13494/2001). The authors thank the technicians of SGIker, Dr. J.P. Chapman and Dr. I. Orue, financed by the National Program for the Promotion of Human Resources within the National Plan of Scientific Research, Development and Innovation, "Ministerio de Ciencia y Tecnología" and "Fondo Social Europeo" (FSE), for the X-ray diffraction and magnetic measurements, respectively. B. Bazán wishes to thank the "Ministerio de Educación y Ciencia" (MAT2004-02071) for funding.

References

[1] G. Ferey, C. R. *Acad. Sci.* Paris 1 (1998) 1.
[2] H. de Sainte Claire Deville, C. R. *Acad. Sci.* Paris 54 (1862) 324.
[3] S.T. Wilson, B.M. Lok, C.A. Messina, T.T. Cannann, E.M. Flanigen, *J. Am. Chem. Soc.* 104 (1982) 1146.
[4] (a)J.V. Smith, *Chem. Rev.* 88 (1988) 149,(b) J.B. Paris, *Inorg. Chem.* 24 (1985) 4312.
[5] J.L. Guth, H. Kessler, R. Wey, *Stud. Surf. Sci. Catal.* 28 (1986) 121.
[6] (a) M. Estermann, L. McCusker, C. Baerlocher, A. Merrouche, H. Kessler, *Nature* 352 (1991) 320),(b) G. Ferey, *J. Fluorine Chem.* 72 (1995) 187.

[7] R.E. Morris, S.J. Weigel, *Chem. Soc. Rev*. 26 (1997) 309.
[8] C.F. Baes, R.E. Mesmer, *"The Hydrolysis of Cations"*, Krieger Publishing Company, Malabar, Florida, EE.UU. (1986).
[9] (a) B. Bazan, J.L. Mesa, J.L. Pizarro, L. Lezama, M.I. Arriortua, T. Rojo, Inorg. Chem. 39 (2000) 6056; (b) S. Ekambaran, S.C. Sevov, *Inorg. Chem*. 39 (2000) 2405.
[10] S. Chakrabarti, S.K. Pati, M.A. Green, S. Natarajan, *Eur. J. Inorg. Chem*. (2004) 3846.
[11] S. Chakrabarti, S. Natarajan, *Angew. Chem. Int.* Ed. 41 (2002) 1224.
[12] Y-C. Liao, S-H Luo, S.L. Wang, H-M. Kao, K-H. Lii, *J. Solid State Chem*. 155 (2000) 37.
[13] (a) S-H. Luo, Y-C. Jiang, S-L. Wang, H-M. Kao, K-H. Lii, *Inorg. Chem*. 40(21) (2001) 5381; (b) B. Bazan, J.L. Mesa, J.L. Pizarro, L. Lezama, A. Peña, M.I. Arriortua, T. Rojo, *J. Solid State Chem*. In press (2006).
[14] S. Chakrabarti, S.K. Pati, M.A. Green, S. Natarajan, *Eur. J. Inorg. Chem*. (2003) 3820.
[15] B. Bazan, J.L. Mesa, J.L. Pizarro, M.I. Arriortua, T. Rojo, *Mat. Res. Bull*. 38 (2003) 1193.
[16] B. Bazan, J.L. Mesa, J.L. Pizarro, A. Peña, M.I. Arriortua, T. Rojo, Z. *Anorg. Allg. Chem*. 631 (2006) 2026.
[17] S. Chakrabarti, M.A. Green, S. Natarajan, *Solid State Sciences* 4 (2002) 405.
[18] (a) B. Bazan, J.L. Mesa, J.L. Pizarro, A. Goñi, L. Lezzama, M.I. Arriortua, T. Rojo, *Inorg. Chem*. 40 (2001) 5691; (b) B. Bazan, J.L. Mesa, J.L. Pizarro, L. Lezama, J.S. Garitaonandia, M.I. Arriortua, T. Rojo, *Solid State Sciences* 5 (2003) 1291.
[19] Y. Tanabe, S. Sugano, *J. Phys. Soc*. Japan, 9 (1954) 753.
[20] (a) J.E. Orgel, *J. Chem. Phys*., 23 (1955) 1004; (b) A.B.P. Lever, *"Inorganic Electronic Spectrosocpy"*, Elsevier Science Publisher B.V., Amsterdam, Netherlands (1984).
[21] (a) G.R. Rossman, Am. Mineral., 59 (1974) 868; (b) D.M. Sherman, N. Vergo, *Am. Mineral*. 73 (1988) 1346.
[22] R.A. Brand, J. Larner, D.M. Herlach, *J. Phys*. F 14 (1984) 555.
[23] H.O. Stumpf, L. Ouahab, Y. Pei, P. Bergerat, O. Khan, *J. Am. Chem. Soc*. 116 (1994) 3886.
[24] (a) H.W. Wijn, L.R. Walker, J.L. Daris, H. Guggenheim, *J. Solid State Commun*. 11 (1972) 803; (b) P.M. Richars, M.B. Salamon, *Phys. Rev*. B9 (1974) 32; (c) A. Escuer, R. Vicente, M.A.S. Goher, F. Mautner, *Inorg. Chem*. 34 (1995) 5707; (d) A. Bencini, D. gatteschi, "EPR of Exchange Coupled Systems, Springer-Verlag, Berlin Heidelberg (1990); (e) T.T.P. Cheung, Z.G. Soos, R.E. Dietz, F.R. Merrit, *Phys. Rev*. B17 (1978) 1266.
[25] M.E. Fisher, *Am. J. Phys*. 32 (1964) 343.
[26] R. Dingle, M.E. Lines, S.L. Holt, *Phys. Rev*. 187 (1969) 643.
[27] G.S. Rushbrooke, P.S. Wood, *Mol. Phys* 1 (1958) 257.
[28] R.L. Carlin, *"Magnetochemistry"*, Springer- Verlag: Berlin Heidelberg (1986).
[29] J.B. Goodenough *"Magnetism and the Chemical Bond"*, Interscience: New York (1963).

In: Progress in Solid State Chemistry Research
Editor: Ronald W. Buckley, pp. 239-284

ISBN 1-60021-313-8

Chapter 6

CRYSTAL STRUCTURES AND GALVANOMAGNETIC PROPERTIES OF EPITAXIAL FILMS IN A NI-S SYSTEM

Hiroshi Nozaki, Mitsuko Onoda and Kousuke Kosuda
National Institute for Materials Science
1-1 Namiki, Tsukuba, Ibaraki, 305-0044, Japan

ABSTRACT

Molecular beam epitaxy (MBE) and X-ray diffraction (XRD) methods were used for preparation and characterization of epitaxial films in a Ni-S system. Epitaxial films of Ni_3S_4, rhombohedral NiS, Ni_7S_6 and Ni_3S_2 were prepared in the first time. Galvanomagnetic properties of the Ni_3S_4 and $Ni_{1-x}S$ films were investigated. Crystal chemistry of the films was discussed together with that of related materials' epitaxial films.

A Ni_3S_4 film with a spinel-type structure was deposited on MgO(001) at about 440 K. It demonstrated out-of-plane epitaxy of [111] || [001]MgO and in-plane alignment of [1,1,-2] || [110]MgO. The diagonal axis of two-dimensional hexagonal lattice within the (111) sulfur layer of the Ni_3S_4 film was aligned along [110]MgO, which was observed previously for hexagonal $Ni_{1-x}S$ epitaxial films on MgO(001) substrates. The Ni_3S_4 films deposited on SnS(010)/MgO(001) and $Ni_{1-x}S(001)$/MgO(001) substrates had identical orientation to that described above.

A nickel sulfide (M1 phase) film was deposited on an $Al_2O_3(001)$ substrate at 595 K; it had a monoclinic lattice. The 0*k*0 (k=1-7) reflection intensities, collected using a θ-2θ scanning XRD pattern, were simulated using a Rietveld profile refinement program. Those results indicated that the monoclinic structure was distorted from an orthorhombic Ni_7S_6-type structure. The in-plane alignment was [001]||[1,-1,0]Al_2O_3. The XRD pattern showed some amounts of M2 phase inclusion, as described below. By increasing substrate temperature, another monoclinic epitaxial film (M2 phase) was formed at 611 k with small amounts of the Ni_7S_6-type phase and rhombohedral NiS (millerite). This film had a crystal structure that was distorted from the rhombohedral Ni_3S_2-type structure. The out-of-plane epitaxy was [010]|| [001]Al_2O_3 and the in-plane alignment was [100]||[1,-1,0]Al_2O_3. After deposition at 543 K, the M1 phase coexisted with millerite with a uniaxial orientation [100]||[001]Al_2O_3 for the hexagonal lattice setting. The *a*-dimension of this M1 phase was larger by 9.3% than that of the above Mi phase. This fact, along with the coexistence of the Ni_7S_6-type phase with the Ni_3S_2 phase or millerite, suggests some homogeneous range of the M1 phase.

Electrical resistivity, Hall coefficient and magnetoresistance were studied using a $Ni_{1-x}S$ film (x is approximately 0) prepared in this study and other previously reported $Ni_{1-x}S$ films. Metal-metal transition temperatures of the films were apparently changed, probably because of constraint resulting from the substrate surface periodicity. Transverse magnetoresistance of the $Ni_{1-x}S$ films showed magnitudes that were greater by four orders magnitude than those estimated from their Hall coefficients and resistivities. The resistivity and Hall coefficient of the Ni_3S_4 film indicated normal metallic p-type conduction.

1. Introduction

Epitaxial growth has some merits from the viewpoint of material synthesis. First, a compound film that is difficult to prepare is producible using epitaxial growth if a substrate appropriate for the compound is selected. Second, epitaxial growth stabilizes unstable structures that differ from that of the bulk compound at room temperature. Third, an unexpected or a new compound can be synthesized using epitaxial growth. In the present study, molecular beam epitaxy (MBE) was used for syntheses of nickel sulfides in the expectation of producing a Ni_3S_4 film, a stabilized $Ni_{1-x}S$ film, and other nickel sulfide films.

In nickel sulfide systems, several phases are known to be stable only at low temperatures: Ni_3S_4 (polydymite), NiS (millerite), Ni_9S_8 (godlevskite), and Ni_3S_2 (heazlewoodite) [1-3]. The compound Ni_3S_4 has a spinel-type structure and is stable below 629 K [1]. A single phase of Ni_3S_4 is difficult to prepare using thermal equilibrium solid-state reaction of the elements using an evacuated silica tube method [1]. The Ni_3S_4 compound has recently been synthesized using precipitation methods [4-6]. A few studies of physical properties for Ni_3S_4 specimens prepared using the precipitated method have been reported [5, 6].

A specimen of Ni_3S_4 in an epitaxial form will be more appropriate to measure its galvanomagnetic properties than that of a specimen in a powder form prepared using the precipitation method. For that reason, we expected that a Ni_3S_4 epitaxial film can be prepared using molecular beam epitaxy with a MgO(001) substrate. Unexpectedly, we obtained an epitaxial film of hexagonal $Ni_{1-x}S$ [7]. As reported in our previous work of epitaxial $Ni_{1-x}S$ films on MgO(001) substrates [7], the $Ni_{1-x}S$ ($x = 0.07$) film was stable at room temperature one year later. It had an ordered arrangement of Ni-vacancy in the structure, which allowed estimation of the film composition using an XRD reflection intensity associated with the Ni-vacancy arrangement. These characteristics differed from those of bulk $Ni_{1-x}S$. That difference promoted us to prepare epitaxial films of other nickel sulfides such as Ni_3S_4, millerite and more nickel-rich sulfides.

The low-temperature phase of millerite is dimorphic with the high-temperature phase of hexagonal $Ni_{1-x}S$ and stable below 652 K. It has a rhombohedral crystal structure [8, 9] and a metallic conductivity [10].

Hexagonal $Ni_{1-x}S$ has a NiAs-type structure and is stable above 652 K for $x = 0$ and above 555 K for $x = 0.07$ [1]. In addition, $Ni_{1-x}S$ with $x = 0$ exhibits a phase transition from a paramagnetic metal to an antiferromagnetic semiconductor at 264 K with decreasing temperature [11]. However, a recent photoemission study of $NiS_{1-x}Se_x$ has shown that the transition is not the metal-semiconductor transition but a metal-anomalous metal transition [12]. The transition accompanies dilatation of the crystal lattice below that temperature [13, 14]. The transition temperature is strongly dependent on composition of 1-x [15]. Physical property investigations have been performed extensively for specimens that are quenched at high temperatures [11 -19]. Some Hall effect data of $Ni_{1-x}S$ have been reported for quenched specimens [16, 17]. However, such specimens degenerate after several months: they change to rhombohedral NiS (millerite) over time. A kinetics study of the change from $Ni_{1-x}S$ to millerite has shown that $Ni_{1-x}S$ (x = 0) changes to millerite by 1.5 wt% during 0.25 h annealing at 343 K and that initial compositions of $Ni_{1-x}S$ play an important role in the kinetics of change [20]. It has also been shown that the change becomes more sluggish for more metal-deficient compositions.

The crystal structure of Ni_9S_8 is orthorhombic [21]. Recently, Raman and photoluminescence spectroscopy have been used to examine Ni_9S_8 nanorods that had been prepared using a precipitation method [22]. The Ni_9S_8 disproportionates at 709 K to $Ni_{1-x}S$ and Ni_7S_6 [3]. Ni_7S_6 has an orthorhombic structure [23] and a homogeneity range is denoted as $Ni_{7\pm x}S_6$ [3]. It is stable above 675 K, but specimens are obtainable even with furnace cooling to room temperature [3]. No physical properties of Ni_7S_6 have been reported.

Nickel sulfide, Ni_3S_2, has a rhombohedral structure [24, 25]. It transforms discontinuously into a high temperature phase, $Ni_{3\pm x}S_2$, with cubic symmetry [1, 26]. Physical properties of Ni_3S_2 single crystals such as DC resistivity, magnetic susceptibility and heat capacity indicate that the material is a good metallic conductor [27]. An electronic structure calculation also indicates that Ni_3S_2 is a conventional metal with marginal electron-electron correlation effect [28]. High-temperature electrical conductivity of Ni_3S_2 until 1100 K has been reported [29].

In this investigation, epitaxial films of Ni_3S_4, $Ni_{1-x}S$, Ni_7S_6, Ni_3S_2 and rhombohedral NiS films were prepared and characterized, respectively, using MBE and X-ray diffraction methods. Galvanomagnetic properties of Ni_3S_4 and $Ni_{1-x}S$ were investigated. Crystal chemistry was discussed for several epitaxial films' characteristics of nickel sulfides and related sulfides.

2. EXPERIMENTAL

2.1. Film Preparation

Nickel sulfide films were prepared using molecular beam epitaxy. A previous study includes a description of the MBE equipment [30]. The Ni powder (5N sponge powder, Soekawa Chemicals Co. Ltd.) was reduced at 1613 K in an alumina crucible under H_2 atmosphere in a chamber of the MBE equipment until the chamber's oxygen content decreased to the background level (less than 6.7×10^{-8} Pa). Intensities of nickel and sulfur molecular beams were calibrated using a naked ionization gauge before and after deposition experiments. The intensity ratio of the molecular beams was fixed at I(S-beam)/I(Ni-beam)=10 during deposition experiments for Ni_3S_4 and $Ni_{1-x}S$ films and I(S-beam)/I(Ni-beam)=3.8-3.9 for Ni_3S_2, Ni_7S_6 and millerite. Changing substrate temperature and the beam intensity ratio controlled composition of the deposited films. Deposition experiments of nickel sulfides were usually performed with deposition rates 2.5-3.5 nm m^{-1} for 60 m. Film thickness was measured using a stylus method.

(a)

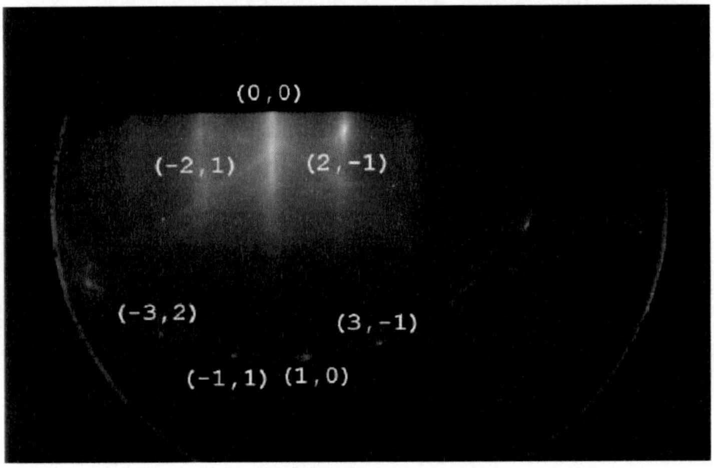

(b)

Figure 1. Typical RHEED patterns of substrates, (a) for MgO(001) with the incident electron beam along the direction parallel to [010]MgO and (b) for Al_2O_3(001) with that along [010]Al_2O_3.

2.2. Substrates

Nickel sulfide films were prepared using various substrates (MgO(001), Al_2O_3(001), ZnO(001) and some alkali halide single crystals). According to characterization of the films using XRD and optical microscopy, we selected MgO(001) and Al_2O_3(001) single crystals as substrates for the nickel sulfide film preparation. In addition to both the oxide substrates, SnS(010) and $Ni_{1-x}S$(001) epitaxial films deposited on MgO substrates were also used as substrates. These oxide substrates were annealed at 1353 K in an O_2 atmosphere for 2 h before film growth experiments to obtain flat surfaces [31, 32]. The substrates were further annealed at about 973 K for 0.5-2 h under gaseous pressure below 1.3×10^{-7} Pa. Then, surfaces of the substrate were examined using reflection high-energy electron diffraction (RHEED) before deposition experiments. Typical RHEED patterns of MgO(001) and Al_2O_3(001) substrates are shown in Figs. 1a and 1b. The former pattern was taken with an incident electron beam along [010]MgO and the latter one was done with that along [010]Al_2O_3. An acceleration voltage of the electron beam was 15 keV. Indexes of the two-dimensional reciprocal lattice for each pattern are shown in the figures.

2.3. XRD Characterization

Crystallographic characteristics of the film and film orientation were measured using an X-ray diffractometer equipped with a line-focused Cu$K\alpha$ source and a curved incident monochrometer. The diffractometer had no conventional divergent incident beam, but rather a converged incident beam that was produced by the monochrometer [33]. The converged beam was cut out by a slit (0.15 mm wide): the incident X-ray beam width at the film surface was about 0.1 mm. The diffractometer had a device to move the film along the direction vertical to the surface and rotate it around the ϕ-axis normal to the surface. The film was positioned along the direction perpendicular to the film surface to obtain the most intense X-ray reflection of the film. The 2θ was calibrated using interplanar spacing of the MgO or Al_2O_3 substrates for each film.

X-Ray reflections for lattice planes parallel to the substrate surface were collected using θ-2θ scanning mode with an ordinary symmetrical arrangement of incident and reflected X-ray beam. An X-ray reflection for a lattice plane that was inclined relative to the substrate surface was measured using an appropriate angle of ϕ-rotation and an offset angle for the θ-axis (Ω-rotation). The offset angle corresponded to an interplanar angle (denoted as α_{ip}) between one lattice plane parallel to the substrate surface and the indicated lattice plane. Therefore, it is necessary to know the unit lattice of the film structure for estimating Ω-rotation and ϕ-rotation positions to measure an indicated lattice reflection. Conversely, we can construct a unit lattice of unknown film structure from measured magnitudes of Bragg angles (θ_B), Ω, ϕ and α_{ip} for reflections. The substrate body intercepted many reflections with $\theta_B<\alpha_{ip}$, but as many X-ray reflections as possible were measured in each film to determine the lattice dimensions precisely. The XRD data of θ_B, Ω, ϕ, α_{ip} and integrated intensity were recorded for every film reflection. These processes were also necessary to determine the lattice symmetry of the film correctly because most X-ray reflections of epitaxial films were not readily visible: they required much searching.

2.3. EPMA, SEM and RHEED Characterizations

Compositional analyses of the film were performed using an electron probe microanalyzer (EPMA, JXA-R500; JEOL) with an acceleration voltage of 5keV. That voltage was used in all cases to avoid characteristic X-rays resulting from the substrate. Scanning electron microscopy (SEM) images or backscattered electron images were taken using the device. The RHEED patterns were obtained using a RHEED gun (VE-052S; Vieetech Japan Co., Ltd.).

2.4. Galvanomagnetic Measurements

Electrical resistivity, the Hall coefficient and magnetoresistance of the film were measured using a five-probe DC method. We used Ag paste for probe contacts. The film specimen was cut to a rectangular form (about 2.5 mm wide and 10 mm length). Linear dependence of Hall voltage on magnetic field strength was always examined for the film specimen at room temperature. Because of the film's flat surface, the strength of contact between Ag paste and the film surface was not sufficiently strong enough for some cases. In such cases, the electrical contact was made using Ag deposition on the film. The temperature dependence of the galvanomagnetic measurement was made using both increasing and decreasing temperatures above 80 K but with only increasing temperature below that temperature.

3. Characterization of Ni-S Films

3.1. Crystal Structure of Ni_3S_4 Epitaxial Films on Mgo(001)

3.1.1. Ni_3S_4 Films on Various Substrates

The Ni_3S_4 epitaxial films were prepared on various substrates such as MgO(001), SnS(010)/MgO(001) and $Ni_{1-x}S$(001)/MgO(001) at 418-439 K because the metal-metal distances of those substrates were approximately equal to the sulfur-sulfur distance within (111) of the cubic spinel-type structure for Ni_3S_4. Epitaxial SnS(010) and $Ni_{1-x}S$(001) films deposited on MgO(001) substrates were produced using this proximity in our previous studies [7, 34], as summarized for the $Ni_{1-x}S$(001)/MgO(001) film in 3.2 section. A spotty but diffuse RHEED pattern of the Ni_3S_4 film deposited on the MgO(001) substrate was obtained, but the pattern was not interpreted because of the pattern's complexity.

The θ-2θ scanning XRD patterns of Ni_3S_4 films deposited on the MgO(001), SnS(010)/MgO(001) and $Ni_{1-x}S$(001)/MgO(001) substrates are shown respectively in Figs. 2a, 2b and 2c. The *lll* reflections of Ni_3S_4 (l=1-5) are readily visible in all patterns. A few rhombohedral NiS (millerite) reflections are visible in Fig. 2c, where the phase is denoted as "r-NiS". Although the XRD pattern of Fig. 2b clearly depicts 0*k*0 reflections of the SnS substrate, the XRD pattern of Fig. 2c shows 002 and 004 reflections of the $Ni_{1-x}S$ substrate in a manner whereby the 222 and 444 reflection profiles of Ni_3S_4 broaden at their right-side base. For the $Ni_{1-x}S$(001)/MgO(001) substrate, the nominal thickness of the $Ni_{1-x}S$ film on MgO(001) was very thin (13 nm). However, a RHEED pattern of the $Ni_{1-x}S$(001)/MgO(001) substrate before the Ni_3S_4 deposition experiment indicated that the MgO(001) surface was covered with a $Ni_{1-x}S$(001) epitaxial film, as shown in Figs. 3a and 3b. The former RHEED pattern was taken with an incident electron beam along the direction at an azimuth of 15° with [100]MgO; the latter pattern was done with that along the direction of [100]MgO. The

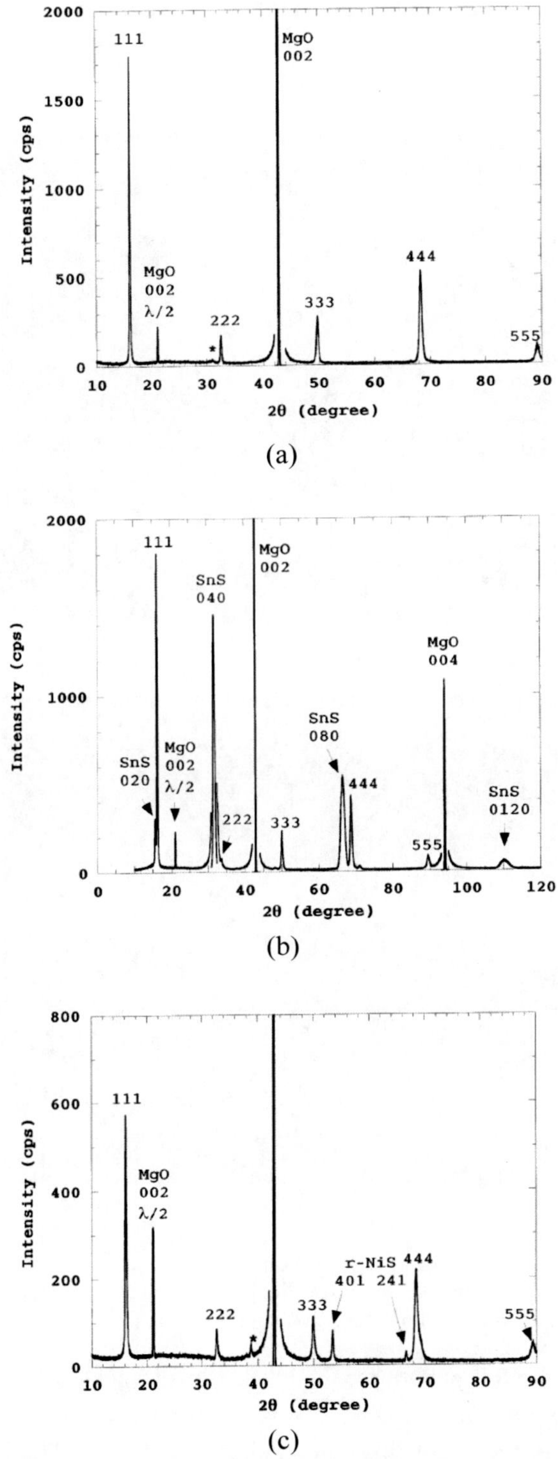

Figure 2. θ-2θ scanning XRD patterns of Ni_3S_4 films. The patterns for Ni_3S_4 films deposited respectively on (a) MgO(001), (b) SnS(010)/MgO(001), and (c) $Ni_{1-x}S$/MgO(001). “r-NiS” in Fig. 2c denotes rhombohedral NiS (millerite). Reflections denoted as * were ascribed to foreign phases.

RHEED patterns were identical to those described in a previous $Ni_{1-x}S$ film study [7]. The relation between the RHEED pattern's reflections and the reciprocal lattice of the $Ni_{1-x}S(001)$ film has been described in that previous study [7].

(a)

(b)

Figure 3. RHEED patterns for a $Ni_{1-x}S$ film deposited on MgO(001) substrate with nominal thickness of 13 nm. The patterns: (a) with an incident electron beam along the direction at an azimuth of 15° with [100]MgO and (b) with an incident electron beam along the direction parallel to [100]MgO.

Each θ-2θ scanning XRD pattern for the Ni_3S_4 films indicates that [111] of the Ni_3S_4 film is perpendicular to the substrate surface. A θ-scanning rocking profile of the 111 reflection intensity for the film on the MgO substrate is shown in Fig. 4. The half-value width of the profile (1.66°) is rather broad. The half-value width of the film on the SnS substrate was 2.2° and that of the film on the $Ni_{1-x}S$ substrate 1.55°. Therefore, the out-of-plane epitaxy of the Ni_3S_4 films was not so sharp, perhaps because of the mismatch between the sulfur-sulfur distance of the film and the metal-metal distance of the substrate in a range from 1.1% for Ni_3S_4/SnS, via 2.1 % for $Ni_3S_4/Ni_{1-x}S$, to 2.7% for Ni_3S_4/MgO.

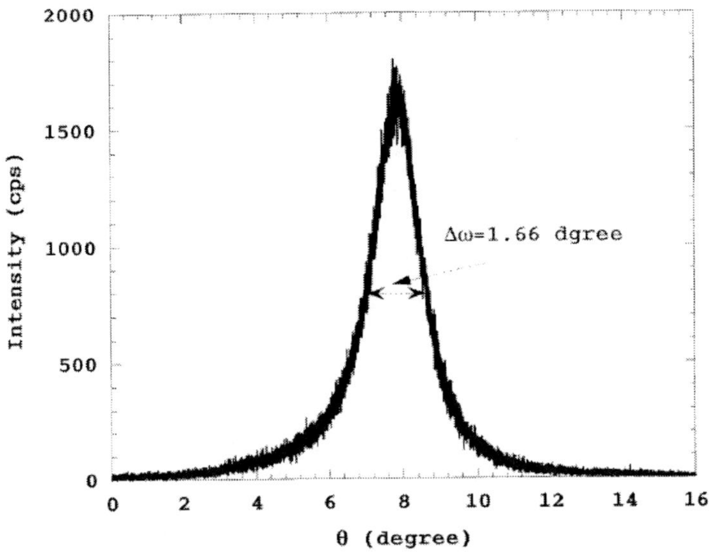

Figure 4. θ-scanning rocking profile of 111 for the Ni_3S_4 film deposited on MgO(001).

3.1.2. In-Plane Alignment of Ni_3S_4 Film

In-plane alignment of the Ni_3S_4 film deposited on the MgO(001) substrate was determined by measuring the ϕ-dependence of 335 reflection intensity. The ϕ-dependence of 335 reflection intensity is shown in Fig. 5. In the figure, the ϕ-dependence of 204 intensity for the MgO substrate is also shown to clarify the film's in-plane alignment. The peak positions of the 335 intensity profile are ϕ=15°, 45°, 75° and 105°. The appearance of this "twelve-fold" rotation symmetry results from the facts that the three-fold rotation axis of the Ni_3S_4 film, that is, [111], coexists with the four-fold rotation axis of the MgO(001) substrate, i.e. [001]MgO, along the same direction with each other. Because the position at ϕ=45° corresponds to [110]MgO, the in-plane alignment of the Ni_3S_4 film on the MgO substrate was [1,1,-2] || [110]MgO, although the alignment was not so sharp, as observed in the broad profile in the figure. Figure 6 shows that the alignment is such that the diagonal axis of the two-dimensional hexagonal sulfur lattice within (111) of the Ni_3S_4 structure is directed along the diagonal axis of two-dimensional Mg cubic lattice within (001) of the MgO substrate. The situation is the same as that for the $Ni_{1-x}S$(001)/MgO(001) epitaxial film [7]. The ϕ-dependence for the Ni_3S_4 film deposited on the SnS substrate was identical to that for the film on the MgO substrate. Although the ϕ-dependence for the Ni_3S_4 film on the $Ni_{1-x}S$(001) substrate was not measured, we inferred from the procedures to measure the interplanar spacing of the film that a similar result of ϕ-dependence for the film was obtained.

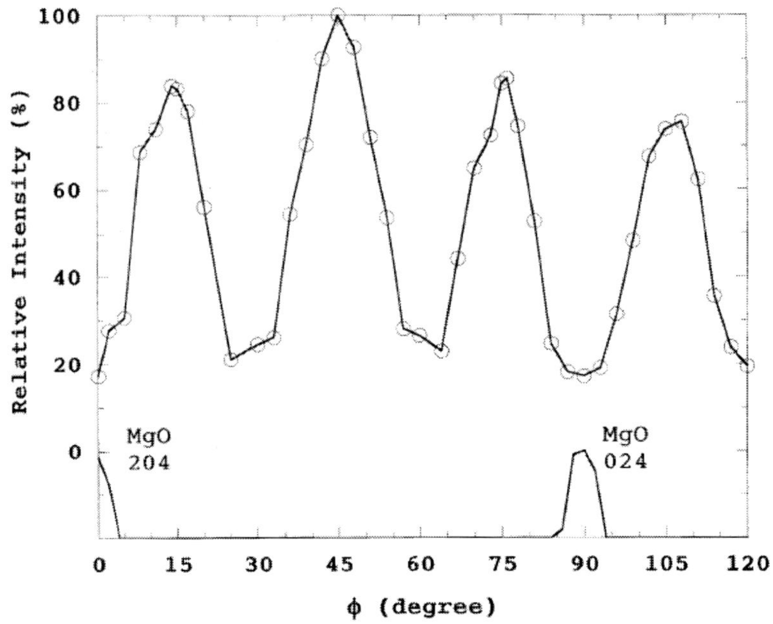

Figure 5. ϕ-dependence profile of 335 reflection intensity for a Ni_3S_4 film deposited on MgO(001). That of 204 reflection intensity for the MgO substrate is also depicted.

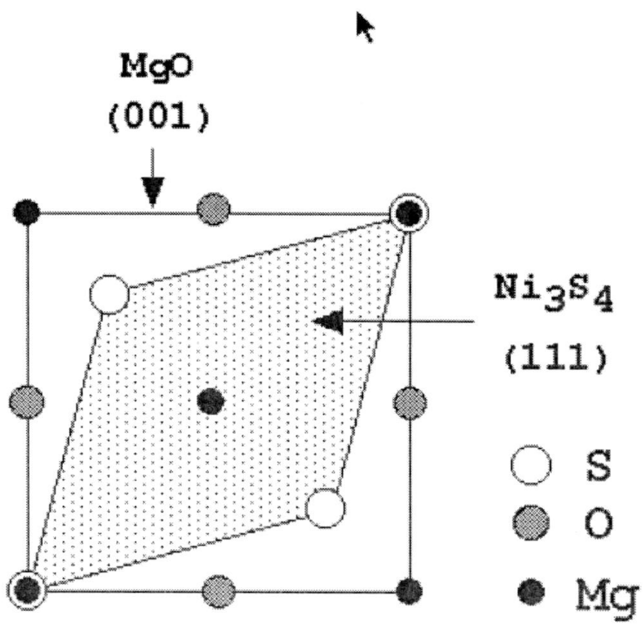

Figure 6. Schematic illustration of in-plane alignment. The diagonal axis of two-dimensional hexagonal sulfur lattice aligns along [110] of the MgO (001) substrate. The sulfur lattice is within the (111) sulfur layer of the Ni_3S_4 structure.

3.1.3. XRD Characterization of Ni_3S_4 Films

Interplanar spacing data of the film deposited on MgO(001) are listed in Table 1, Data for lattice planes inclined to (111) were collected as described in the Experimental section. The film's lattice parameters, in addition to those of the films deposited on the SnS(010)/MgO(001) and $Ni_{1-x}S$(001)/MgO(001) substrates, are presented in the table. Lattice parameters of the films, are slightly larger than the lattice parameter of bulk Ni_3S_4, 0.94761 nm [35] probably because the diagonal length of the substrate two-dimensional lattice (two times metal-metal distance) is 2.7% larger than the diagonal length of the two-dimensional sulfur rhomb within (111) of bulk Ni_3S_4 (0.5803 nm). Lattice parameters of the films are correlated with the diagonal lengths of the substrates, as shown in Fig. 7, where the diagonal length of the SnS substrate was that for the F4 film in reference [34] and the diagonal length of the $Ni_{1-x}S$ substrate was that for Film2 in reference [7].

Table 1. Observed and calculated interplanar spacing and observed intensity of Ni_3S_4 film on MgO(001). The lattice parameters of the three films are also shown.

h k l	d_{obs} (nm)	d_{cal} (nm)	I_{obs} (%)
1 1 1	0.54881	0.54848	100.0
2 2 2	0.27424	0.27424	1.1
4 2 2	0.19375	0.19392	0.3
3 3 3	0.18289	0.18283	3.0
5 3 3	0.14503	0.14487	0.9
4 4 4	0.13714	0.13712	7.4
5 5 3	0.12368	0.12368	0.6
5 5 5	0.10968	0.10970	1.5
7 5 1	0.10969	0.10970	0.3
8 8 0	0.083861	0.083969	0.7

a=0.9500 nm for Ni_3S_4 on MgO
a=0.9481 nm for Ni_3S_4 on SnS
a=0.9489 nm for Ni_3S_4 on $Ni_{1-x}S$

The difference between lattice parameters of the films and bulk Ni_3S_4 should engender strain on the films. The strain should broaden the half-value width of X-ray reflection of the film, particularly that of reflection with a larger θ_B. The half-value width of the reflection with the larger θ_B was rather large, e,g,, about 0.77° for 555, as reflected in $\tan\theta_B$-dependence of the half-value width. The linear $\tan\theta_B$-dependence of the *lll* reflection half-value width (l=1-5), as shown in Fig. 8, indicates that the film is bearing some strain. However, the gradients of the linear dependences of the films were almost identical. That fact suggests that an integrated breadth of strain distribution function is independent of the substrate lattice dimension, and that other factors produce the strain.

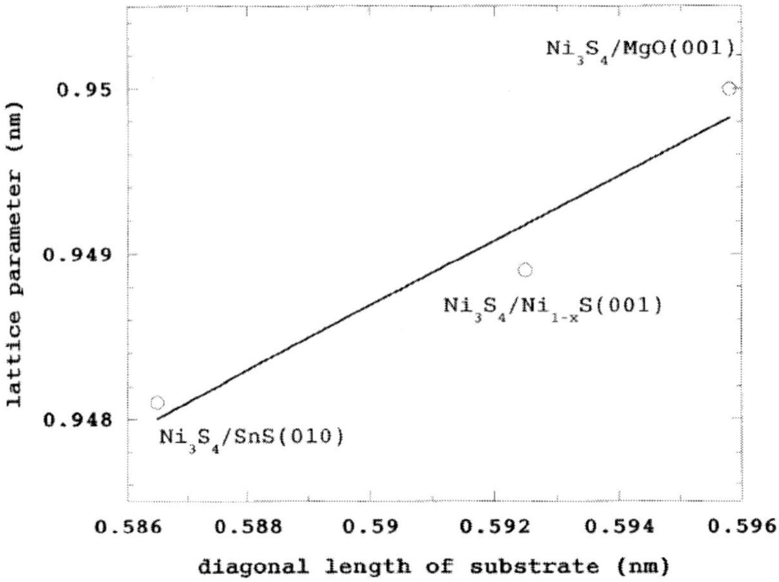

Figure 7. Relation between the lattice parameter of Ni_3S_4 film and the diagonal length of substrate.

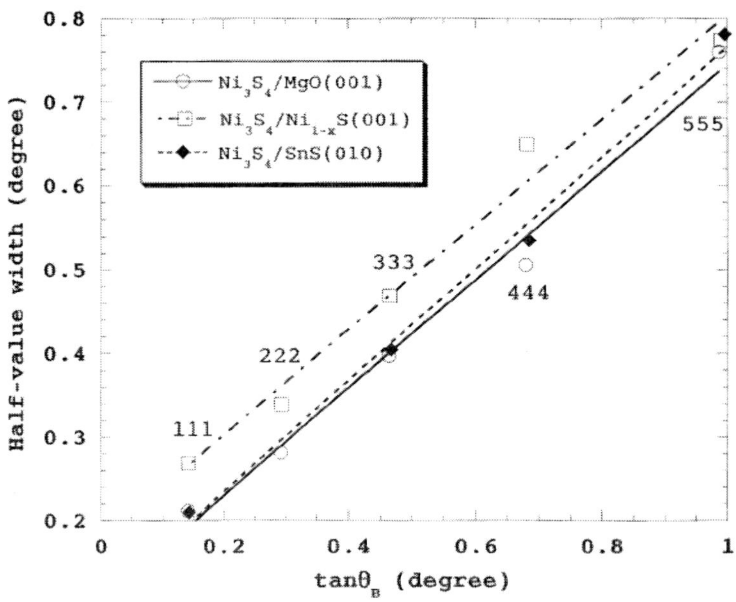

Figure 8. $\tan\theta_B$-dependence of half-value width for the *lll* reflections (l = 1-5) of Ni_3S_4 films.

An SEM image of the Ni_3S_4 film on the MgO substrate is shown in Fig. 9. Pyramidal crystallites of Ni_3S_4 with side sizes of triangle bases smaller than about 300 nm are visible in the figure. The apex of the pyramid is directed along the direction normal to the film surface, which is consistent with the out-of-plane epitaxy observed from the θ-2θ scanning XRD pattern.

The composition of the Ni_3S_4 film on the MgO substrate was examined using EPMA. The composition obtained ($Ni_{3.22}S_{4.00}$, on average, for five points) was in a range between

$Ni_{3.20}S_{4.00}$ and $Ni_{3.23}S_{4.00}$. The result suggested that excess metals occupied another site different from 16*d* and 8*a* sites in an ideal spinell-type structure.

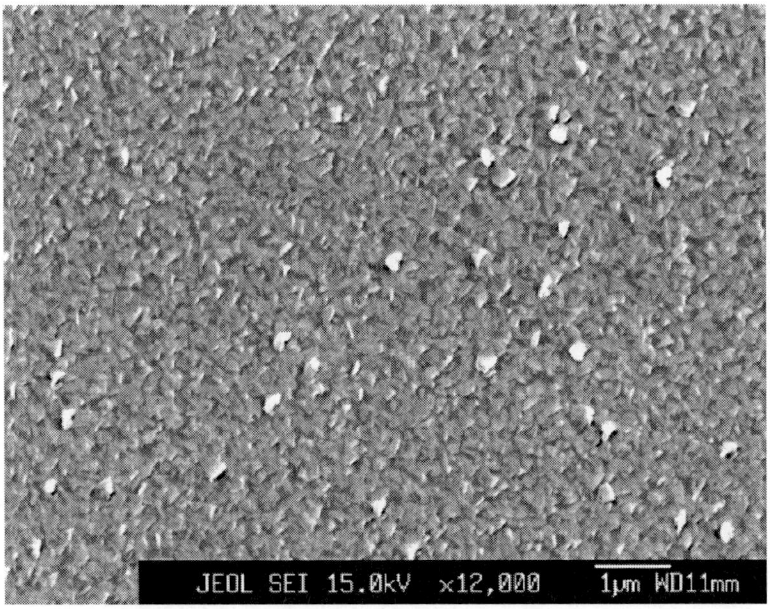

Figure 9. An SEM image of the Ni_3S_4 film deposited on the MgO substrate.

3.1.4. Crystal Structure Simulations of Ni_3S_4 Film

Several *lll* reflections (*l*=1-5) were observed in θ-2θ scanning XRD patterns, as shown in Figs. 2a, 2b and 2c. These reflection intensities depend only on the atomic position along [111], not on the atomic positions within (111). Therefore, we tried to determine the *u* parameter of the spinel-type structure of the Ni_3S_4 film deposited on the MgO(001) substrate by a simulation using a Rietveld profile analysis program (PREMOS by A. Yamamoto) because the *u* parameter is associated with the S atomic position along [111] for the $Fd\bar{3}m$ lattice of the structure. However, the Rietveld simulation result implied an unusually large value of the displacement parameter *B* for Ni at the 8*a* site of the structure and a negative value for Ni at the 16*d* site. The large value of the displacement parameter for Ni at 8*a* suggests that partial Ni atoms at 8*a* move to other positions. Furthermore, the strong intensity of 555 that is apparent in Fig. 2a was inexplicable using the ideal spinel-type structure.

For those reasons, we adopted another model with Ni occupancy that was different from the ideal spinel-type structure. In the model, the Ni occupancy at 8*a* was 33%, that at 16*d* was 55% and that at 16*c* was 33% for the $Fd\bar{3}m$ lattice. Ni does not occupy the latter site in the ideal spinel-type structure. The displacement parameters at the all Ni sites and the sulfur site (32*e*) were B = 1.0 in the model. The result is shown in Fig. 10. Many data blanks occur around 21°, 26.5°, 31°, angles between about 41° and 47°, etc. in the figure. Around the blank 2θ-positions, reflections of Ni_3S_4 arise in a polycrystalline specimen, but should never arise in the present film with [111] out-of-plane epitaxy. Therefore, we deleted the XRD data around these θ-positions in the figure as well as the 002 reflection of the MgO substrate. The intensity calculated with the model agrees well with the observed intensity. In the model, the *u* parameter was $u = 0.380(1)$. The *u* parameter of bulk Ni_3S_4 was reported as $u = 0.385$ [36].

However, the above occupancy of the model made the composition of the film to an unusually Ni-deficient one in comparison with the Ni_3S_4 composition. Furthermore, the Ni-deficient composition was inconsistent with the EPMA result. The number of reflections used in the Rietveld simulation was too small to allow further investigation of the Ni_3S_4 film structure. In spite of the compositional contradiction, the Rietveld simulations suggest that the Ni_3S_4 film structure is modified from the ideal spinel-type structure. One XRD study using single crystals has shown that several natural thiospinells, such as $CuCo_2S_4$, Co_3S_4 and $(Co,Ni)_3S_4$, possess a symmetry that is lower than $Fd\bar{3}m$ [37], but natural single crystals of Ni_3S_4 and Fe_3S_4 were not available. Therefore, we inferred that Ni_3S_4 possess a less than ideal spinel-type structure, as mentioned in the reference of [37]. The deviation of the Ni occupancy in the film structure from that in the spinel-type structure might engender the strain mentioned in the section of 3.1.3.

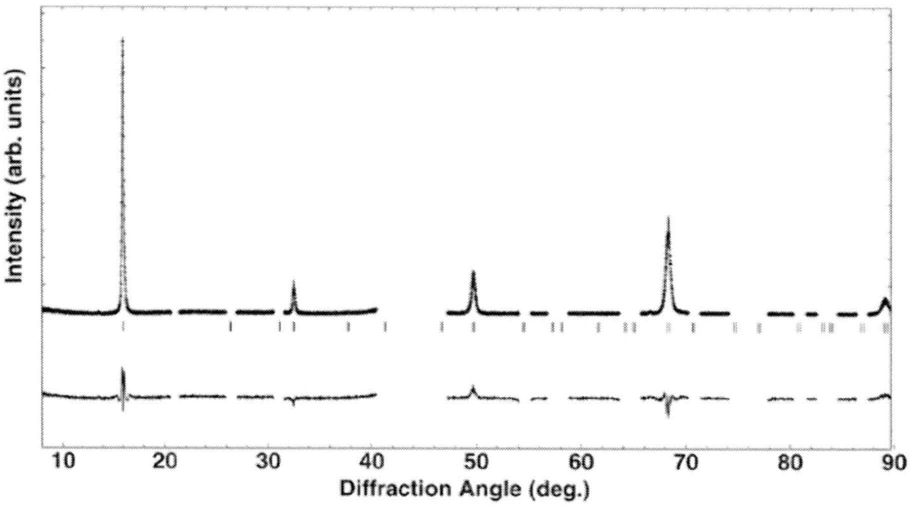

Figure 10. A θ-2θ scanning XRD pattern fitted by the Rietveld profile analysis program. The indexes of five reflections (*lll*, *l* = 1-5) are assigned respectively from 111 at the leftmost side to 555 at the rightmost side.

3.2. Stability of $Ni_{1-x}S$ Films on MgO(001)

The crystallography of epitaxial $Ni_{1-x}S$ film has been described in our previous study [7]. The $Ni_{1-x}S$ films deposited on MgO(001) had the following epitaxial relations; [001] || [001]MgO for the out-of-plane epitaxy and [1,-1,0] || [110]MgO for the in-plane alignment. The film composition of 1-x depended strongly on the MgO substrate temperature. The Ni vacancies of the film were ordered in alternate metal-layers of the NiAs-type structure, but such an ordering of Ni vacancy was mostly not observed in JCPDS diffraction data of bulk $Ni_{1-x}S$. Consequently, the X-ray reflection condition for *P63/mmc* of the NiAs-type structure (l=even for 00l) was broken and the reflection 001 of the structure should appear. The intensity ratio between 001 and 002 reflections yielded an estimation of the Ni vacancy amount: x.

In that previous study, stability of $Ni_{0.93}S$ film (Film2 denoted in the study) was confirmed by reinvestigating the θ-2θ pattern and lattice parameters about one year later. The film composition about 1 year later, as estimated from the intensity ratio between 001 and 002 reflections, was slightly more metal-deficient than the fresh film. The composition 1-x changed from 0.930±0.003 to 0.927±0.003 about 1 year later. Furthermore, trace reflections for 111 of Ni_3S_4 and 101 and 300 of millerite appeared in the θ-2θ pattern together with 444 of Ni_3S_4, which was also observed for the fresh film. The lattice parameters of the film changed from a = 0.3421(1) nm and c = 0.5334(2) nm to a = 0.3420(3) nm and c = 0.5333(5) nm about 1 year later. Because of unchanged optical reflectance for all films in the previous study, and because of the very slight changes in 1-x and the lattice parameters for Film2, we concluded that the film and other films (denoted as Film1, Film3 and Film4 in the previous study) were stable at room temperature. However, the previous study did not reexamine those films using XRD, except Film2.

We measured electrical resistivity and the Hall coefficient of Film4 1.5 years later after preparation. The composition of the fresh Film4 was the least metal-deficient (1-x = 0.01) among Film1-Film4. However, the temperature of the so-called metal-semiconductor transition observed in these measurements was 70 K, which was much lower than the transition temperature of 264 K for bulk $Ni_{1-x}S$ (x = 0) [11]. Therefore, we felt it necessary to confirm the stability of $Ni_{1-x}S$ epitaxial films with the composition near x = 0.

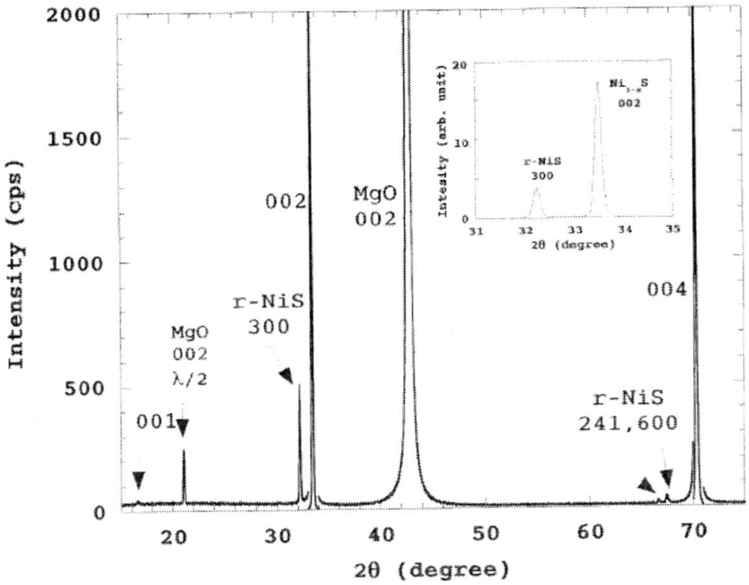

Figure 11. A θ-2θ scanning XRD pattern of fresh Film5. Intensities of 002 and 004 were reduced to about one-tenth. The inset shows a θ-2θ scanning pattern of the Film5 0.4 y later around 300 (millerite) and 002 ($Ni_{1-x}S$) reflections.

We prepared a fresh $Ni_{1-x}S$ epitaxial film/MgO(001) deposited at 603-609 K to investigate the galvanomagnetic properties. The substrate temperature was almost equal to that of Film4 used in the previous study. The RHEED patterns of the fresh film were identical to those in Figs. 3a and 3b. Figure 11 shows a θ-2θ scanning XRD pattern for the film (denoted as Film5 hereafter), which indicates that the film composition was approximately x

= 0 because of very weak 001 reflection. The intensity ratio between 001 and 002 was I_{001}/I_{002} = 0.2±0.08%. Its magnitude was the same as the intensity ratio of Film4 in the previous study. The Film5 composition was $1\text{-}x$ = 0.99, as estimated by XRD. Composition measured using EPMA was $1\text{-}x$=0.99, on average, for four points. The observed interplanar spacing and intensity are listed Table 2. The lattice parameters, which were obtained using the least-squares method, are shown in that table. The *c*-dimension was very close to 0.53482 nm for bulk $Ni_{1-x}S$ with x = 0, but the *a*-dimension was smaller than 0.34398 nm for the bulk $Ni_{1-x}S$ [38].

Table 2. Observed and calculated interplanar spacing and observed intensity of Film5. The lattice parameters are also shown

h k l	d_{obs} (nm)	d_{cal} (nm)	I_{obs} (%)
0 0 1	0.53460	0.53484	0.1
0 0 2	0.26749	0.26742	68.5
0 0 4	0.13370	0.13371	100.0
1 0 4	0.12198	0.12193	3.6
1 1 4	0.10555	0.10546	11.8
1 0 5	0.10071	0.10064	1.6
2 0 4	0.099534	0.099381	1.3
2 0 5	0.086751	0.086805	1.2
2 1 4	0.085986	0.085991	0.8
1 0 6	0.085260	0.085380	5.9
3 0 4	0.079495	0.079582	5.9
3 1 2	0.078734	0.078747	7.2

a=0.34306 nm and c=0.53484 nm

The *a*-dimensions of all films (Film1-Film5) were in the range between 0.3421 nm and 0.3431 nm [7]. This is strong evidence that the in-plane lattice dimension of the $Ni_{1-x}S$ film is constrained using the lattice dimension of the substrate surface. The in-plane alignment of the $Ni_{1-x}S$ films on the MgO(001) substrate was [1,-1,0] || [110]MgO, which resulted from proximity between diagonal length of the sulfur rhomb within (001) of $Ni_{1-x}S$ and that of the metal square within (001) of MgO [7]. Although the former diagonal length for the bulk hexagonal $Ni_{1-x}S$ (0.5958 nm for x = 0) agrees with the latter diagonal length of the cubic MgO substrate (0.5958 nm), the *a*-dimension of the $Ni_{1-x}S$ epitaxial film with $x \approx 0$ does not agree with $0.5958/\sqrt{3}$ nm for the bulk but is in that range mentioned above. This paradox might result from the fact that the surface periodicity of the MgO substrate, on which the $Ni_{1-x}S$ film is deposited, does not necessarily agree with the bulk periodicity by reconstruction of the MgO surface. We inferred that the *c*-dimension of the film was a better measure for the film composition than the *a*-dimension in this epitaxial film. It is noteworthy that the 300 reflection of millerite is readily apparent in the θ-2θ scanning pattern.

A θ-2θ scanning pattern of Film5 about 0.4 years later after preparation of Film5 revealed that the 300 reflection of millerite became very stronger than that of the fresh film, as shown in the inset of Fig. 11. This fact indicates that part of Film5 changed from $Ni_{1-x}S$ to millerite. The composition of $Ni_{1-x}S$ near the part might change to a more metal-deficient composition. For XRD, the 004 reflection came to have a shoulder at the higher 2θ side, which indicated that more metal-deficient $Ni_{1-x}S$ occurred in the film. Furthermore, the metal-semiconductor transition temperature decreased from 180 K for fresh Film5 to 160 K for Film5 0.25 years later, as mentioned in 4.2.1. Therefore, the intensity change of 300 reflection for millerite is considered as a measure for stability of the $Ni_{1-x}S$ epitaxial film in addition to change of the *c*-dimension.

Table 3. Intensity ratios between 300 of millerite (r-NiS) and 002 of $Ni_{1-x}S$ (h-NiS) for the fresh films and those observed later. Intensity ratios of 001 and 002 reflections and 1-*x* determined by the latter ratio are also shown

films	period (year)	I_{001}/I_{002} (h-NiS)	1-*x*	I_{300} / I_{002} (r-NiS) (h-NiS)
Film2	0.0	0.17	0.93	0.0
	1.0	0.19	0.93	0.030
	2.0	0.26	0.92	0.077
Film4	0.0	0.002	0.99	0.011
	1.1	0.001	0.99	0.073
Film5	0.0	0.002	0.99	0.035
	0.3	0.001	0.99	0.245

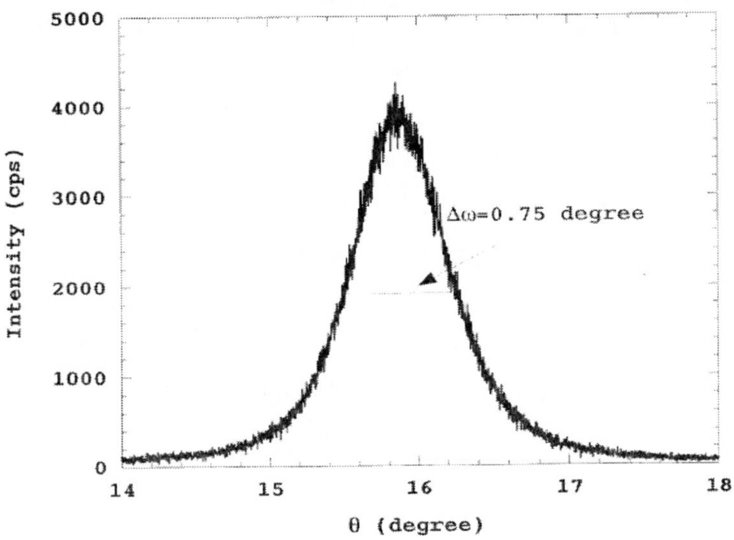

Figure 12. A θ-scanning rocking profile of 300 for millerite segregated from the $Ni_{1-x}S$ in Film5 0.4 y later.

For the fresh films of Film2, Film4 and Film5 and those measured at some time after their preparation, the intensity ratios between the 300 reflection of millerite and the 002 reflection of $Ni_{1-x}S$ are shown in Table 3. The table shows that the $Ni_{1-x}S$ films with 1-*x* = 0.93 is more stable than those with 1-*x* = 0.99. That fact is consistent with the result of kinetic study for the change from bulk $Ni_{1-x}S$ to millerite [20]. It is considered that the film with 1-*x* = 0.99 has greater stress than the film with 1-*x* = 0.93 because the *a*-dimension of the former film differs more from that of the bulk $Ni_{1-x}S$ than that of the latter. The *a*-dimension of Film2 almost agreed with that of bulk $Ni_{1-x}S$ with $x = 0.07$ [7].

(a)

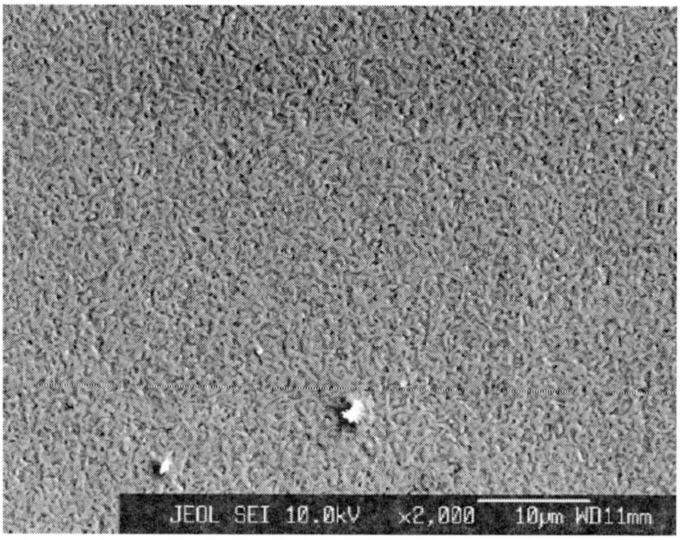

(b)

Figure 13. SEM images of (a) fresh Film5 and (b) Film5 0.4 y later.

Millerite was segregated from the $Ni_{1-x}S$ film with a preferential orientation. A θ-scanning rocking profile of 300 for millerite, as shown in Fig.12, showed that the half-value width of the profile was 0.75°. Therefore, [100] of millerite was directed sharply along [001]MgO, which was parallel to [001]$Ni_{1-x}S$. The segregated millerite had a preferred orientation within the plane parallel to the substrate surface that was [010]millerite || [100]MgO. There exists no natural $Ni_{1-x}S$, but there is some natural millerite. Therefore, the coexistence of $Ni_{1-x}S$ and millerite with clear orientation to each other might be a novel phenomenon.

An SEM image of Film5 measured 0.4 y later differed clearly from that of fresh Film5, as shown in Figs. 13a and 13b, although needle-like crystallites of millerite [39] are not visible in the image shown in Fig. 13b.

3.3. Epitaxial Nickel Sulfide Films on Al_2O_3(001)

3.3.1. XRD Characterization of Epitaxial Ni_7S_6 Film

The MgO(001) substrate was useful for preparation of $Ni_{1-x}S$ and Ni_3S_4 epitaxial films, but it failed to prepare an epitaxial film of millerite. A deposition experiment was performed using an Al_2O_3(001) substrate at 587-595 K with an intensity ratio of sulfur and nickel molecular beams, *I*(S-beam)/*I*(Ni-beam) = 3.8 because a polycrystalline millerite film on a glass substrate was prepared at that temperature with an identical intensity ratio of molecular beams. Consequently, the deposition experiment produced an epitaxial film of unknown nickel sulfide (M1 phase). Results of XRD characterization are described below.

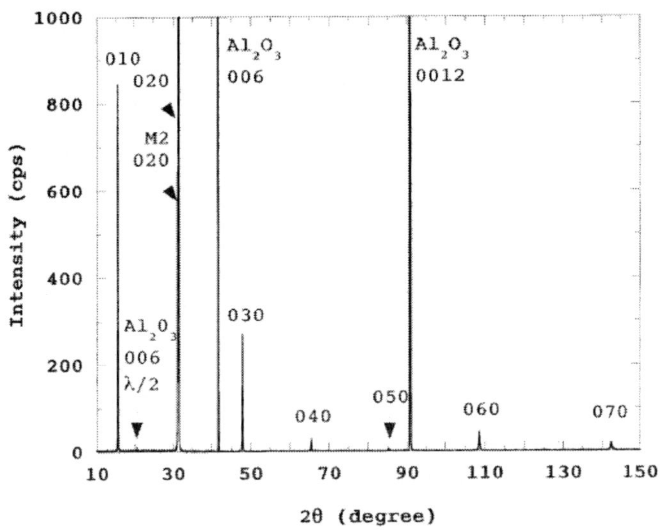

Figure 14. A θ-2θ scanning XRD pattern of M1 phase. Reflections of 0*k*0 (*k* = 1-7) are visible. A reflection denoted as "M2 020" was ascribed to 020 of the M2 phase.

Figure 14 shows a θ-2θ scanning pattern of the film. In that figure, many 0*k*0 (*k* = 1-7) reflections for a new nickel sulfide (M1 phase) are visible, being accompanied by 020 for another new nickel sulfide (M2 phase). The M2 phase is described in the 3.3.3 section. The pattern indicated that (010) of the film was parallel to the substrate surface. Interplanar spacing data of the film deposited on Al_2O_3(001) are listed in Table 4. Data for lattice planes inclined to (010) were collected as described in the Experimental section. The data were assigned with a monoclinic lattice with a = 0.3548 nm, b = 0.5698 nm c = 0.3350 nm, and α = 89.23° using the least-squares method.

Table 4. Observed and calculated interplanar spacing, observed intensity; observed and calculated interplanar angles for the M1 phase

h k l	d_{obs} (nm)	d_{cal} (nm)	I_{obs} (%)	$□_{obs}$ (°)	$□_{cal}$ (°)
0 1 0	0.5693	0.5698	29.0	0.0	0.0
0 2 0	0.2848	0.2849	100.0	0.0	0.0
0 3 0	0.1898	0.1900	21.0	0.0	0.0
0 4 0	0.1424	0.1425	2.4	0.0	0.0
0 4 1	0.1315	0.1316	1.0	23.7	23.2
1 4 0	0.1312	0.1311	0.1	23.1	23.0
0 4-1	0.1296	0.1303	0.9	23.1	23.0
1 4 1	0.1225	0.1225	2.6	31.3	31.2
1 4-1	0.1219	0.1215	<0.1	31.2	30.9
0 5 0	0.1139	0.1140	0.4	0.0	0.0
0 4 2	0.1092	0.1090	1.1	40.4	40.8
2 4 0	0.1085	0.1086	0.3	40.6	40.4
0 4-2	0.1077	0.1075	<0.1	40.2	40.2
0 5 1	0.1086	0.1083	0.4	18.7	18.9
1 5 0	0.1080	0.1079	0.3	18.8	18.8
0 5-1	0.1080	0.1074	0.1	18.9	18.8
1 4 2	0.1036	0.1037	0.5	45.4	44.0
2 4 1	0.1035	0.1035	0.6	44.2	43.7
1 4-2	0.1021	0.1024	3.1	44.5	43.3
0 6 0	0.09492	0.09497	4.8	0.0	0.0
1 5 2	0.09125	0.09110	0.2	38.8	37.6
1 6 1	0.08852	0.08843	<0.1	22.1	22.0
0 6 2	0.08277	0.08263	1.0	30.4	29.8
2 6 0	0.08266	0.08264	<0.1	29.5	29.6
0 7 0	0.08136	0.08141	2.1	0.0	0.0

a=0.3548 nm, b=0.5698 nm, c=0.3350 nm and □=89.23°

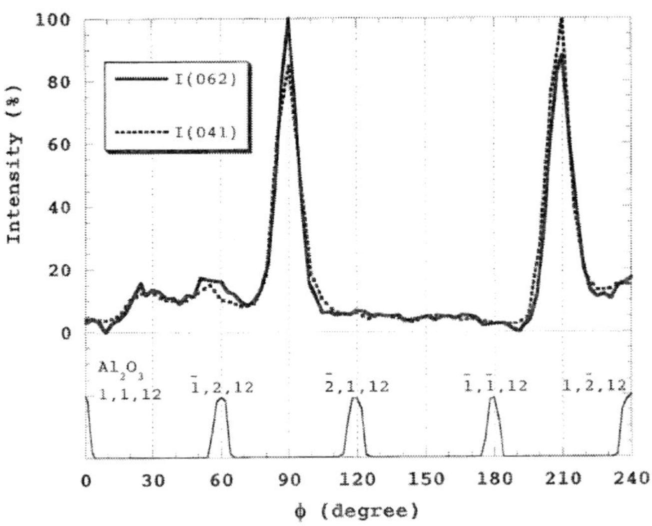

Figure 15 φ-dependencies of 041 and 062 intensities along with that of 1,1,12 for the Al_2O_3 substrate.

Figure 15 shows that the in-plane alignment of the film was examined using ϕ-dependencies of 041 and 062 intensities. In that figure, the ϕ-dependence of 1,1,12 intensity for the Al_2O_3 substrate is also shown to clarify the relation between orientations of the film and the substrate. Both profiles of the 041 and 062 ϕ-dependences agree well with each other in their details. The peak positions of the 041 intensity profile are ϕ=90° and 210°. Consequently, the in-plane alignment of the film is [001] || [1,-1,0]Al_2O_3.

Considering a two-dimensional orthogonal cell (A' = 0.4760 nm and B' = $\sqrt{3}$A' = 0.8245 nm) consisting of Al sites within the (001)Al_2O_3 surface: there are two Al sites and one vacant site are situated along the *B*'-axis per unit periodicity and one Al site along the *A*'-axis. The distance between Al sites including vacant sites along the B'-axis, which is parallel to [001] of the M1 phase, is 0.2748 nm. The distance between Al sites along the A'-axis, which is parallel to [100] of the M1 phase, is 0.4760 nm. Therefore, the periodical relations between lattice dimensions of the film and the substrate are $4a \approx 3A'$ and $5c \approx 2B'$.

3.3.2. Crystal Structure of Ni_7S_6 Film

The θ-2θ pattern of the M1 phase is very different from the pattern of $Ni_{1-x}S$ and that of Ni_3S_4. In the $Ni_{1-x}S$ and Ni_3S_4 epitaxial films, the sulfur layer of hexagonal sulfur stacking for the $Ni_{1-x}S$ structure and that of cubic sulfur stacking for the Ni_3S_4 structure were parallel to the MgO(001) substrate surface. Nickel atoms in these structures occupy octahedral interstices of sulfur stacking in $Ni_{1-x}S$. They occupy octahedral and tetrahedral interstices of sulfur stacking in Ni_3S_4. Similarly, the sulfur layer of the monoclinic structure for the M1 phase is considered to be parallel to the Al_2O_3(001) substrate surface. Therefore, the pattern suggests that sulfur atoms are stacked differently from hexagonal close-packed stacking for $Ni_{1-x}S$ and cubic close-packed one for Ni_3S_4 with a distribution of Ni atoms in interstices of the sulfur stacking. A Ni_7S_6 structure reported by Fleet [23] was considered as a candidate of appropriate crystal structure explaining the present θ-2θ scanning pattern. Therefore, we tried to determine *y*-positional parameters of nickel and sulfur sites along the [010] direction and their occupancies, assuming a Ni_7S_6–type structure.

Table 5. Positional *y*-parameters and occupancies with their standard deviations in parentheses for a bulk Ni_7S_6 structure [23]

Site	Occupancy	y
S1-4(c)	1.000	0.1418(7)
S2-8(f)	0.985(16)	0.1207(4)
S3-8(f)	0.945(16)	0.3858(6)
Ni1-8(f)	0.436(10)	0.2162(6)
Ni2-8(f)	0.919(10)	0.0399(2)
Ni3-8(f)	0.518(10)	0.3164(4)
Ni4-8(f)	0.459(9)	0.4805(5)
Ni5-4(c)	0.965(20)	0.3448(4)

Positional parameters along the [010] direction and occupancies at sulfur and nickel sites of the Ni_7S_6 structure are listed in Table 5; they were reported in the reference of [23]. It was convenient to switch the *b* and *c* parameters relative to the cell described by Fleet to refer to No. 63 in International Tables for Crystallography. Consequently, the space group was changed from *Bmmb* given by Fleet to *Cmcm*. An orthorhombic cell with A' = 0.3274 nm, B' = 1.1359 nm and C' = 1.6157 nm, as reported by Fleet [23], was assumed. Parameters A', B' and C' corresponded to respective parameters of the film, a, $2b$ and $5c$.

As shown in the table, Ni(2) and Ni(5) sites are almost fully occupied, but Ni(1), Ni(3) and Ni(4) sites are about half occupied. Therefore, the number of nickel atoms at Ni(2) in the unit lattice is about 8, but numbers of nickel atoms at the other sites are about 4. Because of the large number of the Ni(2) site, its positional parameter and occupancy are particularly effective to intensity distribution in the θ-2θ scanning pattern.

Figure 16 shows a simulation using the same Rietveld profile analysis program for the θ-2θ scanning pattern as that described in 3.1.4. In the simulation, positional parameters and occupancies at every site except the Ni(2) site were fixed at values shown in the table, but the occupancy at Ni(5) was changed from 0.965 to 1.0. Displacement parameters at the sulfur sites were fixed as B = 1.0 and those at the nickel sites were fixed as B = 5.0. Consequently, magnitudes of y-position and occupancy at the Ni(2) site were changed from magnitudes of 0.0399 and 0.919 in the table to 0.0538 and 0.813, respectively. A rather large value of Rwp = 24.2% was obtained, but the large value resulted from very low background level in the θ-2θ pattern of the epitaxial film. Fig. 16 shows that the calculated profile agreed well with the observed one. The Rietveld simulation strongly suggested that the M1 phase had a Ni_7S_6-type structure with a monoclinic distortion. The film composition was $Ni_{6.73}S_6$ according to the Rietveld simulation. Assuming a monoclinic Ni_7S_6-type structure for the film, lattice parameters of the structure should be changed to be $A' = a$, $B' = 2b$, $C' = 5c$ and $\alpha' = \alpha$.

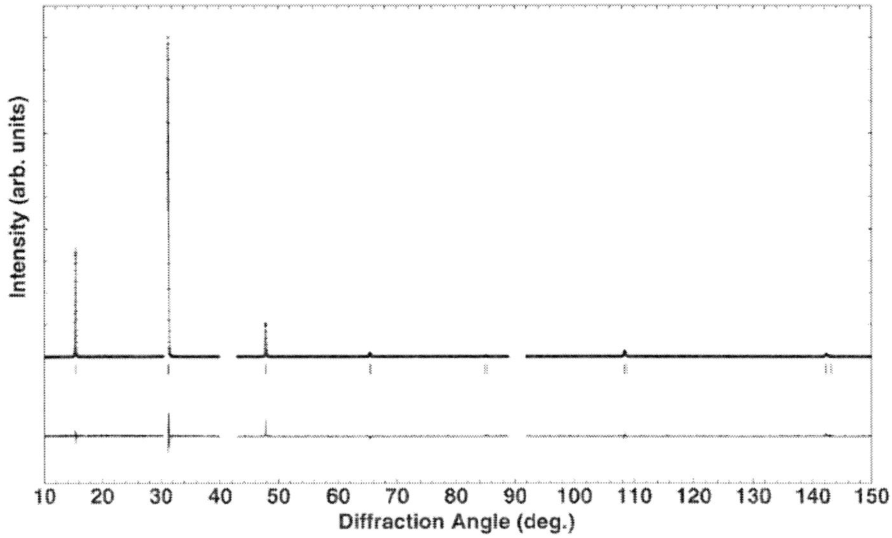

Figure 16. A θ-2θ scanning XRD pattern fitted by the Rietveld profile analysis program. Observed θ-2θ scanning XRD pattern is the one depicted in Fig. 14.

Figure 17 shows an SEM image of the film. Dark crystallites segregated in the matrix appear to align along the direction of the image. The matrix consists of fine crystallites. Composition of the matrix measured using EPMA was $Ni_{6.9}S_{6.0}$, on average, for three points; that for a dark crystallite was $Ni_{3.2}S_{2.0}$. The EPMA and XRD results indicated that the matrix corresponded to monoclinic Ni_7S_6 and the dark crystallite corresponded to the M2 phase with a composition of approximately Ni_3S_2.

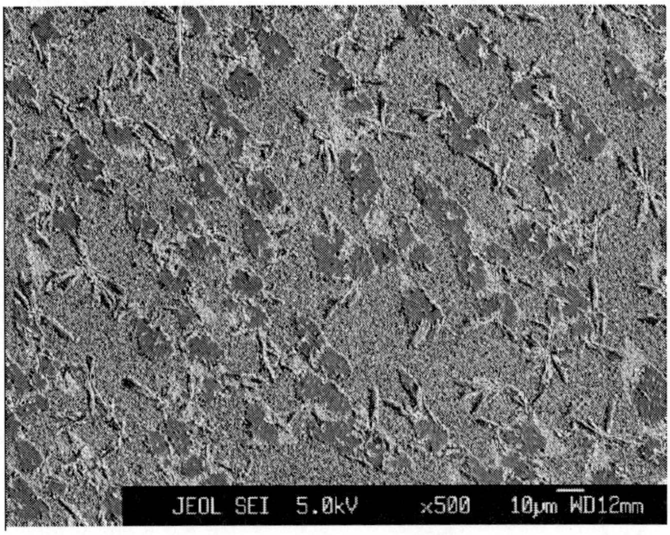

Figure 17. An SEM image of monoclinic Ni_7S_6 (M1 phase) coexisting with M2 phase. The matrix was ascribed to monoclinic Ni_7S_6 and dark crystallites were ascribed to the M2 phase.

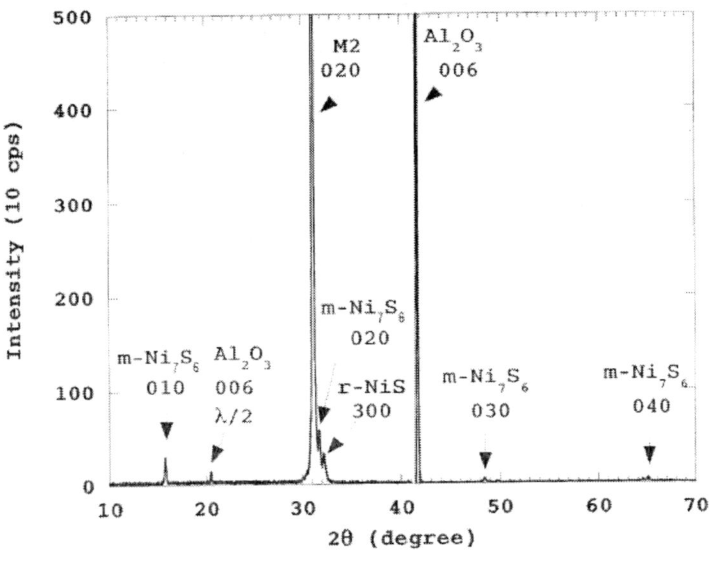

Figure 18. A θ-2θ scanning XRD pattern of M2 phase. Reflections for monoclinic Ni_7S_6 (M1 phase) and millerite are denoted respectively as "m-Ni_7S_6" and "r-NiS".

3.3.3. Crystal Structure of Epitaxial Ni_3S_2 Film

A nickel sulfide film deposited on an $Al_2O_3(001)$ substrate was prepared at an elevated temperature, 601-611 K, using the same intensity ratio of I(S-beam)/I(Ni-beam) as that used in the M1 phase preparation described in 3.3.1. Figure 18 shows a θ-2θ scanning pattern of the film. In that figure, a very strong 020 for a new nickel sulfide (M2 phase) is visible, being accompanied by the 0 *k* 0 reflections (*k*=1-4, denoted as m-Ni_7S_6) of the M1 phase. A weak 300 reflection for millerite is also visible in the figure. Interplanar spacing data of the M2 phase are listed in Table 6. Data for lattice planes inclined to (010) were collected as described in the Experimental section. Data were assigned with a monoclinic lattice with a = 0.2910 nm, b = 0.5735 nm, c = 0.4087 nm, and β = 89.29°.

Table 6. Observed and calculated interplanar spacing, observed intensity; observed and calculated interplanar angles for the M2 phase

h k l	dobs (nm)	dcal (nm)	Iobs (%)	αobs (°)	αcal (°)
0 2 0	0.2867	0.2867	100	0.0	0.0
0 4 0	0.1433	0.1434	0.2	0.0	0.0
0 4 1	0.1353	0.1353	0.7	19.4	19.3
1 4 0	0.1294	0.1286	0.5	26.7	26.3
1 4 1	0.1224	0.1229	2.5	31.4	31.0
0 4 2	0.1174	0.1174	1.2	35.1	35.1
1 4 2	0.1092	0.1091	1.5	40.4	40.4
-1 4 2	0.1085	0.1086	1.3	40.8	40.8
2 4 0	0.1021	0.1921	7.4	44.6	44.6
2 4 1	0.09925	0.09927	0.2	46.2	46.2
-2 4 1	0.09879	0.09886	0.1	46.5	46.4
0 6 0	0.09556	0.09558	1.0	0.0	0.0
1 4 3	0.09383	0.09307	0.7	49.2	49.2
0 6 1	0.09303	0.09307	0.1	13.2	13.2
1 6 0	0.09076	0.09080	0.4	18.2	18.2
1 6 1	0.08873	0.08871	0.1	22.1	21.8
-1 6 1	0.08852	0.08857	0.1	22.1	22.1
0 6 2	0.08657	0.08657	0.4	25.1	25.1
2 6 0	0.07984	0.07988	0.2	33.3	33.3

a = 0.2910 nm, b = 0.5735 nm, c = 0.4087 nm and β = 89.29°

The θ-2θ pattern indicated that (010) of the film was parallel to the (001)Al_2O_3 substrate surface. Figure 19 shows ϕ-dependence for 240 intensity of the film. The ϕ-dependence indicated the in-plane alignment of the film, [100] || [1,-1,0]Al_2O_3.

Consistency of in-plane periodicities between the film and the substrate was not obvious in comparison with that for Ni_3S_4 and $Ni_{1-x}S$ epitaxial films on the MgO(001) substrate. Considering a two-dimensional orthogonal cell, as described in 3.3.1, the distance between Al (or vacancy) sites along the B'-axis ($B'/3$ = 0.2748 nm) is 5.9% shorter than the a-parameter length. Six times the Al-distance along the A'-axis is approximately equal to seven times the c-parameter length.

The lattice parameters of the film were related to those for the rhombohedral Ni_3S_2 (heazlewoodite, A_r = 0.40718(6) nm and α = 89.459(9)°) [25]. Heazlewoodite structure has a pseudo-cubic body centered lattice of sulfur (A_c ≈ 0.407 nm). Therefore, the following relations pertain: $a \approx A_c/\sqrt{2}$, $b \approx \sqrt{2}A_c$ and $c \approx A_c$. Considering these relations, it was concluded that the M2 phase structure was a Ni_3S_2-type structure with a monoclinic distortion. It is noteworthy that (010) corresponds to (110) of the body-centered sulfur lattice, which is the densest sulfur layer. The unit lattice dimension of the film along the *a*-axis might be two times that of the observed magnitude of 0.2910 nm, but it was not confirmed because θ_B data of (10*l*) was not obtainable for the present XRD experimental setup because of $\theta_B < \alpha_{ip}$.

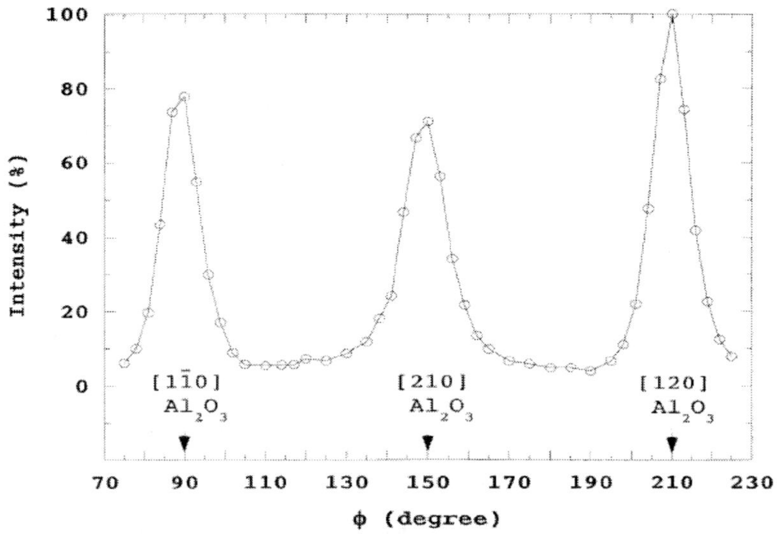

Figure 19. φ-dependence of 240 reflection intensity for the M2 phase. Directions of [1,-1,0]Al_2O_3 and its equivalents are shown on the abscissa.

Figure 20. A backscattered electron image for monoclinic Ni_3S_2 coexisting with monoclinic Ni_7S_6. Light crystallites correspond to monoclinic Ni_3S_2 and intermediately darkened matrix corresponds to monoclinic Ni_7S_6.

Figure 20 shows a backscattered electron image for the film. The image indicates that light crystallites have richer Ni content than the intermediately dark matrix. The light crystallite corresponded to monoclinic Ni_3S_2 according to EPMA results. Composition of the crystallite measured using EPMA was $Ni_{3.0}S_{2.0}$ for two respective points. The intermediately darkened matrix consisted of fine crystallites. Its composition was $Ni_{7.6}S_{6.0}$ in average for three points. The latter composition was intermediate between monoclinic Ni_7S_6 and monoclinic Ni_3S_2. Therefore, the matrix probably includes fine crystallites of both compounds.

3.3.4. Coexistence of Millerite with Ni_7S_6

A nickel sulfide film deposited on an $Al_2O_3(001)$ substrate was prepared at a lower temperature, 533-543 K, using the same intensity ratio of *I*(S-beam)/*I*(Ni-beam) as that used in M1 phase preparation. Figure 21 shows a θ-2θ scanning pattern of the film. In that figure, 0*k*0 (*k* = 1-7) reflections of monoclinic Ni_7S_6 and *h*00 (*h* = 3 , 6 and 9) reflections of millerite are visible. In the figure, the former reflections are denoted as * and the latter ones are shown as "r-NiS". The pattern indicated that the out-of-plane epitaxy for monoclinic Ni_7S_6 was [010] || [001]Al_2O_3 and that for millerite was [100] || [001]Al_2O_3. The lattice planes of (010) for monoclinic Ni_7S_6 and (100) for millerite are the respective basal planes of sulfur of the structures. In-plane alignments of these compounds were not clear, but they were mutually correlated, as shown in ϕ-dependence of 062 intensity for monoclinic Ni_7S_6 and the ϕ-dependence of 410 intensity for millerite (Fig. 22). Figure 23 shows rocking profiles of 020 for monoclinic Ni_7S_6 and 300 for millerite. The rocking profiles indicated that both compounds are oriented sharply along [001]Al_2O_3, although their in-plane alignment was not clearly apparent.

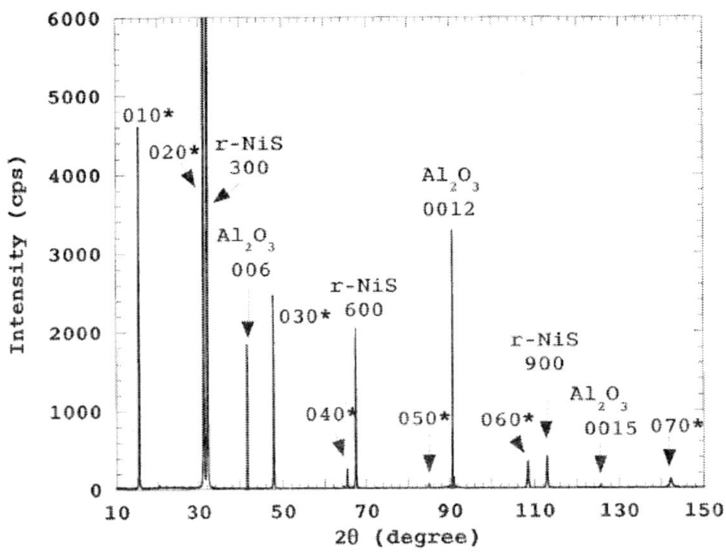

Figure 21. A θ–2θ scanning XRD pattern of monoclinic Ni_7S_6 coexisting with millerite. Reflections of monoclinic Ni_7S_6 are denoted as "0*k*0*" (*k* = 1-7) and those of millerite as "r-NiS *h*00" (*h* = 3, 6 and 9).

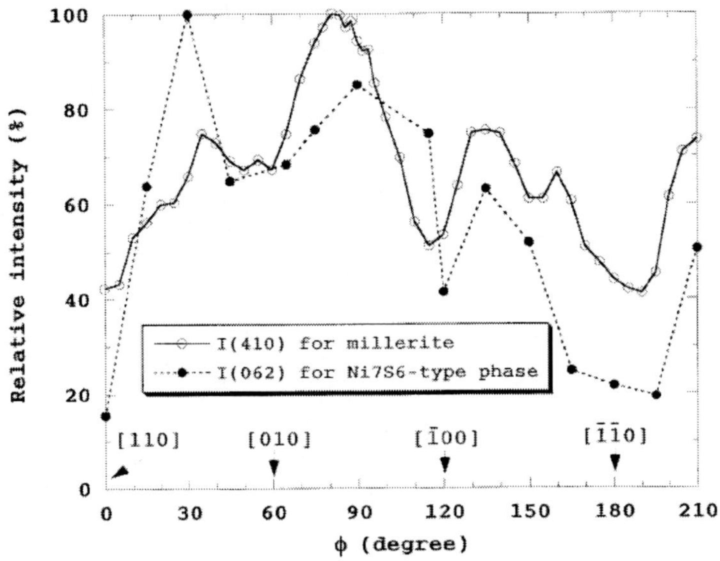

Figure 22. ϕ-dependence of 062 reflection intensity for monoclinic Ni_7S_6 along with that of 410 reflection intensity for millerite. Allows indicate directions of $[110]Al_2O_3$ and its equivalents.

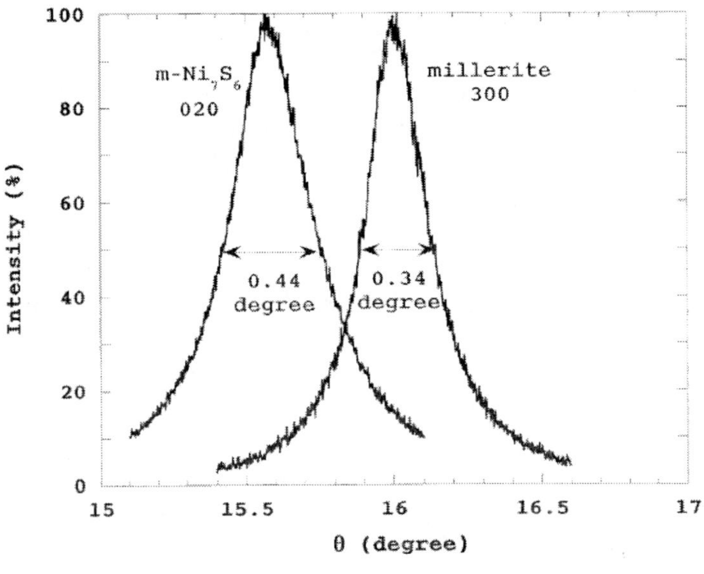

Figure 23. θ-scanning rocking profile of 020 for monoclinic Ni_7S_6 and that of 300 for millerite.

The lattice parameters of monoclinic Ni_7S_6 for the film coexisting with millerite were a = 0.3879 nm, b = 0.5699 nm, c = 0.3348 nm and α = 88.99°. These were obtained using the least squares method with interplanar spacing data for lattice planes parallel and inclined to $(001)Al_2O_3$. It is noteworthy that the a-parameter of the present film was 9.3% larger than that of the film coexisting with monoclinic Ni_3S_2. This fact, along with the coexistence of monoclinic Ni_7S_6 with monoclinic Ni_3S_2 or millerite, suggested that monoclinic Ni_7S_6 had a homogeneity range, which is examined further in 3.3.5.

Table 7 shows interplanar spacing of millerite for the film. The lattice parameters, which were obtained using the least squares method with the interplanar spacing, were a = 0.9611 nm and c = 0.3150 nm for a hexagonal lattice setting. The lattice parameters agreed with those of mineral millerite (a = 0.9620 nm and c = 0.3149 nm) [40].

Table 7. Observed and calculated interplanar spacing, observed intensity; observed and calculated interplanar angles; and observed integrated breadth for millerite in the matrix of monoclinic Ni_7S_6.

h k l	d_{obs} (nm)	d_{cal} (nm)	I_{obs} (%)	α_{obs} (°)	α_{cal} (°)	$\Delta\omega$ (°)
3 0 0	0.27788	0.27745	100.0	0.0	0.0	0.156
4 1 0	0.18161	0.18163	1.7	10.7	10.9	0.168
6 0 0	0.13872	0.13872	9.3	0.0	0.0	0.188
5 2 0	0.13329	0.13328	0.1	15.9	16.1	0.145
4 4 0	0.12019	0.12014	0.2	29.8	30.0	0.201
6 1 1	0.11772	0.11773	0.1	23.0	23.2	0.174
5 3 1	0.11128	0.11124	0.1	29.5	29.7	0.184
7 0 1	0.11128	0.11124	0.1	20.5	20.7	0.193
7 1 0	0.11022	0.11025	0.5	6.4	6.6	0.213
8 0 1	0.098776	0.098793	0.1	18.1	18.3	0.204
9 0 0	0.092325	0.092482	2.6	0.0	0.0	0.246

a=0.9611 nm and c=0.3150 nm

Figure 24a shows a backscattered electron image of the film. Dark areas exist in the light matrix. The image indicates that the light matrix has a richer Ni content than the dark areas. The dark areas appeared to be porous, as shown in an SEM image of Fig. 24b. The composition of the dark areas measured using EPMA was $Ni_{1.12}S$, on average, for five points and that of the light areas $Ni_{1.16}S$, on average, for five points. Considering the XRD result of the film that millerite coexisted with monoclinic Ni_7S_6, the dark areas corresponded to millerite and the light matrix corresponded to monoclinic Ni_7S_6. However, the composition of the dark areas obtained using EPMA was rather metal-rich, even for millerite, with the most metal-rich composition of NiS. The dark area might partly include monoclinic Ni_7S_6; alternatively, porous morphology of the film surface might affect the EPMA result. The composition of monoclinic Ni_7S_6 in the matrix was also rather metal-rich, although it should be more metal-deficient than the composition of monoclinic Ni_7S_6 coexisting with monoclinic Ni_3S_2. The reason for this contradiction is unknown, but some plausible explanations are that the dark area consisted of millerite crystallites with slight inclusions of monoclinic Ni_7S_6 and the present experimantal condition of EPMA indicates slightly larger Ni contents for both the dark area and matrix than their true Ni contents.

Both images showed that aggregates of fine crystallites in the dark areas started to grow along steps on the (001)Al_2O_3 surface and extended from step to step. A similar image was observed in optical photographs for Ag_2S epitaxial thick films (thickness > 85 nm) deposited on cleaved MgO(001) surface [30]. For this case, a two-dimensional body-centered orthogonal cell of sulfur for the Ag_2S film aligned such that one orthogonal axes was parallel to a direction along which steps on the MgO cleaved surface were running [30]. Contrary to the in-plane alignment of the Ag_2S film, the present film had no clear orientation in relation to the step direction.

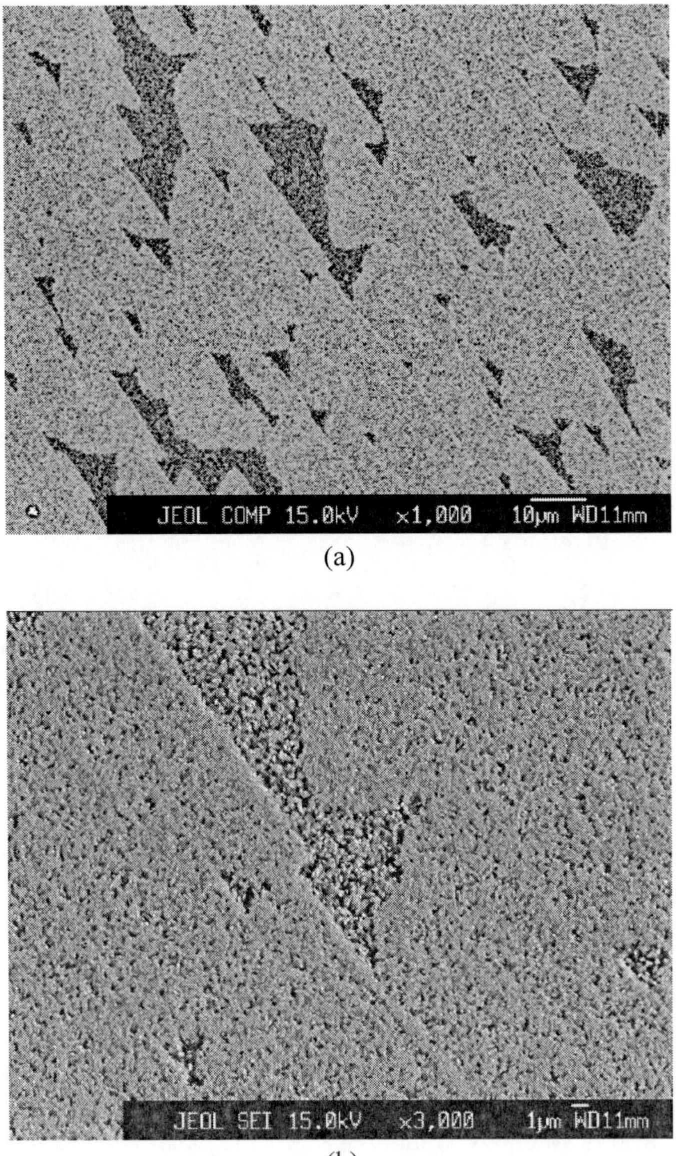

Figure 24. (a) A backscattered electron image of monoclinic Ni_7S_6 coexisting with millerite. Dark areas were ascribed to millerite and the matrix was attributed to monoclinic Ni_7S_6. (b) An SEM image of the same position in the film.

3.3.5. Homogeneity Range of Ni_7S_6

The 0*k*0 intensity of the θ-2θ scanning XRD pattern for the film coexisting with millerite differed slightly from that of the film coexisting with monoclinic Ni_3S_2, which suggested slight changes of the *y*-positional parameter and the occupancy for monoclinic Ni_7S_6

coexisting with millerite. For that reason, a Rietveld profile simulation was performed for the present film. The procedure of the simulation was identical to that for the film coexisting with monoclinic Ni_3S_2, as described in 3.3.2. The simulation gave the result that the *y*-positional parameter at the Ni(2) site was 0.0481 and occupancy at the site was 0.761. The *Rwp* value was 24.4%. The large *Rwp* value was attributable to the reason described in 3.3.2. Figure 25 shows the calculated and observed profiles. The calculated profile agreed with the observed one, as shown in the difference depicted in the figure.

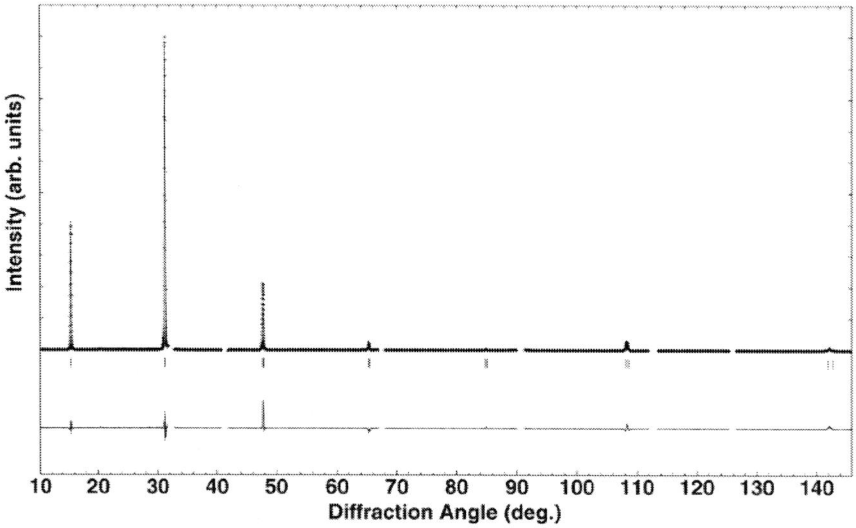

Figure 25. A θ-2θ scanning XRD pattern fitted by the Rietveld profile analysis program. Observed θ-2θ scanning XRD pattern for 0*k*0 (k = 1-7) of monoclinic Ni_7S_6 is that depicted in Fig. 21.

Consequently, the composition of the film coexisting with millerite was $Ni_{6.60}S_6$, which was more metal-deficient than that of the film coexisting with monoclinic Ni_3S_2. Consequently, the Rietveld profile simulations of both films indicated that monoclinic Ni_7S_6 had a homogeneity range, but the composition of the film obtained using the Rietveld profile simulation, which was approximately between $Ni_{6.6}S_6$ and $Ni_{6.7}S_6$, was more metal-deficient than bulk Ni_7S_6. The composition resembled bulk Ni_9S_8 [2, 21]. The crystal structure of Ni_9S_8 (godlevskite) was reported by Fleet [21]. An identical simulation for a Ni_9S_8-type structure using the Rietveld profile analysis program was performed to verify whether the film structure was Ni_9S_8-type or not. The simulation showed that no satisfactory agreement was obtained between the observed θ-2θ scanning pattern and the calculated one.

4. Galvanomagnetic Properties of Nickel Sulfide Films

4.1. Ni_3S_4 Film

Temperature dependencies of electrical resistivity and the Hall coefficient of the Ni_3S_4 epitaxial film deposited on the MgO(001) substrate are shown in Fig. 26. The Hall coefficient is positive throughout the temperature range. Assuming single band conduction, the hole carrier density was 5.4 × 10^{28} m^{-3}, which was estimated from the Hall coefficient magnitude at 285 K. The magnitude corresponded to 1.4 hole carriers per Ni_3S_4 molecule. The resistivity decreased linearly with decreasing temperature above 80 K. In this temperature range, the Hall coefficient decreased very slowly, although the data was rather scattered. These facts indicate that the Ni_3S_4 film is metallic with a hole conduction and electron phonon scattering is dominant for conduction in that temperature range. Below about 80 K, the Hall coefficient decreased rapidly with decreasing temperature, thereby indicating that the film conduction included small amount of electron conduction in addition to the hole conduction, and that the electron conduction increased more rapidly than the hole conduction with decreasing temperature.

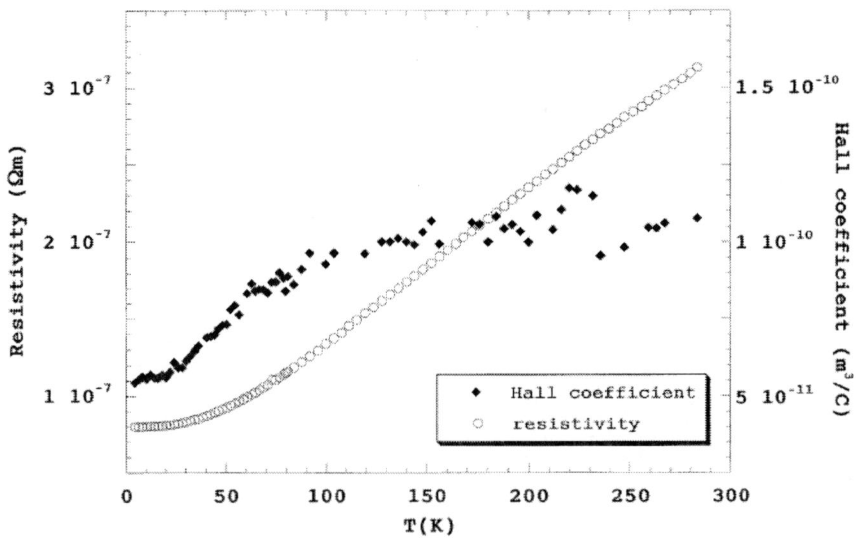

Figure 26. Temperature dependencies of resistivity and Hall coefficient for the Ni_3S_4 film on MgO(001).

We estimated the conduction parameters of the Ni_3S_4 film assuming the two-band model [41]. According to the two-band model, conductivity (σ) and the Hall coefficient (R_H) are given as:

$\sigma = \sigma_e + \sigma_h$, and

$R_H = (\sigma_e^2/n_e - \sigma_h^2/n_h)/q\sigma^2$,

where σ_e and σ_h respectively represent electron and hole conductivities, n_e and n_h are respective electron and hole carrier densities, and q is the electron charge. Considering the temperature dependence of Hall coefficient below 90 K, we performed a parameter-fitting to the temperature dependence of R_H with the following assumption. The carrier densities were constant in the entire measured temperature range and the temperature dependence of σ_e/σ was expressed as an equation, $\sigma_e/\sigma = p_1+p_2/[(T-p_3)^2 + p_4]$, where p_i (i = 1-4) are parameters determined using the best-fitting curve of R_H. The result of the parameter fitting in the temperature range of 4.3-90 K is shown in Fig. 27. The temperature dependence of the conduction rate σ_e/σ is also given in that figure. The carrier densities are n_e=1.99×10^{28} m^{-3} and n_h=6.00×10^{28} m^{-3}. The magnitudes of σ_e/σ are 0.189 at 4.5 K and 0.0237 at 285 K if the fitting temperature range is extended. The latter magnitude gives a value of R_{cal}=1.0×10^{28} m^3 C^{-1} at 285 K which agrees excellently with the observed one at that temperature. However, the estimated magnitudes of n_e and n_h will be varied to some extent according to a choice of functional expression of σ_e/σ.

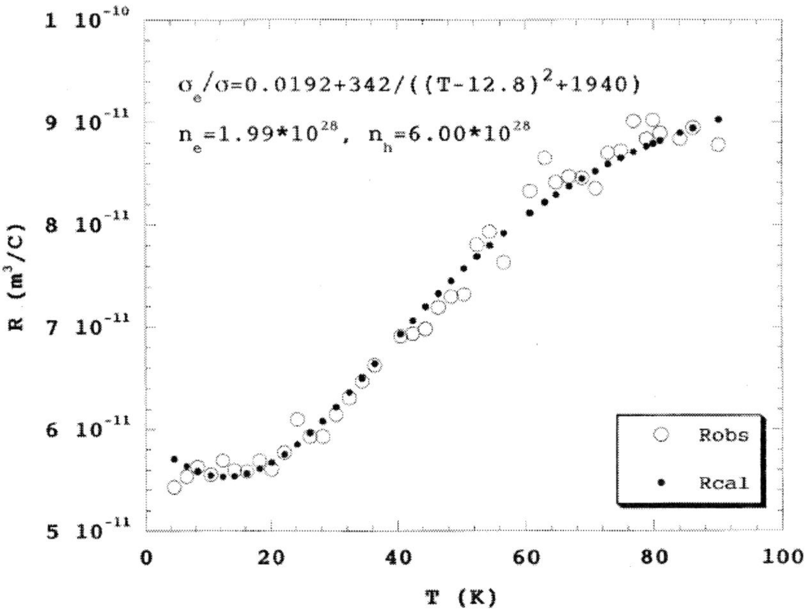

Figure 27. A fitting curve for the temperature dependence of Hall coefficient of the Ni_3S_4 film on MgO(001) using parameters of σ_e/σ, n_e and n_h. An expression of the σ_e/σ temperature dependence is shown in the figure in addition to the magnitudes of n_e and n_h.

It was concluded that the Ni_3S_4 film had hole conduction at 285 K, but 80% hole and 20% electron conduction at 4.5 K. The Hall mobility of the electron ($\mu_e = \sigma_e/(qn_e)$) and that of hole ($\mu_h = \sigma_h/(qn_h)$) are, respectively, 7.3 × 10^{-4} m^2 Ω^{-1} C^{-1} and 1.1 × 10^{-3} m^2 Ω^{-1} C^{-1} at 4.5 K.

Ferrimagnetic magnetization at 5 K was reported for Ni_3S_4 prepared using a PH-controlled homogeneous precipitation method [5]. However, the present Hall coefficient at 4.3 K was independent of applied magnetic field strength below 0.95 T, as shown in a linear field-dependence for Hall resistivity in Fig. 28. From the fact, we infer that no extraordinary Hall effect existed in the Ni_3S_4 film. An extraordinary Hall effect should be observed in a ferrimagnetic or ferromagnetic substance, when the substance is magnetized by an applied magnetic field, as observed in Fe_3S_4 [42]. Furthermore, magnetoresistance was not observed in that field-strength range. Ferrimagnetism should give a large negative magnetoresistance because the applied magnetic field should suppress magnetic scattering in conduction through s-d interaction. The temperature dependence of the resistivity showed no characteristics that suggested a magnetic transition. Therefore, these results indicate that the Ni_3S_4 film was not ferrimagnetic, even at 4.3 K. Figure 26 shows that the electrical resitivity of Ni_3S_4 reported in the previous study [5] is two orders of magnitude larger than that in the present study. A previous study also reported [5] that the Seebeck coefficient of Ni_3S_4 is negative. However, it differs from the hole conduction, as shown in Figure 26. These differences might result from poor crystallinity through an ambient temperature reaction between aqueous nickel chloride and sodium dithionite solutions in the previous study, as pointed out in the reference [6].

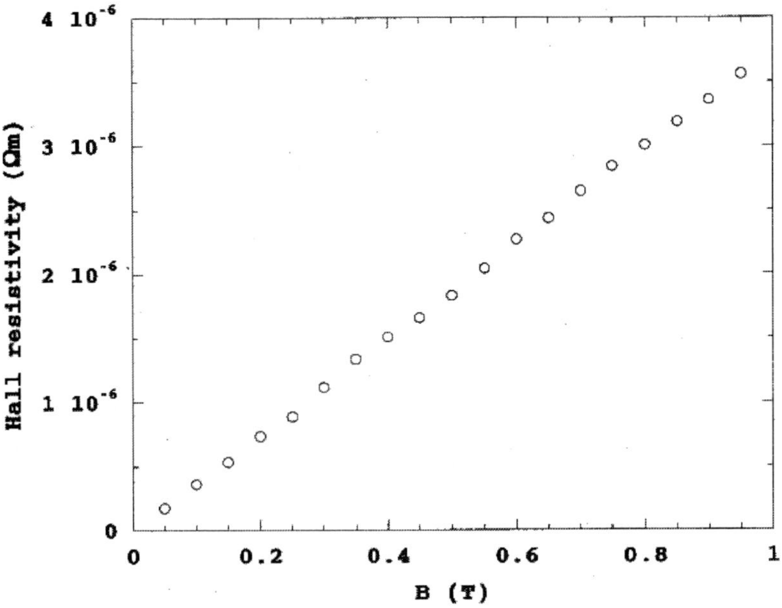

Figure 28. Magnetic field dependence of Hall resistivity at 4.3 K for the Ni_3S_4 film on MgO(001).

4.2. $Ni_{1-x}S$ Films

4.2.1. Electrical Resitivity and Hall Coefficient for $Ni_{1-x}S$ ($x \approx 0$)

Temperature dependence of resistivity for the freshly prepared $Ni_{1-x}S$ epitaxial film (Film5) is shown in Fig. 29 along with the temperature dependence of the film measured about 0.25 y later. The resistivity of the fresh film decreases with a slow linear dependence with decreasing temperature; it increases rapidly at 180 K. The temperature corresponds to the so-called metal-semiconductor transition temperature mentioned in previous studies [11, 16, 17, 19]. For $Ni_{1-x}S$ of bulk material ($x = 0$), the transition temperature has been reported as 264 K [11]. The resistivity increases further with decreasing temperature until 115 K. Subsequently, it begins to decrease below about 100 K. The resistivity increases with increasing temperature until about 130 K. It starts to decrease beyond about 135 K. The transition occurs at 190 K on increasing temperature. Then, it increases linearly with a broad small hump at about 250 K with increasing temperature.

The temperature dependence of the Hall coefficient for the fresh film, in addition to that for the film 0.25 y later, is shown in Fig. 30. The temperature dependence for the fresh film resembles that of the resistivity, but anomalous behavior around 255 K was not so clear in comparison with the hump of resistivity. It is noteworthy that the resistivity and the Hall coefficient for the fresh film decrease clearly below 80 K with decreasing temperature. These temperature dependencies differ from those of bulk $Ni_{1-x}S$ that are almost constant below the transition temperature [16, 17].

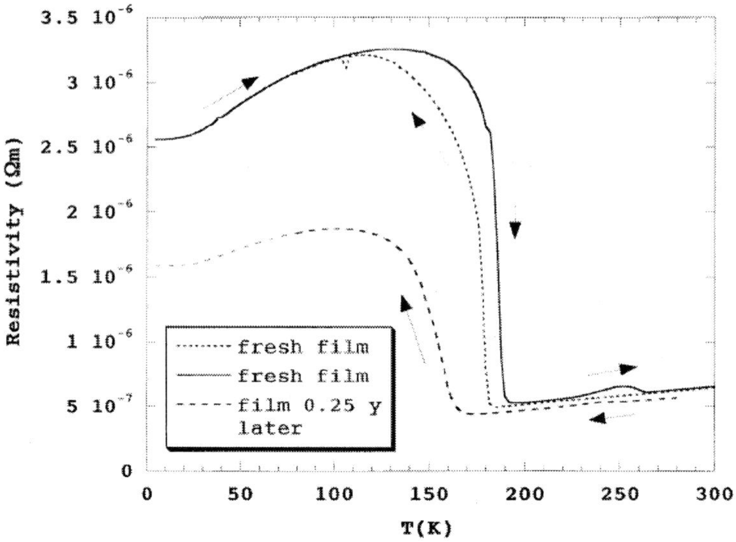

Figure 29. Temperature dependencies of resistivity for fresh Film5 and Film5 0.25 y later. Arrows indicate decreasing and increasing temperatures of measurements. The temperature dependence on increasing temperature shows a hump at ca. 250 K.

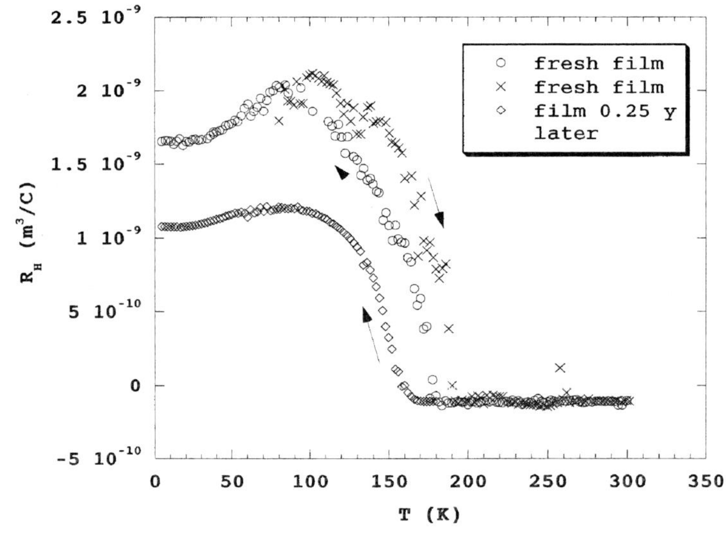

Figure 30. Temperature dependencies of Hall coefficient for fresh Film5 and Film5 0.25 y later. Arrows indicate decreasing and increasing temperatures of measurements. The temperature dependence on increasing temperature shows an anomaly at ca. 255 K.

Temperature dependencies of resitivity and the Hall coefficient for the fresh film indicate several facts and suggestions:

(1). Conduction above the transition temperature is metallic with electron conduction.

(2). The transition temperature decreases greatly in comparison to that of bulk $Ni_{1-x}S$ (x = 0). The volume of bulk $Ni_{1-x}S$ is known to dilate below the transition temperature. It increases by 1.3% at 77 K [14]. Therefore, pressure applied to bulk $Ni_{1-x}S$ suppresses the transition temperature [43, 44]. Similarly, it is considered that the MgO substrate suppresses the dilatation of the epitaxial $Ni_{1-x}S$ film on the transition with decreasing temperature, which decreases the transition temperature.

(3). The decreasing resistivity and Hall coefficient below about 100 K indicate that the conduction in the temperature range is also metallic. Therefore, the respective transitions at 180 K and 190 K with decreasing and increasing temperature are not metal-semiconductor transitions but rather metal-metal transitions. A photoelectron spectroscopy study of bulk $NiS_{1-x}Se_x$ has reported that the transition is not a metal-semiconductor transition but rather a metal-anomalous metal transition [12].

(4). The broad hump in the resistivity on increasing temperature presumably reflects the metal-metal transition that might occur in slight areas within the film at 250 K. These areas might remain free from suppression by the substrate and be caused by dilatation at low temperatures. Therefore, it was considered that the composition in $Ni_{1-x}S$ was approximately $x = 0$ because the transition temperature of 250 K was close to that of bulk $Ni_{1-x}S$ (264 K for $x = 0$).

(5). Hystereses in the temperature dependencies of the resistivity and the Hall coefficient around the transition temperatures showed that the transition was first order, as reported in previous studies [13].

The magnitude of the Hall coefficient at 280 K was -1.11×10^{-10} m^3 C^{-1}. Assuming a single band, the electron carrier density at 280 K was 5.6×10^{28} m^{-3}. The magnitude corresponded to electron carrier density of 3.1 per unit lattice. The Hall mobility of electrons was 1.8×10^{-4} m^2 Ω^{-1} C^{-1}. For the metallic state below the transition temperature, we estimated the conduction parameters of n_e, n_h and σ_e/σ by obtaining a fitting curve to the temperature dependence of the Hall coefficient for temperatures of 20-80 K, assuming that n_e and n_h were constant below 80 K and that σ_e/σ varied similarly to the equation for the Ni_3S_4 film.

Figure 31 shows the result. Both carrier densities of electron and hole were one order of magnitude smaller than the electron carrier density at 280 K. The ratio σ_e/σ increased from 0.08 at 80 K to 0.16 at 20 K. The carrier density of electrons was larger than that of the hole, but the Hall mobility of electrons at 20 K was smaller ($\mu_e = 1.0 \times 10^{-4}$ m^2 Ω^{-1} C^{-1}) than the Hall mobility of the hole ($\mu_h = 7.8 \times 10^{-4}$ m^2 Ω^{-1} C^{-1}) at 20 K. The latter Hall mobility was nearly equivalent to that reported previously [16].

Both transition temperatures of the resistivity and Hall coefficient for the film 0.25 y later were about 160 K with decreasing temperature. This temperature was lower by 20 K than that for the fresh film. As mentioned in the section 3.2, the XRD θ-2θ scanning pattern revealed the strong 300 reflection of millerite. It was therefore considered that the film composition changed from the fresh film after about 0.25 y. The XRD characterization in the section 3.2 indicated a partially heterogeneous structure of the film 0.4 y later. Nevertheless, the film 0.25 y later seemed to be homogeneous because the transition did not occur stepwise, but rather occurred all at once at about 160 K.

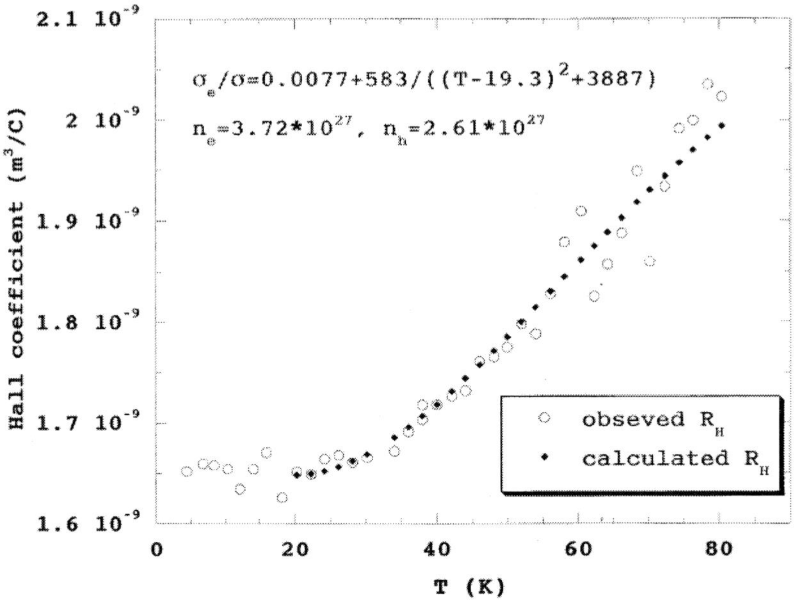

Figure 31. A fitting curve for the temperature dependence of the Hall coefficient of fresh Film5 using parameters of σ_e/σ, n_e and n_h. An expression of the σ_e/σ temperature dependence is shown in the figure in addition to the magnitudes of n_e and n_h.

4.2.2. Electrical Resistivity and Hall Coefficient for $Ni_{1-x}S$ (x=0.07)

The resistivity and Hall coefficient for Film2 ($Ni_{0.93}S$) and the resistivity of Film3 ($Ni_{0.93}S$) were measured 2 y after preparation. Film2 came to include a slight amount of millerite 2 y later, as mentioned in 3.2. We surmised that this change of the film did not affect its galvanomagnetic properties remarkably, except for those associated with the metal-metal transition. Film2 and Film3 showed no transition. Those results are shown in Fig. 32. The Hall coefficient of Film3 was almost equal to that in Fig. 32; it is not included in the figure. A calculated Hall coefficient is shown there. It was fitted to the temperature dependence of Hall coefficient between 20 K and 100K by the method mentioned above. A jump in the resistivity of Film2 at about 230 K is not intrinsic. These films were metallic with the electron conduction. The magnitude of Hall coefficient at 280 K (-5.66×10^{-11} m^3 C^{-1}) corresponds to the electron carrier density of 1.1×10^{29} m^{-3}, that is, three electrons per $Ni_{1-x}S$ molecule. The very small magnitude of the Hall coefficient and the large electron carrier density suggested the coexistence of electron and hole. Although the Hall coefficient in the figure was almost constant above about 120 K, it decreased rapidly below about 80 K with decreasing temperature. This temperature dependence indicated that the electron conduction increased more rapidly than the hole conduction.

Therefore, we estimated the conduction parameters of n_e, n_h and σ_e/σ by obtaining a fitting curve to the temperature dependence of the Hall coefficient in the temperature range of 20 K-100 K, assuming that both carrier densities were constant throughout the measured temperature range and the ratio σ_e/σ varies with temperature, similarly to the equation for the Ni_3S_4 film. The calculated Hall coefficient, including that in an extended temperature range above 80 K, is shown in Fig. 32. The same estimation was performed for temperature dependence of the Hall coefficient for Film3. The carrier densities are $n_e=1.01\times10^{28}$ m^{-3} and $n_h=0.300\times10^{28}$ m^{-3} (0.55 electrons and 0.16 holes per unit lattice, respectively) for Film2.

Those of Film3 are n_e=0.992×10^{28} m^{-3} and n_h=0.312×10^{28} m^{-3}. The temperature dependence of the ratio σ_e/σ for Film2 is shown in the figure. That of Film3 is σ_e/σ = 0.660 + 115/((T - 20.2)^2 + 1074). According to the estimation, the contribution of electron to the conduction was 77% at 20 K and 66% at 280 K for Film2. The Hall mobility of the electron was 7.4 × 10^{-3} m^2 Ω^{-1} C^{-1} at 20 K and 0.64 × 10^{-3} m^2 Ω^{-1} C^{-1} at 280 K, whereas the Hall mobility of hole was 7.1 × 10^{-3} m^2 Ω^{-1} C^{-1} at 20 K and 1.1 × 10^{-3} m^2 Ω^{-1} C^{-1} at 280 K. The electron Hall mobility of Film2 at 20 K was 70 times larger than that of $Ni_{1-x}S$ with $x \approx 0$ at 20 K and the Hall mobility of the hole for Film2 at 20 K was about 10 times larger than that of $Ni_{1-x}S$ ($x \approx$ 0) at 20 K. The Hall mobility of electron at 280 K for Film2 was 3.5 times larger than that of Film5. Consequently, Film2 demonstrated better metallic conduction than Film5 at all measured temperatures.

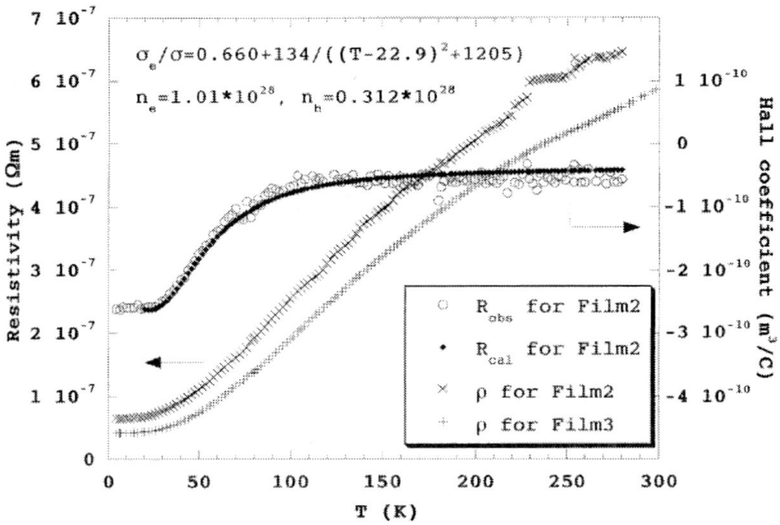

Figure 32. Temperature dependencies of resistivities for Film2 and Film3, and temperature dependence of the Hall coefficient for Film2. A fitting curve for the temperature dependence of the Hall coefficient of Film2 using parameters of σ_e/σ, n_e and n_h. The expression of σ_e/σ is shown in the figure in addition to the magnitudes of n_e and n_h.

4.2.3. Magnetoresistance of $Ni_{1-x}S$ Films

Appreciable transverse magnetoresistance was observed at low temperatures for all $Ni_{1-x}S$ films (Film1-Film5). It has been reported that positive transverse magnetoresistance, 7 × 10^{-4} at magnetic field strength of 1.8 T, is observed at 78 K for $Ni_{0.96}S$ but is not observed even at 4.2 K [16]. The magnetoresistance of Film5 at 4.3 K increases in a quadratic manner with increasing magnetic field. Figure 33 shows the magnetoresistance observed for a magnetic field applied along a direction normal to the film surface was 0.00375 at 0.95 T for Film5, which was in four orders of magnitude larger than that estimated from the two-band model. The magnitude estimated from the model is on the order of $(\sigma/2nq)^2B^2$, i.e., 0.15 × 10^{-6} for n=3 × 10^{27} m^{-3} at B = 0.95 T, where n is the carrier density with the assumption of $n/2 = n_e = n_h$ and B the applied magnetic field. In the figure, the transverse magnetoresistance for the field parallel to the film surface is also shown. It is known that $Ni_{1-x}S$ (x = 0) is antiferromagnetic with a local magnetic moment of 1.66 ± 0.08 μ_B at 4.2 K [45]. The magnetic moments are coupled ferromagnetically within (001) layers, and moments on adjacent layers are coupled antiferromagnetically. The moment directions are aligned along the [001] axis of the structure [45]. The magnetic field, applied along the magnetic spin direction, destabilizes the antiferromagnetic spin-alignment. Therefore, it was expected that resistivity increased in a quadratic manner through increasing the field through s-d interaction [46-48]. However, that explanation contradicted the fact that when the magnetic field was applied along a direction perpendicular to the magnetic spin direction, magnetoresistance also occurred positively, as shown in Fig. 33. In the case of a field perpendicular to the magnetic spin, zero magnetoresistance was expected [46]. Furthermore, the expectation contradicted the fact that Film2 and Film3 showed large magnetoresistance, despite the fact these films were considered to be paramagnetic metals at all measured temperatures because they had no metal-metal transition. The temperature dependencies of magnetoresistance for Film2 and Film5 are shown in Fig. 34. The magnetoresistance decreased until about 55 K with increasing temperature for both Film1 and Film5.

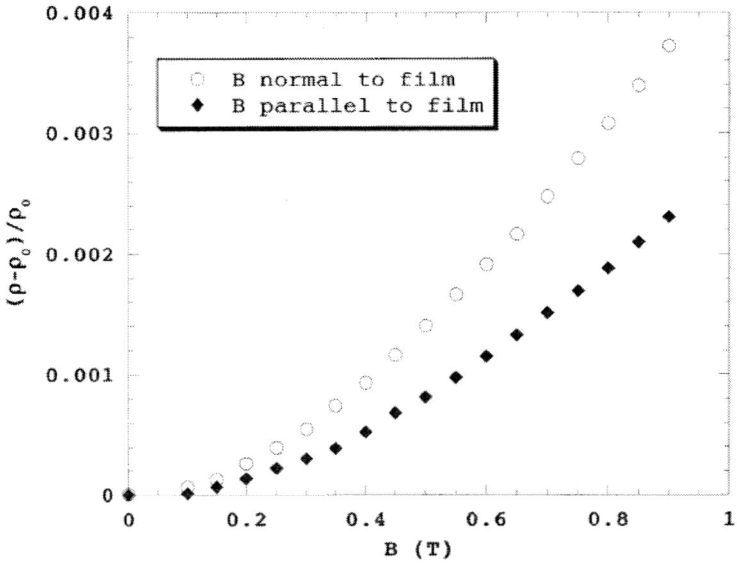

Figure 33. Magnetic field dependencies of transverse magnetoresistance for fresh Film5 at 4.3 K with the magnetic field applied along normal and parallel directions to the film surface.

Negative giant magnetoresistance in NiS around the metal-metal transition temperature has been reported [49]. It has been explained that the large magnetoresistance is attributable to magnetic-field-induced magnetic and electrical transition from antiferromagnetic anomalous-metal phase to paramagnetic metal phase [49]. The present positive magnetoresistance is not associated directly with the negative giant megnetoresistance but it might be associated with the observation that the magnetic field strongly affects the magnetic and electronic state in bulk $Ni_{1-x}S$ ($x = 0$).

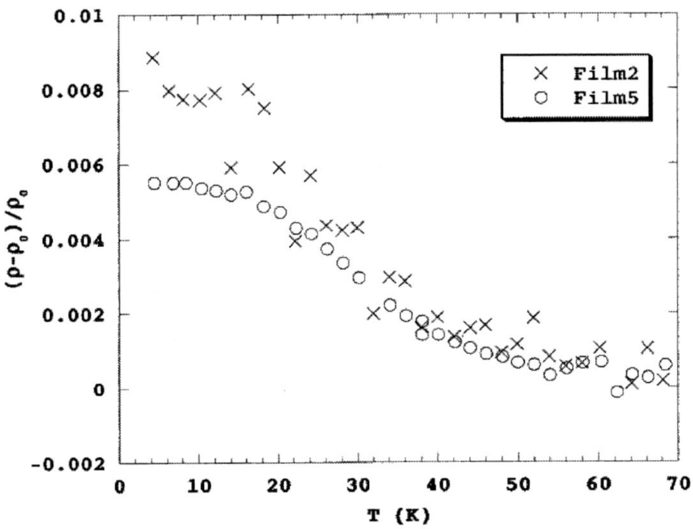

Figure 34. Temperature dependencies of magnetoresistance for Film2 and fresh Film5.

5. Crystal Chemistry of Epitaxial Films for Nickel Sulfides and Related Materials

Nickel sulfide epitaxial films on the MgO and Al_2O_3 substrates, along with epitaxial films of related sulfides (Ag_2S and SnS) on MgO substrates, exhibited characteristic crystal chemistry. Salient characteristics are described below.

(1) Sulfur Layer of Epitaxial Films

The epitaxial films of nickel sulfides possessed a peculiar feature that the densest sulfur layer in the film structure was parallel to the oxide substrate surface, as exemplified by the sulfur layer of (111) for Ni_3S_4, (001) for $Ni_{1-x}S$, (010) for monoclinic Ni_7S_6, (010) for monoclinic Ni_3S_2, and (100) for millerite coexisting with monoclinic N_7S_6. Among the nickel sulfides mentioned above, millerite has a very slender to capillary habit elongated along the [001] direction [39]. That habit, along with the densest sulfur layer of (100), causes the out-of-plane epitaxy of [100] || [001]Al_2O_3.

For Ag_2S epitaxial films deposited on cleaved or polished MgO(001) substrates, (012), (-1,1,2) and (040) of the monoclinic structure at room temperature are parallel to the substrate surface [30, 50]. These lattice planes are the densest sulfur layer because they correspond to (110) of the pseudo-cubic body-centered sulfur lattice and its equivalents. The Ag_2S epitaxial film is another example that the densest sulfur layer is parallel to the substrate surface.

However, SnS(010)/MgO(001) films have an out-of-plane epitaxy that differs from that of the films described above. The compound SnS has an orthorhombic structure that is distorted from a NaCl-type one. The (010) lattice plane corresponds to (100) of the NaCl-type structure, with *a* and *c* parameters that are roughly equivalent to the unit lattice dimension of

MgO. Furthermore, the compound SnS has a layered structure with a cleavage plane of (010), which helps growth of the SnS film to extend within (010). Consequently, the SnS film deposited on MgO(001) had out-of-plane epitaxy of (010) || (001)MgO [34].

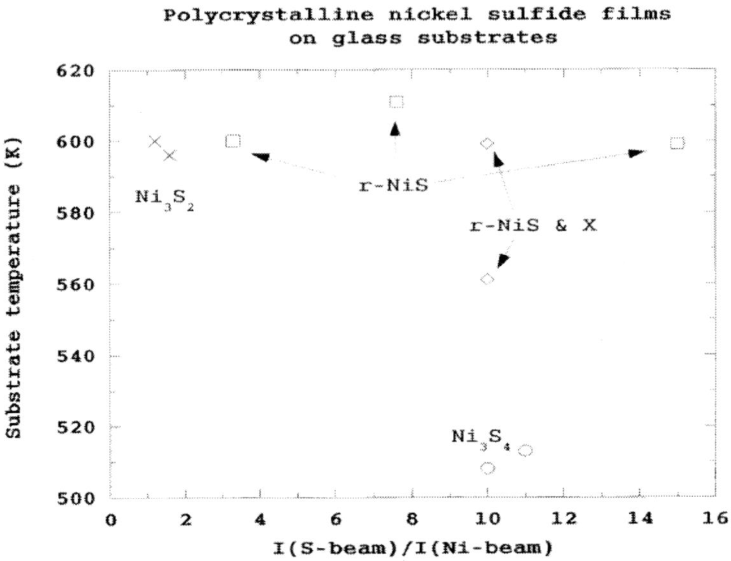

(a)

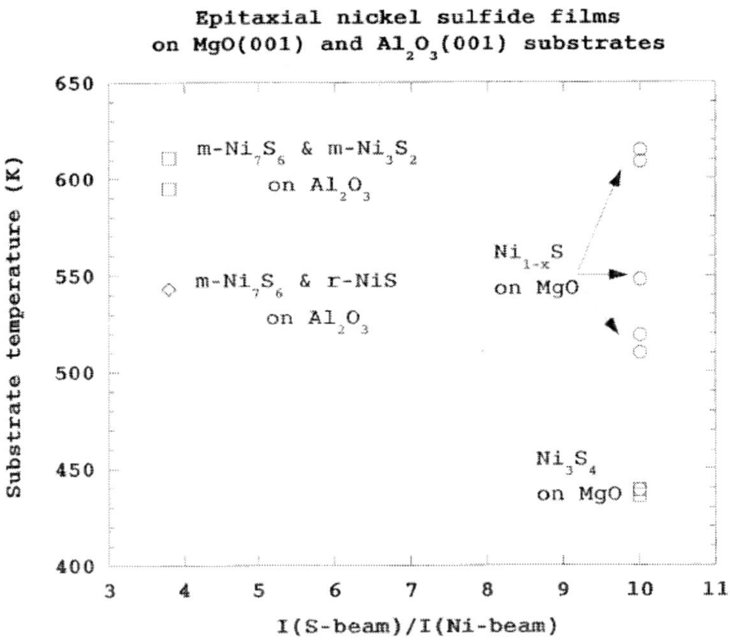

(b)

Figure 35. (a) Phases of films deposited on glass substrates. "m-Ni_3S_2" and "r-NiS" show respective monoclinic Ni_3S_2 and millerite. An unknown phase is denoted as "X". The abscissa shows the intensity ratio between sulfur-molecular beam and Ni-molecular beam. (b) Phases of films deposited on MgO(001) substrates and those on Al_2O_3(001) substrates. "m-Ni_7S_6" shows monoclinic Ni_7S_6 and "m-Ni_3S_2" shows monoclinic Ni_3S_2.

(2) Phases of Epitaxial Films

In phase relations of bulk nickel sulfides, several sulfides exist at 598 K: Ni_3S_2, Ni_9S_8, NiS (millerite), $Ni_{1-x}S$ and Ni_3S_4 with increasing sulfur content [2]. In contrast, nickel sulfide films deposited on glass substrates around 598 K varied from Ni_3S_2 to millerite with increasing *I*(S-beam)/*I*(Ni-beam) from 1.3 to 15, as shown in Fig. 35a. A Ni_3S_4 film appeared at a lower temperature of 508 K with an intensity ratio of *I*(S-beam)/*I*(Ni-beam) = 10. At that fixed intensity ratio, $Ni_{1-x}S$ did not appeared but millerite appeared at substrate temperatures greater than 557 K. Therefore, millerite on glass substrates appeared over wide ranges of substrate temperatures and molecular beam intensity ratios.

In contrast to the glass substrate, use of the MgO(001) substrate engendered the fact that hexagonal $Ni_{1-x}S$ commonly appeared with an intensity ratio of 10 in a temperature range between 510 K and 615 K, as depicted in Fig. 35b. With that intensity ratio, the Ni_3S_4 film grew at a lower substrate temperature of about 438 K for the MgO substrate than 508 K for the glass substrate. These results indicated that millerite grew easily on the glass substrate but did so only slightly on the MgO(001) substrate, and that hexagonal $Ni_{1-x}S$ preferred the MgO(001) substrate because of epitaxial growth on this substrate. For the same reason, monoclinic Ni_7S_6 and monoclinic Ni_3S_2 were considered to prefer an Al_2O_3(001) substrate. Generally speaking, the MgO and Al_2O_3 substrates respectively select particular nickel sulfides preferentially on deposition because they are advantageous for epitaxial growth on those substrates. Consequently, the phases of epitaxial films differ greatly from those of bulk's phase relations.

(3) Relation Between Structures of Epitaxial Film and its Bulk Compound

It is considered that the crystal structure of epitaxial film, which is formed during deposition at a particular substrate temperature, is retained at room temperature for most cases. The film crystal structure that is caused by epitaxy does not necessarily correspond to bulk structure at the substrate temperature but a structure constrained by epitaxy. The film's different structure suggests a possible state of the bulk compound that appears on changing temperature, pressure, and composition of the compound.

The phase stability range of bulk $Ni_{1-x}S$ is $0.04 \leq x \leq 0.07$ at 603 K [51], but a $Ni_{1-x}S$ epitaxial film with $x = 0.01$ is producible at an identical temperature of 603-609 K using molecular beams with *I*(S-beam)/*I*(Ni-beam) = 10. Therefore, the existence of $Ni_{1-x}S$ with $x = 0.01$ at 603-609 K suggests a possible stability range of bulk $Ni_{1-x}S$: the stability range of bulk $Ni_{1-x}S$ is $0 \leq x \leq 0.08$ at 800 K [51].

The SnS epitaxial films have lattice dimensions that differ greatly from those of bulk SnS at room temperature [34]. The lattice dimensions of the SnS films deposited at an almost identical substrate temperature very widely depending on the substrates: polished and annealed MgO(001) surface, cleaved MgO(001) surface, Sn-deposited surface on MgO(001), or glass surface. The *a*-dimension of the SnS film, for example, varied from 0.4329 nm for the glass substrate to 0.4238 nm for the cleaved MgO(001) substrate. The lattice dimension's variation of the SnS films corresponds to that of bulk SnS [34]. The *a*-dimension of bulk SnS, for example, varies greatly with changing temperature from 0.4336(3) nm at 295 K to 0.4242(4) nm at 700 K [52].

(4) Characteristics of Physical Properties for Epitaxial Films

Galvanomagnetic properties of the Ni_3S_4 epitaxial film were measured for the first time. They indicated that a hole carrier was responsible for the conduction, and that Hall mobility of the hole carrier was very low, suggesting that the conduction of the film is not attributable to broad sulfur p-band but to a narrow nickel d-band, although the latter band will hybridize considerably with the former one. This is different from the conduction of CuS. The high Hall mobility of holes for CuS suggests that the sulfur p-band of CuS play an important role in conduction [53].

Galvanomagnetic properties of the $Ni_{1-x}S$ film differed greatly from those of bulk $Ni_{1-x}S$. Galvanomagnetic properties of the $Ni_{1-x}S$ film ($x \approx 0$) indicated that the transition around 180-190 K on increasing and decreasing temperature was not a metal-semiconductor transition but rather a metal-metal one. Suppression on dilatation of the films caused by the substrate decreased the transition temperature greatly, as occurred in bulk $Ni_{1-x}S$ under pressure.

Respective temperature dependencies of resistivity and Hall coefficients for $Ni_{1-x}S$ films (both $x \approx 0$ and $x = 0.07$) indicated that the contribution of electron conduction increased at low temperatures below about 80 K. This phenomenon has not been observed in temperature dependencies of resistivity and the Hall coefficient for bulk $Ni_{1-x}S$. Furthermore, the large positive magnetoresistance of the $Ni_{1-x}S$ films was observed at low temperatures below about 55 K. Large positive magnetoresistance of bulk $Ni_{1-x}S$ has not been reported except that observed at 78 K by Ohtani [16].

6. SUMMARY

Epitaxial films of Ni_3S_4, $Ni_{1-x}S$, monoclinic Ni_7S_6, monoclinic Ni_3S_2 and millerite were prepared in the first time except $Ni_{1-x}S$ using the MBE method. The XRD characterization and galvanomagnetic measurements of the epitaxial films revealed several new or important findings of their crystallography and electrical conduction that are summarized below.

1. N_3S_4 epitaxial films were prepared using substrates of MgO(001), SnS(010)/MgO(001) and $Ni_{1-x}S$(001)/MgO(001). The film orientation was (111)[1,1,-2] || (001)[110]MgO. Lattice parameters of the films were correlated with lattice parameters of the substrates. Broadening of half-value widths for lll (l = 1-5) suggested that strain existed in the crystal structure. Simulations for a θ-2θ scanning pattern using a Rietveld profile analysis program suggested that the crystal structure of the Ni_3S_4 epitaxial film deviated from an ideal spinel-type structure.
2. Stability of $Ni_{1-x}S$ films was examined using XRD for fresh films and films some periods after preparation. $Ni_{1-x}S$ with x = 0.07 was more stable than $Ni_{1-x}S$ with x = 0.01.
3. Epitaxial films of new monoclinic M1 and M2 phases deposited on Al_2O_3(001) substrates were prepared and lattice parameters of those films were obtained. A simulation for a θ-2θ scanning pattern of the M1 phase using Rietveld profile analysis program indicated that the M1 phase was a monoclinic Ni_7S_6-type structure. The monoclinic Ni_7S_6 film orientation was (010)[001] || (001)[1,-1,0]Al_2O_3.

4. Lattice parameters of the M2 phase indicated that the M2 phase had a monoclinic Ni_3S_2-type structure. The orientation of the monoclinic N_3S_2 film was (010)[100] || (001)[1,-1,0]Al_2O_3.
5. Monoclinic Ni_7S_6 coexisted with monoclinic Ni_3S_2 at substrate temperatures of 587 ± 4 K and 606 ± 5 K. Monoclinic Ni_7S_6 coexisted with millerite at the substrate temperature of 528 ± 15 K. The *a*-dimension of monoclinic Ni_7S_6 coexisting with millerite was 9.3% larger than that of monoclinic Ni_7S_6 coexisting with monoclinic Ni_3S_2.
6. A simulation for the θ-2θ scanning pattern of monoclinic Ni_7S_6 coexisting with millerite using the Rietveld profile analysis program indicated that the composition of monoclinic Ni_7S_6 coexisting with millerite was more metal-deficient than that of monoclinic Ni_7S_6 coexisting with monoclinic Ni_3S_2. Therefore, a homogeneity range of monoclinic Ni_7S_6 was suggested.
7. Electrical resistivity and the Hall coefficient of the Ni_3S_4 film were measured. Those measurements indicated that the Ni_3S_4 film was a normal metal with 80% hole and 20% electron conduction at 4.5 K.
8. Electrical resistivity and the Hall coefficient of $Ni_{1-x}S$ ($x \approx 0$) showed that the metal-metal transition occurred at respective 180 and 190 K with decreasing and increasing temperature. The transition temperature was greatly suppressed by substrate constraint to the film. The temperature dependencies of resistivity and Hall coefficient blow the transition temperature differed from those of bulk $Ni_{1-x}S$ ($x = 0$). Carrier densities of electron and hole were estimated using a curve fitting for the temperature dependence of resistivity and the Hall coefficient based on the two-band model.
9. The temperature dependence of resistivity and the Hall coefficient for $Ni_{1-x}S$ ($x = 0.07$) at low temperatures was interpreted using the two-band model; the carrier densities of electron and hole were obtained.
10. Large transverse magnetoresistance of the $Ni_{1-x}S$ films was observed below about 55 K. The origin was not clarified.
11. Crystal chemistry was discussed for several characteristics of epitaxial films for nickel sulfides and related materials.

7. Acknowledgements

The authors thank Dr. H. Nakazawa (NIMS) for valuable comments and Drs. S. Takekawa (NIMS) and Y. Kanke (NIMS) for their kind assistance in sample preparation. Our appreciation is also extended to Mr. K. Natori (NIMS) for his help in galvanomagnetic measurements. One author (H. Nozaki) would like to thank Dr. N. Hirosaki (NIMS) for his support of this investigation.

References

[1] Kullerud, G.; Yund, R. A. *J. Petrol.* 1962, 3, 126-175.
[2] Fleet, M. E. *Can. Mineral.* 1988, 26, 283-291.

[3] Stølen, S.; Fjellvåg, H.; Crønvold, F.; Seim, H.; Westrum, E. F., Jr. *J. Chem. Thermodynamics* 1994, 26, 987-1000.

[4] Shimizu, Y.; Yano, T. *Chem. Lett.* 2001, No. 10, 1028-1029.

[5] Manthiram, A.; Jeong, Y. U. *J. Solid State Chem.* 1999, 147, 679-681.

[6] Chen, X.; Wang, Z.; Wang, X.; Wan, J.; Liu, J.; Qian, Y. *Chem. Lett.* 2004, 33, 1294-1295.

[7] Nozaki,H.; Aizawa, T.; Kosuda, K. *J. Solid State Chem.* 2005, 178, 2001-2007.

[8] Rajamani, V.; Prewitt, C. T. *Can. Mineral.* 1974, 12, 253-257.

[9] Grice, J. D.; Ferguson, R. B. 1974, 12, 248-252.

[10] Krishnakumar, S. R.; Shanthi,N.; Sarma, D. D. *Phys. Rev.* 2002, B66, 115105 (1-6).

[11] Sparks, J. T.; Komoto, T. *Phys. Lett.* 1967, 25A, 398-399.

[12] Sarma, D. D.; Krishnakumar, S. R.; Chandrasekharan, N.; Weschke, E.; Schüβler-Langeheine, C.; Killan, L.; Kaindl, G. *Phys. Rev. Lett.* 1998, 80, 1284-1287.

[13] Sparks, J. T.; Komoto, T. *J. Appl. Phys.* 1963, 34, 1191-1192.

[14] Trahan, J. T.; Goodrich, R. G.; Watkins, S. F. *Phys. Rev.* 1970, B2, 2859-2863.

[15] Koehler, R. F., Jr.; Feigelson, R. S.; Swarts, H. W.; White, R. L. *J. Appl. Phys.* 1972, 43, 3127-3131.

[16] Ohtani, T. J. *Phys. Soc. Japan* 1974, 37, 701-710.

[17] Barthelemy, E.; Chavant, C.; Collin, G.; Gorochov, O. *J. Physique Colloque* 1976, 37, c4.17-c4.22.

[18] Anzai, S.; Matoba, M.; Hatori, M.; Sakamoto, H. *J. Phys. Soc. Jpn.* 1986, 55, 2531-2534.

[19] Ikoma, H.; Matoba, M.; Mikami, M.; Anzai, S. *J. Phys. Soc. Japan* 1995, 64, 2600-2608.

[20] Wang, H.; Pring, A.; Ngothai, Y.; O'Neill, B. *Am. Mineral.* 2006, 91, 171-181.

[21] Fleet, M. E. *Acta Cryst.* 1987, C43, 2255-2257.

[22] Shen, G.; Chen, D.; Tang, K.; An, C.; Yang, Q.; Qian, Y. *J. Solid State Chem.* 2003, 173, 227-231.

[23] Fleet, M. E. *Acta Cryst.* 1972, B28, 1237-1241.

[24] Fleet, M. E. *Am. Mineral.* 1977, 62, 341-345.

[25] Parise, J. B. *Acta Cryst.* 1980, B36, 1179-1180.

[26] Fjellvåg, H.; Andersen, A. *Acta Chem. Scand.* 1994, 48, 290-293.

[27] Metcalf, P. A.; Fanwick, P.; Kakol, Z.; Honig, J. *M. J. Solid State Chem.* 1993, 104, 81-87.

[28] Lu, Z. W.; Klein, B. M.; Singh, D. J. *Phys. Rev.* 1996, B54, 13542-13545.

[29] Dharmaprakash, S. M. *Cryst. Res. Technol.* 1996, 31, K49-K53.

[30] Nozaki, H.; Onoda, M.; Kurashima, K.: Yao, T. *J. Solid State Chem.* 2001, 157, 86-93.

[31] Kim, S.; Baik, S. J. *Vac. Sci. Technol.* 1995, 13A, 95-100.

[32] Guo, L. W.; Hanada, T.; Ko, H. J.; Chen, Y. F.; Makino, H.; Yao, T. Surf. Sci. 2000, 445, 151-158.

[33] Yukino, K.; Okamura, F. P.; Nozaki, H.; Kobayashi, Y.; Yamada, Y. In Advances in X-ray Analysis; Barret, C. S. et al.; Eds.; Proceedings of the Combined First Pacific-International Congress on X-ray Analytical Methods (PICXAM) and Fourth Annual Conference on Applications of X-ray Analysis; Plenum Press, New York, 1992; Vol. 35B, pp. 1275-1283.

[34] Nozaki, H.; Onoda, M.; Sekita, M.; Kosuda, K.; Wada, T. *J. Solid State Chem.* 2005, 178, 245-252.

[35] JCPDS-International Center for Diffraction Data, No. 47-1739.

[36] Wyckoff, R. W. G. *Crystal Structures;* Second edition; Interscience Publisher; New York London Sydney, 1965; Vol. 3, pp 81.

[37] Higgins, J. B.; Speer, J. A.; Craig, J. R. *Phillos. Mag.* 1975, 32, 683-685.

[38] JCPDS-International Center for Diffraction Data, No. 89-7141.

[39] Dana's System of Mineralogy; Palache, C.; Berman, H.; Frondel,C.; Eds.; John Wiley and Sons, Inc; New York and London, 1963; Vol. 1, pp 239-242.

[40] JCPDS-International Center for Diffraction Data, No. 02-0041.

[41] J.–P. Yan In Solid State Physics; Seitz, F.; Turnbal, D.; Eds.; Academic Press Inc., Publishers; New York, 1957; vol. 5, pp. 1-96.

[42] Nozaki, H. *J. appl. Phys.* 1980, 51, 486-489.

[43] Anzai, S; Ozawa, K. J. *Phys. Soc. Japan* 1968, 24, 271-274.

[44] McWhan, D. B.; Marezio, M.; Remeika,J. P.; Dernier, D. *Phys. Rev.* 1972, B5, 2552-2555.

[45] Sparks, J. T.; Komoto, T. *Rev. Modern Phys.* 1968, 40, 752-754.

[46] Yamada, H.; Takada, S. *J. Phys. Soc. Japan,* 1973, 34, 51-57.

[47] Usami, K. J. Phys. Soc. Japan, 1978, 45, 466-475.

[48] Nozaki, H.; Ishizawa, Y. *Phys. Lett.* 1977, 63A, 131-132.

[49] Chen, P.; Du, Y. W. *J. Phys. Soc.* Japan, 2001, 70, 209-211.

[50] Nozaki,H.; Onoda, M.; Yukino, K.; Kurashima, K.; Kosuda, K.: Maki, H.; Hishita, S. *J. Solid State Chem.* 2004, 177, 1165-1172.

[51] Grønvold, F.; Stølen, S. *Thermochim. Acta* 1995, 266, 213-229.

[52] Chattopadhyay, T; Pannetier, J; Schnering, H. G. v. *J. Phys. Chem. Solids* 1986, 47, 879-885.

[53] Nozaki, H.; Shibata, K.; Ohhashi, N. *J. Solid State Chem.* 1991, 91, 306-311.

In: Progress in Solid State Chemistry Research
Editor: Ronald W. Buckley, pp. 285-299
ISBN 1-60021-313-8

Chapter 7

MID- AND NEAR-INFRARED SPECTROSCOPY OF HYDROTHERMALLY SYNTHESIZED 2:1 PHYLLOSILICATES IN THE SYSTEM NA_2O-AL_2O_3-SIO_2-H_2O

J. Theo Kloprogge[1*], Loc V. Duong[1], Wayde N. Martens[1], Ad van der Eerden[2] and Ray L. Frost[1]

[1] Inorganic Materials Research Program, Queensland University of Technology, 2 George Street, GPO Box 2434, Brisbane Q 4001, Australia

[2] Department of Inorganic Chemistry, University of Utrecht, P.O. Box 80083, 3508 TB Utrecht, The Netherlands

ABSTRACT

The near-infrared spectra of hydrothermally synthesized beidellites with increasing layer charge, pyrophyllite and paragonite have been compared and interpreted based on their associated mid-infrared spectra. The 6000-7500 cm^{-1} region for beidellite is dominated by the first overtone of the adsorbed H_2O stretching mode around 6900 cm^{-1}, while the 7100 and 7150 cm^{-1} bands are the first overtones of the two Al-OH stretching modes. Pyrophyllite is characterised by a sharp band at 7172 cm^{-1} while paragonite shows a weaker and broader band at 7147 cm^{-1}. The 5000-6000 cm^{-1} region is characterised by combination bands of interlayer H_2O stretching modes with the adsorbed H_2O bending mode around 5240 and 5140 cm^{-1}. Furthermore, the combination modes are broad as expected based on the broad nature of the H_2O bands in the mid-infrared region. Minor bands in the pyrophyllite spectrum are due to the presence of a small amount of beidellite. Bands in the 4000-5500 cm^{-1} region at 4545 and 4595 cm^{-1} are attributed to the combination bands of the stretching and libration modes of the Al_2OH units in the beidellite structure. Pyrophyllite exhibits a single sharp band around 4615 cm^{-1}. A similar but broader band is observed for paragonite at 4575 cm^{-1}.

* Corresponding author: phone +61 7 3864 2184, fax +61 7 3864 1804, E-mail t.kloprogge@qut.edu.au

Key words: beidellite, hydrothermal synthesis, mid-infrared spectroscopy, near-infrared spectroscopy, paragonite, pyrophyllite

1. INTRODUCTION

Phyllosilicates form an important group of silicate minerals with a layered structure composed of polymeric sheets of $(Si,Al)O_4$ tetrahedra linked to sheets of $(Mg, Fe, Al)(O,OH)_6$ octahedra. The 2:1 clay minerals and especially the smectites form an important subgroup within the phyllosilicates together with the micas. In the 2:1 phyllosilicates, isomorphous substitution of cations having different valences can lead to charge imbalances within a sheet. This may be partly compensated by the opposite type of charge imbalance in the adjacent sheet (e.g., a positively charged octahedral sheet versus a negatively charged tetrahedral sheet). This overall negative layer charge is balanced by interlayer cations (e.g., Na^+, K^+, Ca^{2+}, and Mg^{2+}) that are coordinated to the basal surfaces of the tetrahedral sheets from adjacent layers. Tetrahedral substitution (as in beidellite) leads to a more localized charge distribution at the surfaces of the layer and to a higher three-dimensional order (Suquet et al. 1975, Brindley 1984).

In pyrophyllite, $Al_4Si_8O_{20}(OH)_4$, all tetrahedral sites are filled with Si^{4+}. Two-thirds of the octahedral sites in pyrophyllite are filled with Al^{3+} and the remaining one-third of the sites are vacant (dioctahedral). The electrically neutral sheets in pyrophyllite are bonded together by relatively weak dipolar and van der Waals forces (Giese 1975). Paragonite, $Na_2Al_4[Si_6Al_2]O_{20}(OH)_4$, belongs to the group of the so called dioctahedral true micas together with other phyllosilicate minerals as muscovite, phengite, illite, celadonite and glauconite (Bailey 1988). Paragonite has a structure relatively similar to muscovite but the chemical composition differs from that of muscovite in that Na^+ replaces K^+ as the interlayer cation (Deer et al. 1996). The interlayer Na^+ ion is positioned in a very irregular cavity due to the offset in the adjacent 2:1 layers by approximately 0.23 Å (Lin and Bailey 1984). The most abundant polytype found in nature is $2M_1$ (Chatterjee 1974, Sidorenko et al. 1977b, Lin and Bailey 1984), although the 3T polytype (Sidorenko et al. 1977a) has also been observed. The 1M polytype has only been observed in synthetic paragonites.

Despite their importance, the 2:1 clay minerals form a group of minerals that are difficult to study due to their small size, variable structural composition, and relatively slow kinetics of formation and alteration. New information about the formation mechanisms, variation in surface properties with structure, composition, and solution composition, and the long-term stability of clay minerals in surficial and hydrothermal environments may be gained from studies of synthetic clay minerals. Although a number of clay minerals including kaolins and various smectites have been synthesized during the past 50 years, most of this synthetic work has been empirical in nature (Güven 1988). The most common synthetic approach has been to treat starting materials (mostly solid phases like other aluminosilicates, glasses, gels) having a composition close to that of the desired clay by heating in water or under hydrothermal conditions. A major difficulty is that the reactions rarely go to completion unless hydrothermal conditions are employed and, even then, the reaction products quite often consist of a mixture of products rather than of pure clay minerals.

In this study we have selected a number of phase pure 2:1 phyllosilicate samples in the system Na_2O-Al_2O_3-SiO_2-H_2O, i.e. pyrophyllite, beidellite and paragonite, for spectroscopic characterization. Mid-infrared spectroscopy is more commonly used for identification, characterization and quantification, than near- and far-infrared spectroscopy. In this study a description will be given of the near-infrared spectra of these phyllosilicates and for their interpretation they will be compared to their corresponding mid-infrared spectra.

2. Experimental Methods

2.1. Phyllosilicate Synthesis

As starting materials Na-Si-Al gels with appropriate molar ratios were prepared according to the method of Hamilton and Henderson (1968). Fifty milligrams of gel together with 70 μl water were placed together in a gold capsule and treated at pressures between 200 bar and 5 kbar and temperatures between 200 and 500°C for 10 to14 days in a Tuttle-type, externally heated cold-seal pressure vessel (Tuttle 1949). Details of this synthesis method have been published earlier by Kloprogge and coworkers (Kloprogge 1992, Kloprogge et al. 1993, 1996, 1999a). The final reaction products were characterised by powder X-ray diffraction (XRD) on a Philips PW 1050/25 diffractometer using CuKα radiation. The chemical compositions and morphology were determined using a JEOL JXA-8600 electron microprobe. The sodium analyses were generally lower then expected from the Si and Al analyses, presumably due to evaporation of Na during exposure to the electron beam in the microprobe (Kloprogge et al. 1990).

2.2. Characterisation

As phyllosilicate reaction products pyrophyllite, beidellite and paragonite were obtained. A review of the synthesis products and their reaction conditions is given in Table 1. Figure 1 displays a phase diagram determined in detail by Kloprogge et al. (1993). Beidellite forms rather big flakes up to 100 μm and a thickness of approximately 20 μm. A basal spacing of approximately 12.4 Å is observed in the XRD pattern, which expanded to 16.7 Å upon expansion with ethylene glycol. Kloprogge et al. (1993) have observed that above 300°C and a sodium content lower than 0.7 pyrophyllite is formed, while at higher sodium concentrations paragonite is formed. Two polytypes of pyrophyllite are known: 1*Tc* and 2*M*. Eberl (1979) has suggested that the 1*Tc* polytype is formed at high temperatures whereas the 2*M* polytype is formed as the low-temperature form. The XRD data of the experiments described here indicate the formation of the 2*M* polytype in agreement with Eberl's remark (Kloprogge et al. 1996). The paragonite is characterised by a sodium content and Si/Al ratio close to the expected values. The XRD pattern is characterised by a basal (002) reflection around 9.6 Å. Both the pyrophyllite and paragonite do not expand upon exposure to glycerol or ethylene glycol. A minor amount of amorphous material is present in all samples but the amount is estimated to be less then 5 vol% based on SEM observations and will not interfere with the spectroscopic study reported below.

Table 1. Synthesis conditions and compositions of the final products

Sample	Tetrahedral substitution x in starting gel[1]	P (bar)	T (°C)	Time (days)	Products	Chemical analysis	XRD Basal spacing (Å)	Expandable in ethylene glycol
$B_{0.7-1}$	0.7	400	350	15	Beidellite	$Na_{0.61}Al_{4.69}Si_{7.31}O_{20}(OH)_4.nH_2O$	12.4	yes
Par	2.0	1000	450	14	Paragonite	$Na_{1.97}Al_{5.98}Si_{6.02}O_{20}(OH)_4$	9.8	no
$B_{0.6}$	0.6	1000	350	7	Beidellite	$Na_{0.60}Al_{4.70}Si_{7.30}O_{20}(OH)_4.nH_2O$	12.3	yes
$B_{0.7-2}$	0.7	1000	350	7	Beidellite	$Na_{0.72}Al_{4.75}Si_{7.25}O_{20}(OH)_4.nH_2O$	12.4	yes
Pyr	0.0	1000	450	7	Pyrophyllite	$Al_{3.98}Si_{8.02}O_{20}(OH)_4$	9.4	no
$B_{1.5}$	1.5	1000	275	7	Beidellite	$Na_{1.67}Al_{5.68}Si_{6.32}O_{20}(OH)_4.nH_2O$	12.5	yes
$B_{2.0}$	2.0	1000	275	7	Beidellite	$Na_{1.77}Al_{5.90}Si_{6.10}O_{20}(OH)_4.nH_2O$	12.4	yes

[1] x based on $Na_xAl_4(Si_{8-x}Al_x)O_{20}(OH)_4.nH_2O$

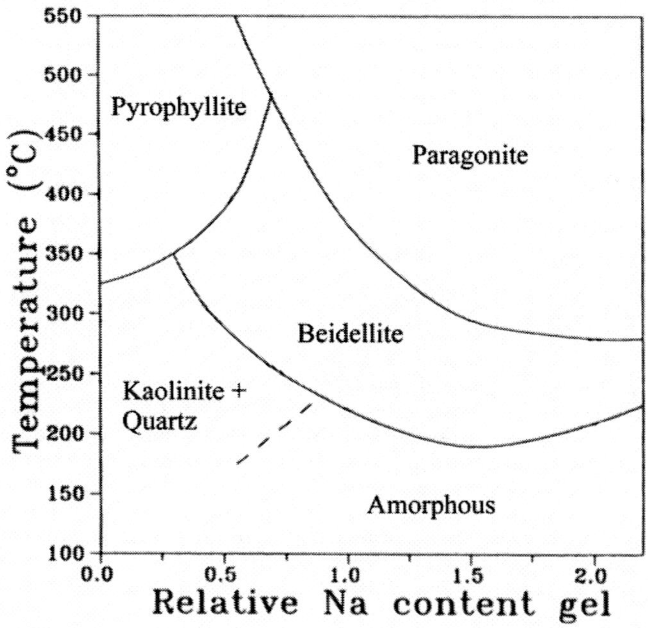

Figure 1. Composition-Temperature phase diagram at 1 kbar (modified after (Kloprogge et al. 1993)) showing the conditions under which the samples used in this study were synthesised. The relative Na content in the gel is based on the formula $Na_xAl_4(Si_{8-x}Al_x)O_{20}(OH)_4.nH_2O$.

2.3. Mid-Infrared (IR) and Near-Infrared (NIR) Spectroscopy

The samples were finely ground for one minute, mixed with oven dried spectroscopic grade KBr (containing approximately 1 wt% sample) and pressed into a pellet under vacuum. The transmittance spectrum of the sample was recorded in triplicate by accumulating 512 scans at 4 cm^{-1} resolution between 400 cm^{-1} and 4000 cm^{-1} using a Perkin-Elmer 1600 series Fourier transform mid-infrared spectrometer equipped with a LITA detector.

The NIR transmittance spectroscopic analyses were performed on a Perkin Elmer System 2000 NIR-FT Raman spectrometer. For the samples 32 scans were accumulated at a spectral resolution of 16 cm^{-1} between 4000 and 10,000 cm^{-1} using a mirror velocity of 0.2 cm^{-1}/s to get an acceptable signal/noise ratio.

Spectral manipulation such as baseline adjustment, smoothing and normalisation were performed using the Spectracalc software package GRAMS (Galactic Industries Corporation, NH, USA). Band component analysis was undertaken using the Jandel 'Peakfit' software package, which enabled the type of fitting function to be selected and allows specific parameters to be fixed or varied accordingly. Band fitting was done using a Lorentz-Gauss cross-product function with the minimum number of component bands used for the fitting process. The Gauss-Lorentz ratio was maintained at values greater than 0.7 and fitting was undertaken until reproducible results were obtained with regression coefficients r^2 greater than 0.995.

3. Results And Discussion

The near-infrared (NIR) spectral region has been defined to extend from roughly 700 to 3500 nm (14285 to 2860 cm^{-1}) (Rossman 1988, Kloprogge et al. 2000c). The only vibrations in the NIR region between 4000 and 8000 cm^{-1} are those associated with hydrogen atoms associated with hydroxyl groups in the case of micas and with hydroxyl groups and water in the case of clay minerals. Rossman (Rossman 1988) reported average band positions for OH-groups and H_2O in the NIR region around 4500 cm^{-1} as being due to M-OH motions, 5200 cm^{-1} as the H_2O combination mode (bend + stretch) and around 7100 cm^{-1} as the first OH stretch overtone. This means that NIR spectroscopy is a very suitable technique to study phyllosilicates.

In order to come to a detailed interpretation of the NIR spectra of the synthetic micas and clays it is necessary to have a more detailed understanding of the associated bands in the mid-infrared spectra of these minerals. Figure 2 shows the mid-infrared low frequency region between 400 and 1900 cm^{-1} and the OH-stretching region between 2900 and 4000 cm^{-1}. Table 2 reports the band centres for the structural OH-bending and OH-stretching vibrations shown in Figure 2.

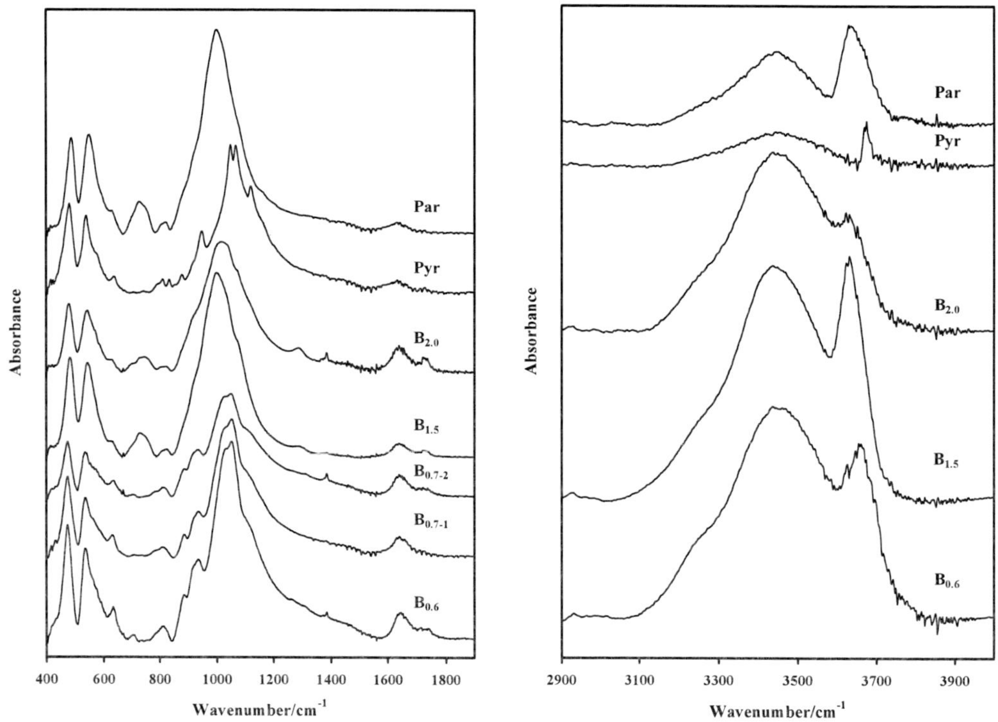

Figure 2. Mid-infrared spectra of synthetic beidellite ($B_{0.6}$, $B_{1.5}$ and $B_{2.0}$), pyrophyllite (Pyr) and paragonite (Par) in (a) the low frequency region between 400 and 1900 cm^{-1} and (b) the OH-stretching region between 2900 and 4000 cm^{-1}.

Kloprogge et al. (2000b) have reported the effect of increasing layer charge in synthetic beidellites on their mid-infrared spectra. They showed that an increase in sodium and aluminium content together with a decrease in silicon content in the starting gel resulted in an increase in the layer charge of the beidellite run products (Kloprogge et al. 1993). However, they found that the resulting clay layer charge could deviate substantially from the layer charge calculated from the chemical composition of the initial gel. In this study the difference between the starting gel and the reaction products for the low charge beidellite was within approximately 15 %. For the high layer charge beidellites synthesised from $Na_{1.5}$ gel the layer charge varied between 1.68 and 1.87 and for the $Na_{2.0}$ the layer charge was 1.89 for the beidellite and 2.0 for the paragonite.

Table 2. Mid-infrared spectral analysis (in cm^{-1}) of the OH band centres of synthetic pyrophyllite, beidellite and paragonite (assignment after (Farmer 1974))

Sample	x		Al-OH lib	Al-OH lib	Al-OH lib	Al-OH stretch	Al-OH stretch
$B_{0.6}$	0.6	beidellite	882	913	934	3668	
$B_{0.7-2}$	0.7	beidellite	882	912	933	3662	
$B_{0.7-1}$	0.7	beidellite	882	915	936	3663	
$B_{1.5}$	1.5	beidellite	880	919	966	3638	
$B_{2.0}$	2.0	beidellite	878	913	965	3634	
Par	2.0	paragonite	883	917	-	3627	3650
Pyr	0.0	pyrophyllite	879	-	948		3675

The mid-infrared spectra of the lower charge beidellites (0.6 and 0.7) are very similar in the Si-O region around 1000 cm^{-1}. The high charge beidellite spectra are much more similar to the paragonite spectrum. In the low frequency region the OH-librational modes associated with the octahedral Al-OH groups in the beidellite structures are observed around, 878-882, 913-920 and 935-965 cm^{-1}. Similar bands were reported by Farmer (Farmer 1974) around 915-950 cm^{-1} for the Al-OH in-plane librational modes in all dioctahedral species and around 880 cm^{-1} for the out-of-plane librational mode in pyrophyllite. Only the last band shows a clear shift with increasing layer charge and has been interpreted as the in-plane vibration in an earlier paper (Kloprogge et al. 2000b). For paragonite the shoulders around 917 and 883 cm^{-1} have been ascribed to the in-plane and the out-of-plane Al_2OH bending modes. The bands around 819 and 803 cm^{-1} have then been ascribed to the Al_2OH bending vibration and the Al-O vibration of the tetrahedral aluminium substituting silicium (Kloprogge and Frost 1999a). Small amounts of other silica species such as amorphous material or quartz, which were observed in the SEM and XRD, are indicated by the presence of a minor band at ~810 cm^{-1} (Madejova et al. 2000a). In an earlier paper Kloprogge and Frost (1999b) have described the mid-infrared spectrum of synthetic pyrophyllite. In accordance with their observations the Al_2OH in-plane and out-of-plane librational modes were observed at 948, 879 and 854 cm^{-1}, respectively. The band around 795 cm^{-1} has been interpreted as another Al_2OH mode, probably due to the presence of local areas with a high number of defects in the pyrophyllite structure. Madejova et al. (2000a) interpreted a band at 804 cm^{-1} for Li-smectites as being due to the AlMgLi-OH bending mode. The assignment of these bands to Al_2OH modes is also supported by earlier results of infrared emission spectroscopy of beidellite and other clay minerals, pyrophyllite, paragonite and rectorite (Kloprogge and Frost 1999a, 1999b, 2000a, 2000b, Kloprogge et al. 1999b, 2000a).

For beidellite the OH-stretching vibrations are affected by the configuration and charge distribution in the neighbouring tetrahedral layers. Vedder and McDonald (1963) have shown that in dioctahedral clay minerals such as beidellite the OH-transition moment is tilted by a small angle out of the 001 plane. A consequence of this orientation is that the proton of the OH-group lies closely to two of the six oxygen atoms at the apices of the tetrahedra forming a hexagonal cavity in the tetrahedral layer. In the relatively low-charge beidellite structure this results in two OH-stretching modes, one associated with hydroxyl protons located near charge-bearing oxygen atoms of the Al substituted hexagonal rings and one associated with neutral oxygen atoms of non-substituted hexagonal rings. Increasing the layer charge will consequently be reflected in the intensity ratio between these two OH-stretching modes. For paragonite the hydroxyl stretching region is characterised by a silanol band visible as a shoulder around 3745 cm^{-1} associated with Si-OH groups present on the crystal edges, two overlapping Al-OH stretching bands around 3650 and 3627 cm^{-1} and an absorbed water band around 3450 cm^{-1} (Kloprogge and Frost 1999a). The silanol band at 3745 cm^{-1} agrees well with similar bands observed for Si-OH stretching modes in silica gel around 3740 cm^{-1} (Anderson and Wickersheim 1964). In comparison, pyrophyllite shows a strong sharp band associated with the Al_2OH stretching vibration around 3675 cm^{-1} in the mid-infrared region and a comparable Raman band around 3668 cm^{-1}. A broad and relatively weak band associated with the absorbed water is observed around 3454 cm^{-1} in the mid-infrared region combined with a Raman band at slightly higher wavenumber around 3626 cm^{-1}. The bending mode of H_2O is observed around 1632 cm^{-1} (Kloprogge and Frost 1999b). Very dominantly present in the mid-infrared spectra of the beidellites are two bands due to interlayer water. These interlayer water bands are absent in the paragonite and pyrophyllite spectra, although these spectra do show the presence of some absorbed water on the external surfaces of the crystals. The interlayer water in the beidellite samples is characterized by a H_2O bending vibration at 1635 cm^{-1} plus the associated OH-stretching vibration around 3450 cm^{-1}. This band is most likely associated with water molecules in the outer hydration sphere of the interlayer sodium cation (Bishop et al. 1994). The shoulder around 3240 cm^{-1} can be ascribed to interlayer H_2O chemically bound or coordinated to the interlayer cation with an ice-like structure (van der Marel and Beutelspacher 1974). An alternative explanation for this band is that it is the first overtone of the H2O bending vibration enhanced by resonance interaction with the broad stretching band around 3450 cm^{-1}. Examples have been found in e.g. beryl and hemimorphite (pers. Comm. Eugen Libowitzky). Clearly only one Al_2OH band can be observed around 3668 cm^{-1}. The small band at 3620 cm^{-1} is thought to be characteristic for the asymmetric water OH-stretching vibration associated with the inner hydration shell of the interlayer cation (Bishop et al. 1994). However, due to the presence of strong water bands overlapping the 3668 and 3620 cm^{-1} bands, the positions and intensities of these bands can deviate significantly and obscure the second Al_2OH band.

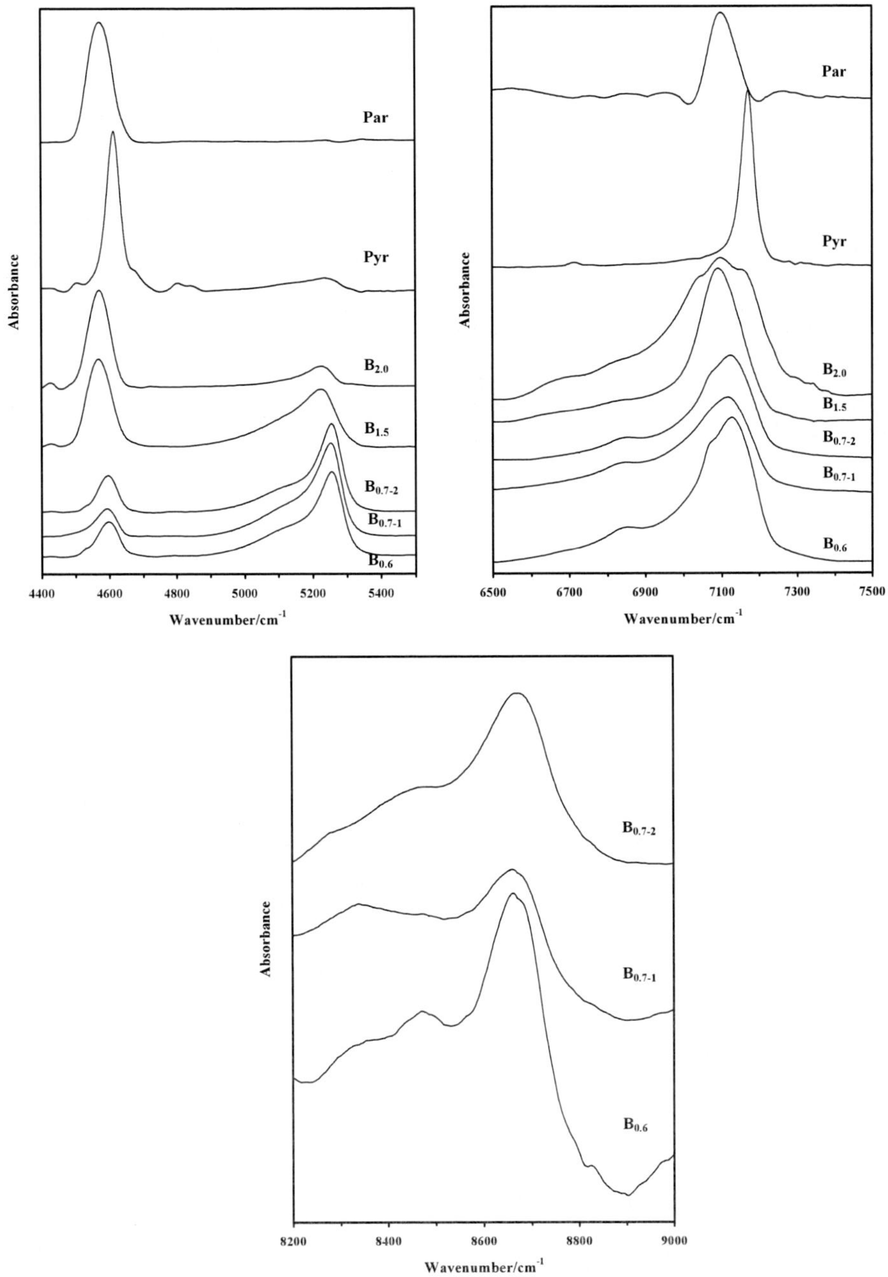

Figure 3. Near-infrared spectra of synthetic beidellite ($B_{0.7-1}$, $B_{0.6}$, $B_{0.7-2}$, $B_{1.5}$ and $B_{2.0}$), pyrophyllite (Pyr) and paragonite (Par) in the region between (a) 4400 and 5500 cm^{-1}, (b) between 6500 and 7500 cm^{-1} and (c) between 8200 and 9000 cm^{-1} for the low charge beidellites.

The NIR spectra of the synthetic beidellites, pyrophyllite and paragonite are shown in Figure 3, while Figure 4 shows the band component analysis of a low charge beidellite, pyrophyllite and paragonite. A detailed list of the band centers for all samples is given in Table 3. As for the mid-infrared spectra the NIR spectra of the low charge beidellites are very similar. Band component analysis indicates the presence of two bands around 4545 (weak) and 4595 (strong) cm^{-1}. This split disappears in the spectra of the high charge beidellites. The range around 4500 cm^{-1} is characteristic for the combination bands of the Al-OH-stretching modes and the two Al-OH libration modes around 3660 cm^{-1} and 880 and 915 cm^{-1} respectively (Cariati et al. 1981, 1983a, 1983b, Bishop et al. 1993, Madejova et al. 2000b). Pyrophyllite shows a strong sharp band around 4615 cm^{-1} accompanied by two weak bands at both slightly higher and lower frequencies. The presence of the extra band around 4548 cm^{-1} may indicate the presence of a very small amount of beidellite, not observed in the XRD, corresponding with the band around 854 cm^{-1} in the mid-infrared spectrum. Kloprogge et al. (1999b) observed similar bands for both natural and synthetic pyrophyllites. Clark et al. (1990) reported a band at 4617 cm^{-1}, which agrees very well with the band observed in this study at 4615 cm^{-1}, and a weaker band at 4794 cm^{-1} with a clear shoulder at 4850 cm^{-1}. This second, weaker band with its shoulder is identical to the two weak bands observed in this study around 4805 and 4844 cm^{-1}

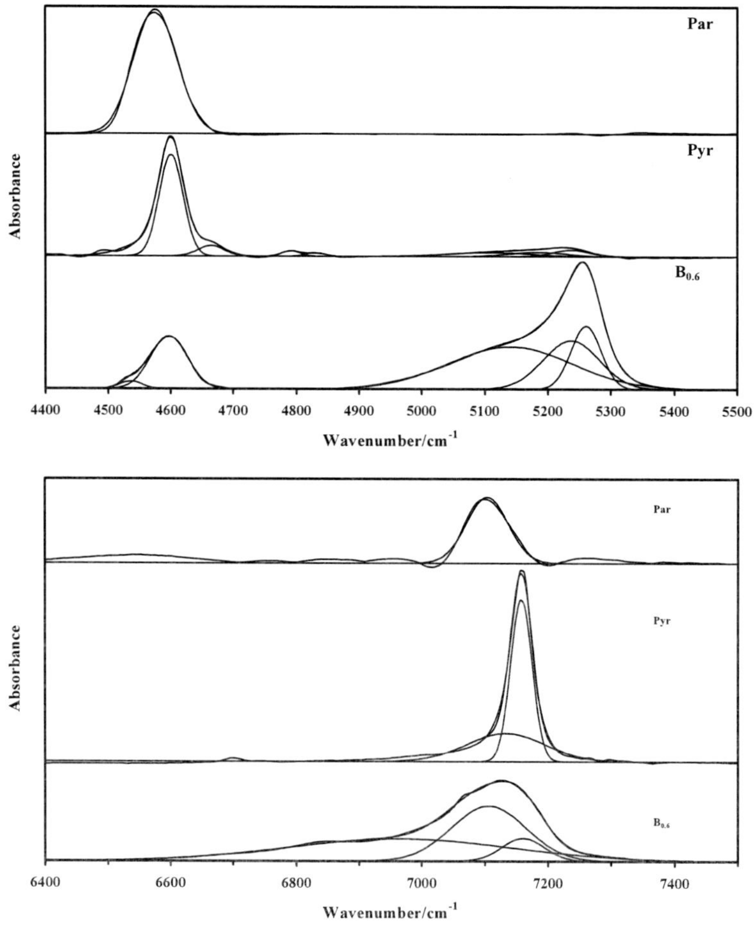

Figure 4. Band component analysis of the near-infrared spectra of synthetic beidellite ($B_{0.6}$), pyrophyllite (Pyr) and paragonite (Par) in the region between (a) 4400 and 5500 cm^{-1} and (b) 6400 and 7500 cm^{-1}.

Table 3. Near-infrared band centres (in cm^{-1}) of synthetic pyrophyllite, beidellite and paragonite

Sample	x									
$B_{0.6}$	0.6	4536	4597	-	5140	5237	5260	6963	7106	7160
$B_{0.7\text{-}2}$	0.7	4553	4596	-	5139	5233	5258	6939	7073	7140
$B_{0.7\text{-}1}$	0.7	4543	4596	-	5144	5241	5259	6999	7098	7157
$B_{1.5}$	1.5	-	4571	-	5075	5172	5233	6938	7099	-
$B_{2.0}$	2.0	-	4570	-	5158	-	5221	6866	7094	-
Par	2.0	-	4575	-	-	-	-	-	7103	-
Pyr	0.0	4598	4615	4678	5120	5198	5251	-	7147	7172

Three bands can be observed in beidellite around 5140, 5240 and 5260 cm^{-1}. Isolated water molecules show an absorption band near 5333 cm^{-1}, due to the combination mode of the stretching and bending OH-mode of the water molecule (Hunt and Salisbury 1970). The presence of interlayer water in smectitic clay minerals will be observed as broad bands close to this position, due to the poorly ordered sites in the interlayers of a smectitic clay (Clark et al. 1990). In the case of the synthetic beidellites this means that the first two bands around 5140 and 5240 cm^{-1} can be assigned to the combination modes of interlayer water modes (Bishop et al. 1993). Similar bands were observed for montmorillonite around 5130 and 5250 cm^{-1} by Sposito et al. (1983). Band component analysis indicates that the relatively broad band around 5250 actually contains two components around 5240 and 5260 cm^{-1}, which can probably be explained by the fact that a part of the interlayer water is coordinated to the interlayer sodium cations, while the remaining water molecules are less tightly bond to the clay structure, in agreement with dehydration studies that indicate two dehydration stages (Koster van Groos and Guggenheim 1984, 1986, 1987, Kloprogge et al. 1992). The strong intensities of these bands are due to the presence of a relatively large amount of interlayer water. Clearly visible is a decrease in intensity with increasing layer charge. This increase in layer charge is compensated by the incorporation of more sodium interlayer cations leaving less interlayer space for water. Similar bands, although very weak compared to the beidellites, are observed for the synthetic pyrophyllite in good agreement with the earlier observation that a small amount of beidellite is present in this sample. No combination bands, as expected, associated with absorbed or interlayer water are observed for the synthetic paragonite.

The bands observed in the high wavenumber region can be attributed to the overtones and combination modes associated with fundamental stretching and bending modes observed in the mid-infrared region. For the beidellites this region is relatively complex showing multiple overlapping bands. Two relatively sharp bands occur around 7100 and 7150 cm^{-1} plus a much broader band around 6900 cm^{-1}. This broad band deceases in intensity with increasing layer charge of the beidellite and is absent in the spectra of the pyrophyllite and paragonite. Therefore, plus the fact that the free water molecule has an absorbance near 6878 cm^{-1} (Hunt and Salisbury 1970, Sposito et al. 1983), this band must be assigned to the first overtone of the hydrogen bonded interlayer water OH-stretching mode (Madejova et al. 2000b). The 7100 cm^{-1} band is characteristic for the first overtone of the Al-OH stretching modes of the beidellites. The origin of the second band around 7150 cm^{-1} is less clear though. Earlier infrared emission spectroscopy work on synthetic beidellites has shown however that the broad band around 3660 cm^{-1} actually consists of two overlapping Al-OH stretching modes around 3615 and 3650 cm^{-1} (Kloprogge et al. 1999b), while montmorillonite SAz shows also two bands around 3622 and 3668 cm^{-1} after heating to 110°C (Madejova et al. 2000a). This means that even though the two Al-OH stretching modes cannot always be identified in the mid-infrared spectra, both bands do show up in the NIR spectra. For pyrophyllite a strong sharp band is observed at 7172 cm^{-1}, together with a much broader, weaker band around 7147 cm^{-1}. This last band is comparable to the strongest band observed for the synthetic beidellite, supporting the presence of a minor amount of beidellite already inferred from the band around 4598 cm^{-1}. The band at 7172 cm^{-1} agrees well with the band observed by Clark et al. (Clark et al. 1990) at 7177 cm^{-1}. Paragonite only shows a relatively broad band around 7103 cm^{-1} associated with the first overtone of the Al-OH stretching mode.

In the very high wavenumber region the low charge synthetic beidellites show a very weak band around 8670 cm^{-1}. For the high-charge beidellite this band is shifted to 8631 cm^{-1}. Madejova et al. (2000b) described a similar band around 8661 cm^{-1} plus a band at 10550 cm^{-1} for montmorillonite Saz. Pyrophyllite exhibits a weak absorption band at 8122 cm^{-1}, close to the value of 8110 cm^{-1} reported by Clark et al. (1990) for natural pyrophyllite. The interpretation for the second band around 10550 cm^{-1} given by Madejova et al. disagrees with the observations of Clark et al. (1990) for pyrophyllite, which has no interlayer water, with a similar band at 10500 cm^{-1}. A better alternative explanation for this weak band is that it is a combination band of Al-OH in the phyllosilicate structure. No corresponding band was observed for the synthetic paragonite in the 8000-10000 cm^{-1} region.

In conclusion it can be said that near-infrared spectroscopy of phyllosilicates is a strong technique to identify different phases such as beidellite, pyrophyllite and paragonite in the system Na_2O-Al_2O_3-SiO_2-H_2O. These spectra are dominated by overtone and combination modes of the Al-OH mid-infrared vibrational modes together with bands due to the combination and overtone modes involving vibrational modes of absorbed and interlayer water.

Acknowledgments

The financial and infra-structural support of the Queensland University of Technology, Centre for Instrumental and Developmental Chemistry is gratefully acknowledged. The authors are gratefull for the critical insights offered by Eugen Libowitzky, which strongly improved this manuscript.

References

Anderson, J.H. and Wickersheim, K.A., 1964. Near infrared characterization of water and hydroxyl groups on silica surfaces. *Surface Science* 2: 252-260.

Bailey, S.W., 1988. *Hydrous phyllosilicates* (exclusive of micas). Mineralogical Society of America, (Washington D.C.). 725 pp.

Bishop, J.L., Pieters, C.M. and Burns, R.G., 1993. Reflectance and Mössbauer spectroscopy of ferrihydrite-montmorillonite assemblages as Mars soil analogues. *Geochimica et Cosmochimica Acta* 57: 4583-4595.

Bishop, J.L., Pieters, C.M. and Edwards, J.O., 1994. Infrared spectroscopic analyses on the nature of water in montmorillonite. *Clays and Clay Minerals* 42: 702-716.

Brindley, G.W. (1984) Order-disorder in clay mineral structures. In G.W. Brindley, and G. Brown, Eds. *Crystal Structures of Clay Minerals and their X-ray Identification*, p. 125-189. Mineralogical Society, (London).

Cariati, F., Erre, L., Micera, G., Piu, P. and Gessa, C., 1981. Water molecules and hydroxyl groups in montmorillonites as studied by near infrared spectroscopy. *Clays and Clay Minerals* 29: 157-159.

Cariati, F., Erre, L., Micera, G., Piu, P. and Gessa, C., 1983a. Effect of layer charge on the near-infrared spectra of water molecules in smectites and vermiculites. *Clays and Clay Minerals* 31: 447-449.

Cariati, F., Erre, L., Micera, G., Piu, P. and Gessa, C., 1983b. Polarization of water molecules in phyllosilicates in relation to exchange cations as studied by near infrared spectroscopy. *Clays and Clay Minerals* 31: 155-157.

Chatterjee, S.D., 1974. X-Ray Powder Pattern and Molar Volume of Synthetic 2M-Paragonite; a refinement. *Contr. Miner. Petrol.* 43: 25-28.

Clark, R.N., King, T.V.V., Klejwa, M. and Swayze, G.A., 1990. High spectral resolution reflectance spectroscopy of minerals. Journal of Geophysical Research 95: 12653-12680.

Deer, W.A., Howie, R.A. and Zussman, J., 1996. *An introduction to the Rock-Forming Minerals*. Addison Wesley Longman Ltd., (Harlow). 324-326 pp.

Eberl, D., 1979. Synthesis of pyrophyllite polytypes and mixed layers. *Amer. Mineral* 64: 1091-1096.

Farmer, V., 1974. *The Infrared Spectra of Minerals*. Mineralogical Society, (England)

Frost, R.L., Kristof, J., Kloprogge, J.T. and Horvath, E., 2000. Rehydration of potassium acetate-intercalated kaolinite at 298 K. *Langmuir* 16: 5402-5408.

Giese, R.F., Jr., 1975. Interlayer bonding in talc and pyrophyllite. *Clays and Clay Minerals* 23: 165-166.

Güven, N. (1988) Smectites. In S.W. Bailey, Ed. *Hydrous phyllosilicates* (exclusive of micas), 19, p. 497-560. Mineralogical Society of America, (Washington D.C.).

Hunt, G.R. and Salisbury, J.W., 1970. Visible and near-infrared spectra of minerals and rocks: I silicate minerals. *Modern Geology* 1: 283-300.

Kloprogge, J.T.. (1992) Pillared clays. Preparation and characterization of clay minerals and aluminium-based pillaring agents. *Chemistry,* p. 349. Universtiy of Utrecht, The Netherlands.

Kloprogge, J.T. and Frost, R.L., 1999a. Infrared emission spectroscopic study of some natural and synthetic paragonites. *Applied Spectroscopy* 53: 1071-1077.

Kloprogge, J.T. and Frost, R.L., 1999b. An infrared emission spectroscopy study of synthetic and natural pyrophyllite. N. Jb. *Miner.* Mh: 62-74.

Kloprogge, J.T. and Frost, R.L., 2000a. Infrared emission spectroscopic study of the dehydroxylation of 10 Å halloysite from a Neogene cryptokarst of South Belgium. *Geologica Belgica* 2: 213-220.

Kloprogge, J.T. and Frost, R.L., 2000b. Study of the thermal behaviour fo rectorite by in-situ infrared emission spectroscopy. *Neues Jahrbuch für Mineralogie*, Monatshefte: 145-157.

Kloprogge, J.T., Frost, R.L. and Hickey, L., 2000a. Infrared emission spectroscopic study of the dehydroxylation of some hectorites. *Thermochimica Acta* 345: 145-156.

Kloprogge, J.T., Hickey, L. and Frost, R.L., 2000b. The effect of increasing layer charge on the infrared absorption spectra of synthetic beidellites. *Journal of Materials Science Letters* 19: 1131-1134.

Kloprogge, J.T., Jansen, J.B.H. and Geus, J.W., 1990. Characterization of synthetic sodium-beidellite. *Clays and Clay Minerals* 38: 409-414.

Kloprogge, J.T., Jansen, J.B.H., Schuiling, R.D. and Geus, J.W., 1992. The interlayer collapse during dehydration of synthetic $Na_{0.7}$-beidellite: A ^{23}Na solid state Magic-Angle spinning NMR study. *Clays and Clay Minerals* 40: 561-566.

Kloprogge, J.T., Komarneni, S. and Amonette, J.E., 1999a. Synthesis of smectite clay minerals: A critical review. *Clays and Clay Minerals* 47: 529-554.

Kloprogge, J.T., Komarneni, S., Yanagisawa, K., Fry, R. and Frost, R.L., 1999b. Infrared emission spectroscopic study of the dehydroxylation via surface silanol groups of synthetic and natural beidellite. *J. Colloid and Interface Science* 212: 562-569.

Kloprogge, J.T., Ruan, H. and Frost, R.L., 2000c. Near-infrared spectroscopic study of synthetic and natural pyrophyllite. *Neues Jahrbuch für Mineralogie,* Monatshefte: 337-347.

Kloprogge, J.T., van der Eerden, A.M.J., Jansen, J.B.H., Geus, J.W. and Schuiling, R.D., 1993. Synthesis and paragenesis of Na-beidellite as a function of temperature , water pressure, and sodium activity. *Clays and Clay Minerals* 41: 423-430.

Kloprogge, J.T., Vogels, R.J.M.J. and van der Eerden, A.M.J., 1996. Hydrothermal synthesis of pyrophyllite from Si-Al gels and salt solutions. N. Jb. *Miner.* Mh 3: 135-144.

Koster van Groos, A.F. and Guggenheim, S., 1984. The effect of pressure on the dehydration reaction of interlayer water in. *Am. Mineral* 69: 872-879.

Koster van Groos, A.F. and Guggenheim, S., 1986. Dehydration of K-exchanged montmorillonite at elevated temperatures and. *Clays and Clay Minerals* 34: 281-286.

Koster van Groos, A.F. and Guggenheim, S., 1987. Dehydration of a Ca- and a Mg-exchanged montmorillonite (SWy-1) at elevated pressures. *Am. Mineral* 72: 292-298.

Lin, C. and Bailey, S.W., 1984. The crystal structure of paragonite-2M1. American *Mineralogist* 69: 122-127.

Madejova, J., Bujdak, J., Petit, S. and Komadel, P., 2000a. Effects of chemical composition and temperature of heating on the infrared spectra of Li-saturated dioctahedral smectites. (I) Mid-infrared region. *Clay Minerals* 35: 739-751.

Madejova, J., Bujdak, J., Petit, S. and Komadel, P., 2000b. Effects of chemical composition and temperature of heating on the infrared spectra of Li-saturated dioctahedral smectites. (II) Near-infrared region. *Clay Minerals* 35: 753-761.

Rossman, G.R. (1988) Vibrational spectroscopy of hydrous components. In F.C. Hawthorne, Ed. *Spectroscopic methods in mineralogy and geology*, 18, p. 193-206. Mineralogical Society of America, (Washington D.C.).

Sidorenko, O.V., Zvyagin, B.B. and Soboleva, S.V., 1977a. The crystal structure of 3T paragonite. *Sov. Phys. Crystallogr*. 22: 557-559.

Sidorenko, O.V., Zvyagin, B.B. and Soboleva, S.V., 1977b. Refinement of the crystal structure of 2M1 paragonite by the method of high-voltage electron diffraction. *Sov. Phys. Crystallogr*. 22: 554-556.

Sposito, G., Prost, R. and Gaultier, J.-P., 1983. Infrared spectroscopic study of adsorbed water on reduced-charge Na/Li-montmorillonites. *Clays and Clay Minerals* 31: 9-16.

Suquet, H., de la Calle, C. and Pezerat, H., 1975. Swelling and structural organization of saponite. *Clays and Clay Minerals* 23: 1-9.

Tuttle, O.F., 1949. Two pressure vessels for silicate-water studies. *Geol. Soc. Amer. Bull.* 60: 1727-1729.

van der Marel, H.W. and Beutelspacher, H., 1974. *Atlas of infrared spectroscopy of clay minerals and their admixtures.* Elsevier, (Amsterdam). 396 pp.

Vedder, W. and McDonald, R.S., 1963. Vibrations of the OH ions in muscovite. *the Journal of Chemical Physics* 38: 1583-1590.

INDEX

A

B

F

G

H

M

N

O

P

Q

R

S

T

U

V

W

X

Y

Z